PHYSICS

FOR

HIGH SCHOOL STUDENTS

BY

HENRY S. CARHART, LL.D.

PROFESSOR OF PHYSICS IN THE UNIVERSITY OF MICHIGAN

AND

HORATIO N. CHUTE, M.S.

INSTRUCTOR IN PHYSICS IN THE ANN ARBOR HIGH SCHOOL

ALLYN AND BACON

Boston and Chicago

Norwood Press:
Berwick & Smith, Norwood, Mass., U.S.A.

PREFACE.

THE present volume has been written with the same purpose that guided the authors in the preparation of their "Elements of Physics," a book which has long been received with marked favor. The advances in physics are so rapid, and the point of view from which many topics are now considered is so different from that of half a dozen years ago, that it seemed best to make an entirely new book rather than a revision of the former one. The order of the general subdivisions has been changed to that employed by one of the authors in his "University Physics," an order which has been approved by experience.

The authors hold firmly to the opinion that the class-room and the laboratory should be provided with separate books. The material which must be included in any text-book for the class-room makes by itself an ample volume; none of it can be omitted without serious sacrifice. The addition of quantitative experiments for laboratory practice must either overload the volume devoted primarily to the exposition of principles, or it will result in an incomplete and unsatisfactory treatment of the directions for the laboratory. A class-book and a hand-book for the laboratory should be prepared from different points of view. The former should be devoted to a clear exposition of principles by qualitative experiments and by precise didactic statements; the latter should contain somewhat minute directions for the quantitative experiments of the laboratory.

The authors' thanks are due to the publishers for their generous aid in furnishing so many excellent woodcuts to illustrate the book. These are not only faithful representations of actual apparatus, but they also possess artistic merit and make the book attractive.

H. S. C.
H. N. C.

ANN ARBOR, MICHIGAN,
February, 1902.

CONTENTS.

HIGH SCHOOL PHYSICS.

INTRODUCTION.

WHEN we examine attentively the facts of nature about us, we study what are called physical as distinguished from mental phenomena. In this study it is necessary to assume the reality of this outer world of nature, and to assert that external objects exist apart from and independently of the mind of any one observing them. It is true that we become acquainted with the physical universe solely by means of our senses, but these alone do not enable us to decide whether the outer material world has a real existence or only the appearance of it. The final test of physical reality, or that the material world is as real as the mental one which gives us thoughts and feelings, is the fact that the physical world remains unchanged in quantity, or fixed in amount, however it may be measured. Tried by this test, there are only two *classes* of things in the physical world — *matter and energy*.

1. Matter. — It is only a colorless definition of matter to say that it occupies space. It is better described by its properties, to which the next chapter is devoted. Science is not yet able to tell what matter *is*, but the balance has demonstrated that it is invariable in amount, whatever form it may be made to assume.

A limited portion of matter is a *body*, and different kinds of matter, having distinct properties, are called *sub-*

stances. A gold coin, a drop of water, air enclosed in a vessel, are bodies. Each is also a substance, since it has properties distinct from the others.

2. Energy. — It is a fact of common observation that a body in motion can impart motion to another body, either by direct collision or otherwise. It is customary to say in such a case that the first body *does work* on the second. One body may also impart motion to another by virtue of its relative position. Thus, the weight of a clock when wound up gives motion to the pendulum and keeps it swinging against the resistance of the air and of friction. Whenever one body changes the motion or relative position of another, against a resistance opposing the change, the first body is said to do work on the second. Energy may be defined as the capacity of doing work. It is a grand doctrine of modern science that the energy of the physical universe is *conserved*, or is invariable in amount. This principle of the Conservation of Energy will become clearer when we have studied the various forms which energy may assume, and its conversion from one of these forms into another.

3. Physics. — In its most general aspect Physics may be defined as *the science of matter and energy*. It is desirable, however, to restrict this definition so as to exclude problems involving celestial bodies and their motions (Astronomy), discussions relating to the nature and interaction of different forms of matter (Chemistry), and the laws of matter and energy in living things (Biology). The distinction between Physics and Chemistry is a somewhat artificial one, and is becoming less clearly defined as science advances.

4. Physical Phenomena, Theory, Law. — Any change taking place in matter and not altering its composition is a *physical phenomenon*. The fall of a raindrop, the freezing of water, the melting of lead, a flash of lightning, the colors of the rainbow, the setting of the sun, the ringing of a bell, are all physical phenomena.

An *hypothesis* is a supposition advanced to explain phenomena. The more varied the phenomena it explains, the greater the probability of its truth. When the evidence in its support becomes large, it is raised to the rank of a *theory;* and, finally, it becomes a *law* when its truth is fully established.

Physical law expresses the constant connection between related phenomena. For example, it is found that if a quantity of gas, as air, be compressed at a constant temperature, its volume will decrease in the same ratio as the pressure is increased. This is the law of the compressibility of gases.

5. An Experiment consists in producing a phenomenon under conditions chosen and controlled by the experimenter. By making definite changes in the conditions one by one, the results may enable him, by a process of exclusion, to fix the conditions necessary to the phenomenon in question, and so to discover the law embracing it. The basis of all experimentation is the belief in the *constancy of nature*, as enunciated in the statement that "*under the same physical conditions the same physical results will always be produced, irrespective altogether of time and place.*"

CHAPTER I.

PROPERTIES OF MATTER.

6. The Properties of Matter are its peculiar qualities which serve to describe it and to define it provisionally. They are, in general, the different ways in which it presents itself to our senses. These properties are either *general*, that is, those common to all kinds of matter in whatever state or physical condition they may be; or *special*, those distinctive of some kinds of matter and conspicuously absent in others. Thus, all matter has *extension*, or occupies space; but a piece of window glass lets light pass through it, or is *transparent;* while a piece of slate does not transmit light, or is *opaque*. A watch spring recovers its shape after bending, or is *elastic;* while a strip of lead does not possess this property, or is *inelastic*. Extension is a general property of matter, while transparency and elasticity are special properties.

7. Extension. — All matter occupies space or has dimensions. This general property is known as *extension*. The dimensions of a body are its length, breadth, and thickness, or all bodies occupy space of *three dimensions*. A sheet of writing paper or of gold-leaf appears, at first thought, to have only two dimensions, length and breadth; but its third dimension is relatively small, and if its thickness should actually become zero, it would cease to be a sheet of paper or a piece of gold-leaf.

8. Measurement of Extension. — The measurement of extension is made in terms of some unit of length, which is chosen arbitrarily. The units of area and of volume are the square and the cube respectively of the unit of length.

The two systems of units in common use are the *English* and the *Metric*. The former is generally used in Great Britain, Canada, and the United States, while the latter is almost exclusively used on the continent of Europe. The standard of length in the English system is the *Imperial Yard*. It was defined by Act of Parliament in 1855 as the distance between the transverse lines in two gold plugs in a certain bronze bar at 62° Fahrenheit deposited in the Office of the Exchequer. One-third of the yard is the *Foot*, and one thirty-sixth is the *Inch*.

In the Metric system the *Metre* is the standard unit of length. It is defined for the United States by the length of a certain platinum-iridium bar, at the temperature of melting ice, or 0° C. (Centigrade scale). This bar is preserved in the National Bureau of Standards.

The metre as a standard length is in no way superior to the yard. Its great advantage lies in the fact that its multiples and submultiples are all in the decimal system. It is therefore much more convenient to use than the yard, with its irrational subdivisions. Its universal adoption by civilized nations is only a question of time.

The metre (m.) is divided into 10 decimetres (dm.), the decimetre into 10 centimetres (cm.), and the centimetre into 10 millimetres (mm.). The only multiple of the metre of practical use is the kilometre (km.), equal to 1000 metres. It is the unit employed on the continent of Europe for such distances as we express in miles. One kilometre equals 0.6214 mile, or one mile equals 1.6093 kilometres.

The United States *Gallon* (231 cubic inches), and the

Litre (1000 cubic centimetres), are legal units of volume for liquid measure. Tables for the conversion of quantities from one system of units into the other will be found in the Appendix.

By Act of Congress in 1866, the use of the metric system of weights and measures became lawful throughout the United States, and the weights and measures in common use were defined in terms of the units of the metric system. By this same act the legal value of the metre in the United States is 39.37 inches, and the yard is now defined as being $\frac{3600}{3937}$ of a metre. When approximate values only are needed, the metre may be taken to be $39\frac{1}{3}$ inches, the centimetre as $\frac{2}{5}$ inch, the kilometre as $\frac{5}{8}$ mile, and the litre as $2\frac{1}{9}$ pints.

9. Mass and Weight. — The *mass* of a body is the *quantity of matter* in it. The distinction between mass and weight will be explained in Chapter II. It may suffice here to say that the weight of a body depends on its location on the earth, or with respect to some other body; but mass is independent of gravitation. The mass of a ball of iron would be the same if the ball could be taken to the north pole, to the centre of the earth, or to the sun. But its weight at the earth's centre would be zero, and at the surface of the sun nearly twenty-eight times as great as at the earth's surface.

10. Measurement of Mass. — Two systems of measuring mass are in legal use in Great Britain and the United States. The standard units of mass are the *Avoirdupois Pound* for the British system, and the *Kilogramme* for the metric system. The *Ton* of 2000 pounds is the chief multiple of the pound in the United States; its submultiples are the *Ounce* and the *Grain*. The avoirdupois pound is equal to 16 ounces, and to 7000 grains. The coinage of the United States is regulated by the "troy pound of the mint," containing 5760 grains.

For the United States the standard unit of mass in the metric system is the mass of a certain piece of platinum-iridium preserved in the National Bureau of Standards in Washington. This standard kilogramme and the standard metre in the same Bureau were prepared by an International Committee, and they are called "National Prototypes." The real prototypes are kept in the national archives in Paris. The kilogramme was originally designed to represent the mass of a cubic decimetre of pure water at 4° C., the temperature of the greatest density of water. The gramme is the thousandth part of a kilogramme, and is very nearly equivalent to the mass of a cubic centimetre of pure water at 4° C. The gramme (gm.) is divided into 10 decigrammes (dgm.), or into 1000 milligrammes (mgm.).

11. Impenetrability. — Matter not only occupies space, but one portion of matter appears to occupy a portion of space to the entire exclusion of all other matter. This property of *impenetrability* means that no two portions of matter, however small, can occupy the same space at the same time. The volume of an irregular solid is sometimes measured by measuring the volume of liquid displaced when the solid is completely immersed in it. The method is based on this property of impenetrability.

12. Porosity. — Impenetrability does not preclude interpenetration of matter. Thus, a sponge or a piece of sandstone may absorb much water without change of volume.

All matter is more or less spongelike or porous in structure, and the property corresponding to this structure is called *porosity*. These pores, whether visible or invisible, are in reality not a part of the space occupied by the material of the body. They may therefore be filled by some other material.

Experiment.—Into a long test-tube pour 27 cm^3. (cubic centimetres) of water. Add carefully 23 cm^3. of strong alcohol, tipping the test-tube so that the liquid flows down its walls. Mark the position of the surface of the alcohol, and then mix the two liquids thoroughly by shaking. The volume will shrink to 48.8 cm^3.

The porosity of some metals is shown by the fact that gases pass through them. Palladium has the capacity of absorbing a large quantity of hydrogen, and carbon dioxide passes quite freely through red-hot cast iron. It is claimed by some that this apparent absorption by metals is not simple absorption, but a chemical union. The salt glaze on earthenware is porous and is penetrated by some liquids. Agate, though extremely hard, is still porous; advantage is taken of this property to color it, some layers being more porous than others. Glass and other vitreous bodies are not known to be porous.

13. Inertia.—The most characteristic property exhibited by matter is *inertia*. Inertia is the persistence of matter in whatever state of rest or of motion it may chance to be, and its resistance to any attempt to change that state. If a moving body be stopped, its arrest is always due to some influence outside of itself; and if a body at rest be set in motion, the motion must be imparted to it by some other body.

Many familiar phenomena are due to inertia: for example, the onward motion of a rider when his horse suddenly shies or stops; the oscillation of water in a pail when carried; the water flying from a revolving grindstone; the persistence with which the axis of rotation of a spinning top maintains its direction. The violent jar to a water pipe on suddenly closing the faucet is due to the inertia of the stream. Tall columns and detached chimneys are sometimes twisted around on their base by sudden gyratory earthquake movements. The sudden twist of the earth under the column leaves it behind, and the slower return motion of the ground carries the column with it. Figure 1 is the picture of such a monument in India twisted on its base in this way by an earthquake.

Fig. 1.

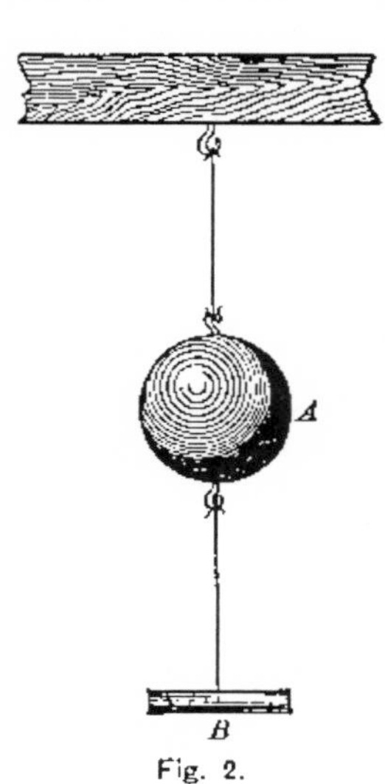

Fig. 2.

Experiment. — Suspend a heavy weight A (Fig. 2) by a string, and to the under side of the weight attach a small bar B by a piece of the same string. If we pull *steadily* downward on B, the string will break above A. The tension in the upper string is greater than in the lower one because it has to support the weight A and resist the pull applied at B. If, however, we pull downward *suddenly* on B, the string will break below A. On account of the inertia of the heavy weight, the lower string breaks before the sudden pull reaches the upper string.

14. Elasticity. — Apply pressure to a rubber ball, stretch a common rubber band, bend a piece of watch spring, twist a stout cotton cord. In each case the form or the volume has been changed, and the body has been *strained.* A *strain* means either a change of size or a change of shape. As soon as the distorting force, or *stress*, has been withdrawn, these bodies recover their initial volume and dimensions. This property exhibited by matter of recovery from a strain on the removal of the stress is called *elasticity.* It is called *elasticity of form* when the body recovers its form on the removal of the force of distortion; and *elasticity of volume* when the temporary distortion is one of volume. Gases, vapors, and liquids possess perfect elasticity of volume; that is, they recover their initial volume when the initial pressure is restored; but they have no elasticity of form. In the case of solids the recovery from distortion is incomplete if the distortion is pushed beyond a limit which is different for different substances. When permanent alteration is about to take place, the body has reached its *elastic limit.* Some solids, when subjected to long-continued forces, suffer elastic fatigue, yield slowly, and never recover their original form. Shoemaker's wax is an example.

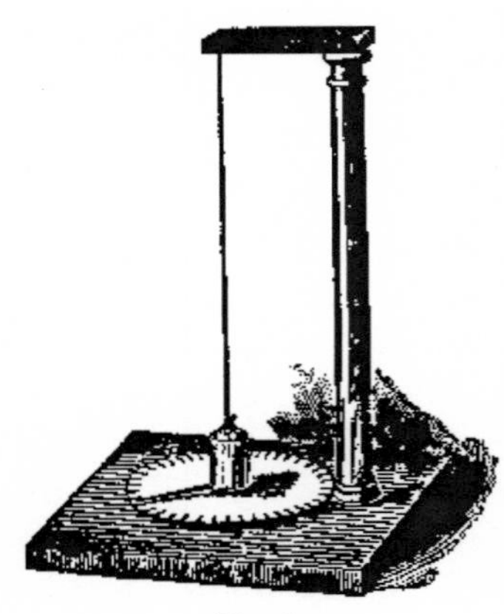

Fig. 3.

The elasticity of a body may be called forth by pressure, by stretching, by bending, or by twisting. The bounding ball and the common popgun are illustrations of the first; rubber bands are familiar examples of the second; bows and springs are instances of the third; the torsional pendulum (Fig. 3) and the stretched spiral spring exemplify the fourth.[1]

15. Hooke's Law. — The law of the distortion of elastic bodies was first announced by Hooke in 1676. It may be briefly expressed by saying that the stress of restitution is proportional to the strain; or, in other words, the force of restitution is proportional to the change of form. Illustrations will make the meaning of the law clearer. Clamp a flat steel bar by one end in a vise, the flat side horizontal. Load the free end with weights by means of a light scale pan, and observe the bending of the bar. The vertical deflection of the end of the bar should be noted with some convenient weight in the pan. Then double the weight and note the new deflection. It will be double the first one. The amount of the bending or distortion of the bar is proportional to the weight. Force is required to bend an inelastic bar like lead, but there is no elastic force of recovery of form when the bending force is removed.

Experiment. — Turn the weight of Fig. 3 through 15 degrees and release it, noting the period of the torsional vibration by observing the intervals in seconds between the successive passages of the pointer through the position of rest. Then turn the weight through 30 degrees and observe the period of vibration again. Repeat with a

---

[1] The *coefficient of elasticity of volume* is the quotient of the pressure applied by the compression produced by it. By pressure is meant the intensity of pressure, or the pressure on unit area; and by compression, the compression suffered by unit volume. Let p be the increase of pressure on unit area, and let the original volume V become $V - v$. Then $\frac{v}{V}$ is the compression for a unit of volume, and the coefficient of elasticity is $p \div \frac{v}{V} = \frac{pV}{v}$. A general definition of the coefficient of elasticity is the quotient of the stress called out in the body by the strain. A stress is a force and a strain is a distortion. Elasticity is measured by the ratio of the *internal* stress to the strain.

twist of 45 degrees, and so on. It will be found that the period of torsional vibration is the same whatever the twist of the weight and wire. The force tending to bring the weight back to its rest position is proportional to the angle of twist or to the distortion, and therefore the period is unaffected by the angular twist. With double the angle to swing through, there is double the force to cause the weight to swing in the same time.

Experiments of similar import may be made by stretching heavy rubber bands or strips. The amount of stretching will be proportional to the force applied.

16. Plasticity. — *Plasticity* is the inability of a body to recover from distortion produced by a stress. In so far as bodies are not *elastic* they are *plastic*, and elastic bodies are plastic beyond the limits where they cease to be perfectly elastic. Plastic bodies require force to change their shape, but they do not require the continued application of force to maintain the change. Elastic bodies spring back when the force of distortion is removed; plastic bodies do not. Bodies are classed as elastic if they have a large limit of elasticity, and plastic if their limit of elasticity is small. A bar of lead will vibrate when struck, showing that it is elastic; but its limit of elasticity is small, and it is therefore classed as a plastic body.

17. Cohesion. — Molecules are the smallest portions of matter with which Physics has to deal. *Cohesion* is the attraction between the molecules of a body. It unites these molecules or particles together throughout the mass, whether the molecules be like or unlike. *Adhesion* is the molecular attraction uniting bodies by their adjacent surfaces. Cohesion holds together the molecules of a substance so as to form a larger mass than a molecule. It resists a force tending to break or crush a body. When two clean surfaces of white-hot wrought iron are brought into intimate contact by hammering, they will cohere. If a clean glass rod be dipped into water and then withdrawn, a drop will adhere to it. Graphite is ground to a very fine powder and cleaned chemically by boiling with nitric acid and chlorate of potash. It is then washed, dried, and compressed in moulds by a powerful hydraulic press. The black powder is by this means converted into a solid slab which may be sawed into thin strips for lead pencils. The pressure causes the clean particles to cohere.

Experiment.—Suspend from one of the scales of a beam balance a perfectly clean glass disk by means of threads cemented at three points (Fig. 4). After counterpoising the disk, place below it a vessel of water and adjust the apparatus so that the disk touches the surface of the water when the beam of the balance is horizontal. Now add weights to the pan till the disk is separated from the water. An examination of the under surface of the glass plate will show that a film of water has been pulled away with the glass. The force of adhesion between the glass and the water is greater than the force of cohesion in the water.

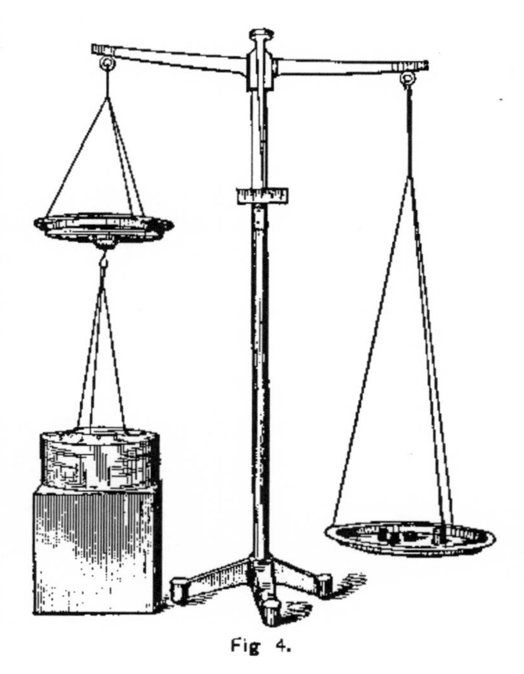

Fig 4.

18. Molecular Attraction affected by Distance.—Little cubes of metal, with plane faces polished so smooth that they will adhere by slightly pressing them together, are known as Barton's cubes. A string of a dozen may be made to cling together, and may be supported by the mutual attraction between the molecules of adjacent surfaces.

Experiment. — Press firmly together two small lead disks, giving to them a slight twisting motion. The surfaces in contact must be flat and bright. They will adhere quite firmly together. Two lead bullets, cut so as to present clean flat surfaces, may be used instead of the disks.

This experiment shows that molecular forces act only through insensible distances. They diminish rapidly as the distance between the molecules increases, and vanish at a range something like a twenty-thousandth of a millimetre.

19. Crystallization. — **Experiment.** — Dissolve 100 gm. of alum in a litre of hot water. Hang some strings in the solution and set aside in a quiet place for several hours. The strings will be found covered with beautiful transparent bodies of regular form and similar in shape.

Most bodies, when they pass slowly from the liquid to the solid state under conditions that allow freedom of motion to the molecules, assume regular geometrical forms called *crystals*. Such is the case when substances pass from the liquid state, either as a molten fluid or in solution, to that of a solid. In the former the crystallization is said to take place by the *dry way*; in the latter, by the *moist way*. Snowflakes are beautiful crystals of myriad forms, but all of them hexagonal in outline. Cast zinc when broken shows a crystalline fracture. Substances which exhibit no plan in the grouping of the molecules are said to be *amorphous*. Glass, glue, and paraffin are examples of amorphous bodies.

When crystalline bodies are broken apart, the fracture consists in separating crystals from the face of other crystals. This separation is most easily effected along certain definite planes called *planes of cleavage.*

20. Tenacity. — *Tenacity* is the ability of a body to resist a force tending to tear it asunder. It is measured by the greatest force the body can bear per unit area of cross-section without breaking.

Wrought iron has greater tenacity than cast iron; the most tenacious metal is steel pianoforte wire. Lead has the least tenacity of all metals. Phosphor-bronze and aluminum-bronze have high tenacity, and they are therefore suitable for telegraph wires, where long spans are necessary.

21. Malleability. — *Malleability* is the property possessed by some bodies of being beaten or rolled into thin sheets without breaking. Pure gold is more malleable than any other substance. By hammering between pieces of gold-beater's skin, it has been reduced to sheets so thin that 300,000 of them, placed one upon another, make but an inch in thickness. Platinum, silver, lead, and tin possess the same property, but to a less degree. Zinc is malleable when heated. It can then be hammered, rolled, or bent without fracture.

Articles of cast iron may be made somewhat malleable, or less brittle, by heating them for several days in the presence of a substance, like the black oxide of iron, which removes from them some of the carbon of the cast iron, and gives to them something of the properties of wrought iron.

22. Ductility. — *Ductility* is the property of a metal which permits of drawing it into wire through a draw-plate. Gold, platinum, copper, and silver are highly ductile. Substances of great malleability are usually also highly ductile ; lead is an exception.

Other substances become highly ductile when heated to a high temperature. Thus, glass has been spun into such fine threads that a mile would weigh only a third of a grain ; and Professor Boys has made quartz fibres so fine, by attaching the white-hot quartz to an arrow shot from a bow, that they are invisible even under a microscope of high power. They are invisible because their diameter is less than a wave length of light.

23. Hardness. — *Hardness* is the relative resistance which a body offers to scratching or abrasion by other bodies. The relative hardness of two bodies is ascertained by finding which of them will scratch the other. For example, glass is harder than copper because it scratches copper.

Most substances, if suddenly cooled from a high temperature, become very hard. A few, like copper and bronze, are softened by sudden cooling. The process of giving to a body a suitable degree of hardness is called *tempering;* that of making it as soft as possible at ordinary temperatures is called *annealing.* Both processes consist in raising the temperature of the body and then cooling, either suddenly or slowly, according to the result desired.

Fig. 5.

24. Absorption. — **Experiment.** — Fill a wide test-tube with dried ammonia gas by displacement over mercury. The gas may be obtained by boiling strong ammonia water in a flask and passing the gas through a tube filled with small lumps of unslacked lime.

Insert beneath the mouth of the test-tube a piece of freshly heated charcoal; the mercury will immediately begin to rise in the tube, because the charcoal absorbs the ammonia (Fig. 5).

Gases are condensed to a greater or less degree on the surface of all solids; and since porous bodies, like charcoal, present an immense surface in their pores, they may absorb large quantities of gas. One cm^3. of boxwood charcoal will absorb 35 cm^3. of carbonic acid gas and 90 cm^3. of ammonia. It should be noted that gases which are most readily liquefied are, as a rule, absorbed by porous bodies most greedily. Finely divided platinum, in the form called platinum sponge, absorbs hydrogen from a small jet so rapidly that the heat generated by the condensation raises the sponge to a red heat and ignites the hydrogen. It will also ignite illuminating gas.

Air condensed on the surface of glass adheres to it with great persistence. In filling a barometer tube (§ 148), it is necessary to boil the mercury in order to expel the air. The last trace of air can be removed only by repeated boiling.

25. Diffusion. — **Experiment.** — Fill a large test-tube on foot two-thirds full of water, colored blue with litmus. Introduce a few drops of sulphuric acid into the liquid at the bottom of the tube by means of a thistle-tube (Fig. 6). A reddish color will appear at the bottom, and, if the liquids are not disturbed for several hours, this change of color will move slowly toward the top.

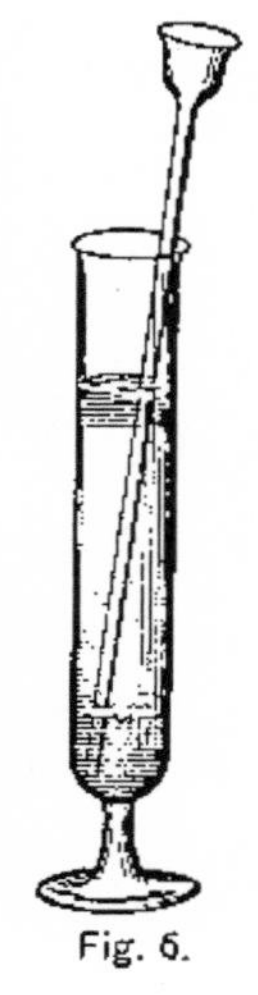
Fig. 6.

Experiment. — Fill two glass cylindrical jars with oxygen and hydrogen respectively, and cover each with a glass plate. Invert the jar of hydrogen and place it over the jar of oxygen. Remove the glass plates and let the jars stand for about 20 minutes. Then apply a lighted taper to each jar; an explosion will follow, showing that the gases have mixed.

In both of these experiments the heavier fluid moves up through the lighter, and the lighter down into the heavier. This intermingling of fluids in contact is called *diffusion*. Liquids diffuse very slowly, but gases more rapidly. Each gas expands and fills both vessels as if the other gas were not present. In time, therefore, both vessels contain a uniform mixture. The presence of a second gas only increases the time required for a uniform distribution.

Fig. 7.

26. Effusion. — **Experiment.** — Cement a small porous battery cup to a funnel tube. Connect the latter to a Florence flask provided with a jet-tube, as shown in Fig. 7. Over the porous cup invert a large glass beaker, into which passes a stream of hydrogen through a glass tube. If all the joints are tight, water will issue from the jet-tube as a small fountain. The hydrogen passes freely through the walls of the porous cup and produces gas pressure in the flask.

When the pores of a solid are exceedingly fine, the passage of a gas through them is by a process called *effusion*. The rate of effusion of different gases is very nearly inversely proportional to the square root of their relative weights. Hydrogen, for example, which is one-sixteenth as heavy as oxygen, passes through very small openings nearly four times as fast as oxygen.

Gases may be separated from one another more or less completely by taking advantage of the different rates at which they pass, by effusion, through fine porous earthenware or unglazed porcelain.

27. Osmosis. — Experiment. — Remove the bottom from a small bottle, and tie over it a dampened piece of parchment paper. Through a stopper fitted to the mouth of the bottle, pass a glass tube terminating in a small funnel at its upper end. A scale of equal parts may be attached to the tube (Fig. 8). Pour into the tube a concentrated solution of copper sulphate. Support the apparatus so that the bottle dips into a jar of water just far enough to bring the two liquids to the same level. If the apparatus be left standing, we shall find that the liquid within the tube is slowly rising, while the water outside is acquiring a bluish tint. The two liquids pass through the membrane, but the greater flow is toward the denser liquid.

Fig. 8.

The passage of liquids at different rates through a porous membrane is called *osmosis*, and the pressure in the tube containing the salt solution is known as *osmotic pressure*. The osmotic pressures, due to solutions of different substances at several concentrations, have been measured by the use of a membrane permeable to water but not permeable to the dissolved substance. Such a membrane is called a semi-permeable membrane. The pressure required to maintain the liquid in the tube at the same level as the water outside the tube is the measure of the osmotic pressure exerted by the dissolved substance.

28. States of Matter. — The three states in which matter exists may be illustrated by water, which may assume either the solid, the liquid, or the gaseous state.

Ice is a solid. A *solid* has a definite shape, and resists any stress tending to change its shape or size.

Water is a liquid. A *liquid* has no shape of its own, but is mobile and conforms to the shape of the containing vessel. It offers but slight resistance to a stress producing relative movement of its parts, so long as the volume remains unchanged.

Steam or water vapor is a gas. A *gas* offers no resistance to stress tending to change its shape. It has neither shape nor size of its own, but completely fills any vessel containing it.

In brief: —

Solids have a definite mass and both size and shape.

Liquids have a definite mass and size, but not shape.

Gases have a definite mass, but neither size nor shape.

Problems.

1. Why is a person, leaping to the ground from a rapidly moving carriage, likely to fall in the direction in which the carriage is going?

2. Why can the head of a hammer be driven on the handle by striking the end of the latter against an anvil?

3. A circus rider leaps vertically in order to go through a hoop, but he alights on the horse's back. How do you explain this?

4. Carbon dioxide is considerably heavier than air. Why does it not collect at the bottom of inhabited rooms?

5. How many metres in 4 rods?

6. If the mercury in a barometer is 30 in. high, what would be its height on a centimetre scale?

7. How many kilometres in 10 mi.?

8. Express the earth's equatorial diameter, 7926.592 mi., in kilometres.

9. At 45° latitude a body falls 16.083 ft. in one second. Express this distance in centimetres.

10. The diameter of a circular plot of ground is 100 ft. Find its area in square metres.

11. What is the perimeter in feet of a square lot whose area is 600 m^2.?

12. The volume of a cylinder is one gallon. Find its dimensions in centimetres, its height being twice its diameter.

Ans. 26.82 cm.; 13.41 cm.

13. Find by calculation the number of cubic centimetres in a pint.

14. Calculate the ratio of a quart to a litre.

15. If the mass of a cubic centimetre of water is one gramme, how many pounds in a cubic foot of water?

16. If a man weighs 150 lb., what will be his weight expressed in kilogrammes?

17. What is the price per kilogramme when the price per pound is 50 ct.?

18. How many litres will a 10-qt. pail hold?

19. How long a piece must be cut from a $\frac{1}{2}$-in. rod of brass in order that the volume may be 10 cm^3.?

20. At 10 ct. a pound, what will 3.5 kgm. of rice cost?

CHAPTER II.

MECHANICS OF SOLIDS.

I. MOTION AND VELOCITY.

29. Mechanics.—The term *Mechanics* is applied to that portion of Physics which deals with the effects of force on matter. It includes the principles applied in the construction of machines.

By *force* we may understand muscular exertion, or whatever else produces the same effects. The making of a muscular effort to overcome resistance give us our primitive idea of force. Whenever any inanimate agency produces effects precisely similar to those due to muscular exertion, it is said to exert force. Thus, a steam engine exerts force in pulling a train, turning a dynamo, or driving a mill; air compressed in a rubber balloon exerts force against the resistance of the stretched rubber; the gases due to the explosion of gunpowder exert force on the ball while it is passing from the breech to the muzzle of a rifle.

The most obvious effects of force on matter are (1) to produce change of motion, and (2) change of size or shape.

30. Motion.—A body moves when it is in different positions at different times. *Motion* is the change in the relative position of a body with respect to some point or place of reference. It involves, therefore, both time and space. The path of a moving body must be continuous, that is, the body must pass in succession through every point of its path.

All rest and motion are relative, since there are no fixed points in space to which absolute motion may be referred. When a passenger walks on the deck of a ship, his motion is relative to the vessel; the motion of the ship across the ocean is relative to the earth's surface; the diurnal motion of the earth's surface is relative to its axis of revolution; while the motion of the earth's centre is relative to the sun.

When a body moves along a straight line, its motion is said to be *rectilinear;* when it moves along a curved line, its motion is *curvilinear.* In the latter case, its direction of motion at any point of its path is that of the straight line drawn tangent to the curve at the point.

31. Velocity. — When a body moves over equal spaces in successive equal times, its motion is said to be *uniform;* if it traverses unequal spaces in successive equal times, its motion is *variable.* For example, the tip of the minute hand of a watch has a uniform motion, though the direction of its motion is constantly changing. So also the apparent motion of a star across the field of a fixed telescope is an instance of uniform motion. On the other hand, the motion of a falling body is variable, for it moves faster and faster as it descends.

Velocity is the time-rate of motion of a body. By the time-rate of any change is meant the whole change taking place in a given time divided by that time. If the motion is *uniform,* the velocity is *constant.* Uniform velocity is measured by the distance a body goes in a unit of time. The word *speed* is often used to denote *rate* of motion without reference to its *direction.* The term *velocity* then includes both the rate of motion and its direction.

When the motion is variable, the velocity at any instant is the distance the body would move in the next unit of time if left wholly to itself, that is, without help or hindrance. For example: the velocity of a shell on leaving the mouth of a gun, called the muzzle velocity, is the space it would pass over in the next second if it should continue to move undisturbed at the same rate; the velocity of a falling stone at any moment is the distance it would fall during the following second, if the attraction of the earth and the resistance of the air could both be withdrawn.

32. Formulæ for Uniform Motion.—Let v be the constant velocity of a body with uniform motion. Then in t units of time the space s passed over will be t times the velocity v, or

$$s = vt. \tag{1}$$

From this relation we have also $v = \frac{s}{t}$, and $t = \frac{s}{v}$.

Even though the motion were not uniform, if s be the space passed over in time t, then the mean or average velocity for the whole time would be $v = \frac{s}{t}$. If both the space and the time be reduced to indefinitely small quantities, then this mean velocity becomes the actual velocity at the instant.

33. Acceleration.—*Acceleration is the time-rate of change of velocity.* If the change of velocity be the same from second to second, the motion is *uniformly accelerated.* If the velocity increases, the acceleration is positive; if it decreases, the acceleration is negative. A falling body has a positive acceleration; the acceleration of a body thrown upward is negative. When a heavy body falls its gain in velocity per second is 9.8 m. for every second of time. Its acceleration is, therefore, 9.8 m. per second per second; in other words, an increase in velocity of 9.8 m. per second is acquired in a second of time.

34. Formulæ for Uniformly Accelerated Motion. — Let a be the acceleration, or the gain in velocity per second acquired in a second of time. (Unless otherwise stated, the unit of time used will be the second of common or mean solar time, of which there are 86,400 in a mean solar day.) Then in t seconds the velocity acquired will be

$$v = at. \tag{2}$$

Since the gain in velocity is uniform, if the body starts from rest, the average velocity for t seconds is $\frac{1}{2}(0 + at)$, or $\frac{1}{2}at$. The distance passed over in t seconds is then $\frac{1}{2}at \times t$, or $\frac{1}{2}at^2$. Hence

$$s = \frac{1}{2}at^2. \tag{3}$$

From (3)

$$a = \frac{2s}{t^2},$$

and

$$t = \sqrt{\frac{2s}{a}}.$$

Combining (2) and (3)

$$s = \frac{a^2t^2}{2a} = \frac{v^2}{2a},$$

or

$$v = \sqrt{2as} \tag{4}$$

Problems.

1. If a train has a speed of 60 mi. an hour, what is its speed in feet per second?

2. A man walks 13 km. in 3 hr.; how long will it take him to walk 25 km.?

3. Roemer found that light takes 16 min. 38 sec. to cross the diameter of the earth's orbit. The mean distance of the earth from the sun is 92.897 million mi. What is the speed of light per second?

Ans. 186,166 mi.

4. If a man walks 38 mi. in 10 hr., what is his average speed in kilometres per hour?

5. A railroad train, 120 yd. long, passes over a bridge 80 ft. long, at the rate of 30 mi. an hour. How long does the train take to pass completely over the bridge?

6. A body moving uniformly in a circular orbit of 10 m. radius makes ten complete revolutions in 7 sec. Find the speed per second.

Ans. 89.76 m.

7. A railroad train, 300 m. long, passes over a bridge 150 m. long at the rate of 48 km. an hour. How long does the train take to pass completely over the bridge?

8. A body starting from rest moves with an acceleration of 20 ft. per second per second. What space does it pass over in 6 sec., and what is its velocity at the end of that time?

9. A train, starting from rest, moves with a uniform acceleration, and takes 5 min. to pass over the first mile. What is its acceleration per second per second, and what is its final velocity?

Ans. 0.1173 ft.; 35.2 ft.

10. A body starting from rest moves with a uniformly accelerated motion for 10 sec., and passes over 100 km. Find the acceleration and the velocity at the end of the tenth second.

Ans. 2 km. per second per second; 20 km. per second.

11. A railroad train, whose motion is uniformly retarded, is moving at the rate of 30 km. per hour, and comes to rest in 10 min. Find the acceleration per minute per minute and the distance the train ran.

Ans. 50 m.; 2500 m.

12. In the same problem, what is the acceleration per minute per second, and what is it per second per second?

13. An electric car, running at the rate of 20 mi. an hour, is uniformly retarded by the application of the air-brake, and is brought to rest after running a distance of 100 ft. Find the time required to stop the car.

14. A freight train on a straight road starts from rest with a uniform acceleration of 3.5 ft. per second per second; find the time required to increase its velocity to 20 mi. an hour.

15. A boy rolls a ball on the grass; it rolls for 5 sec. before stopping and goes a distance of 40 m. Calculate the retardation caused by the grass and the air, and the velocity with which the ball left the hand.

16. In an experiment made with the Westinghouse air-brake, it was found that a train going at the rate of 60 mi. an hour was stopped in 440 yd. Calculate the retardation due to the brake.

35. Graphic Representation of a Velocity. — A velocity has both *magnitude* and *direction* (§ 31). The number of units in the length of a straight line may represent the magnitude of a velocity, and the direction in which the line is drawn may represent the direction of the motion. Thus the line AB (Fig. 9), 4 cm. long, may represent a velocity of 20 m. a second in a direction parallel to the top of the page of this book, if the length of one cm. represents a velocity of 5 m. a second.

A ———————— B

Fig. 9.

36. Composition of Velocities. — When several motions are given to a body at the same time, its actual motion is a compromise between them. The motions are said to be *compounded*, the compromise path being the *resultant*. At the Paris Exposition of 1900 a continuous moving sidewalk carried visitors around the grounds. A person walking on this platform had a velocity with respect to the ground made up of the velocity of the sidewalk relative to the ground and the velocity of the person relative to the moving walk. The several velocities entering into the resultant are known as the *component velocities.*

37. Cases of Composition. — In the composition of uniform velocities three distinct cases arise: —

FIRST.— *When a body has two simultaneous velocities along the same straight line, the resultant velocity is the algebraic sum of the two components; and, if the components are of opposite sign, its direction is that of the greater.*

For example, a boat, which would have in still water a velocity of 5 mi. an hour, will have what velocity against a current of 3 mi. an hour?

The resultant velocity will be $5 - 3 = 2$ mi. an hour up stream. With the current it would be $5 + 3 = 8$ mi. an hour.

SECOND. — *When the two simultaneous velocities are not in the same direction, the magnitude and direction of the resultant are represented by the diagonal of the parallelogram constructed on the two lines representing the component velocities as adjacent sides, and drawn through their intersection.*

This law is known as the *parallelogram of velocities.* To illustrate: If a boat can be rowed in still water at the uniform rate of 5 mi. an hour, what will be the actual velocity if it be rowed at right angles to a current running 3 mi. an hour?

Let the line AB (Fig. 10) represent in length and direction the velocity of 5 mi. an hour across the stream, and the line AC, at right angles to AB, the velocity of the current, 3 mi. an hour, both on a scale of 8 mm. to the mile. Complete the parallelogram $ABCD$, and draw the diagonal AD through the point A common to the two component velocities.

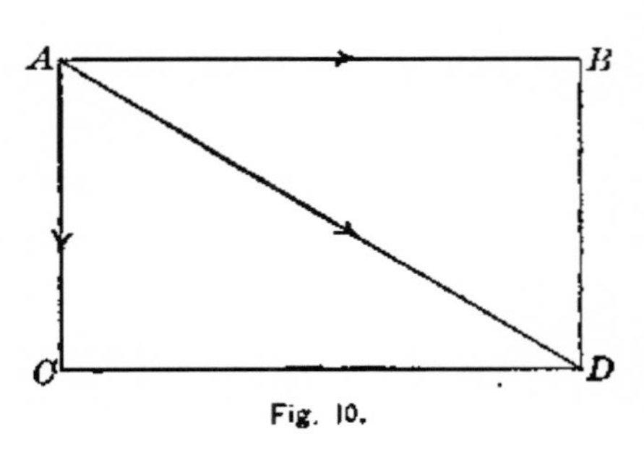

Fig. 10.

AD will represent the actual velocity; for while the boat is rowed 5 mi. in the direction *AB*, it is carried by the current 3 mi. in the direction *AC*. Hence at the end of an hour, instead of arriving at *B*, it will be at *D*, and will have travelled along the line *AD* with uniform motion. The length of the line *AD* on the same scale as the other lines will be 5.83. The resultant velocity is therefore 5.83 miles an hour in the direction *AD*.

When the angle between the components is a right angle, as in the present case, the diagonal *AD* is the hypotenuse of the right-angled triangle *ABD*. Its square is therefore the sum of the squares of 5 and 3, or

$$AD = \sqrt{5^2 + 3^2} = \sqrt{34} = 5.83.$$

When the angle *A* is not a right angle, the resultant must be found by a graphic process of measurement or by the principles of plane trigonometry.

THIRD. — *The resultant of three or more simultaneous velocities may be found by repeated applications of the parallelogram law. The method is known as the polygon of velocities.*

For example, if a body has a velocity eastward of 120 ft. a second, northward of 80 ft. a second, and northwestward of 100 ft. a second, what is the resultant velocity?

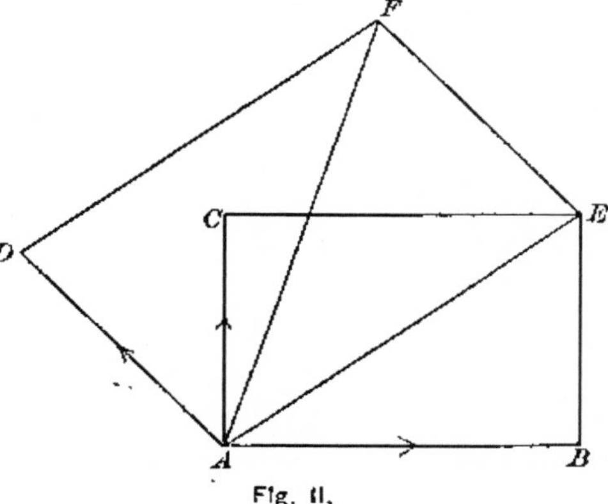

Fig. 11.

Let *AB*, *AC*, and *AD* (Fig. 11) represent these velocities on a scale of 40 ft. to 1 cm. Then *AE* is the resultant of *AB* and *AC* (Why?); and *AF*, of *AE* and AD. The length of AF in centimetres may be found by the aid of a centimetre scale and a pair of dividers. Multiply it by 40, and the product will be the resultant velocity in feet per second.

The same result may be reached by drawing AB eastward to represent 120 ft. a second; BE northward, 80 ft. a second; and EF northwestward, 100 ft. a second. The fourth side AF, required to close the polygon, represents the resultant velocity both in relative direction and in feet per second.

38. Resolution of a Velocity. — Resolution is the converse of composition. The process of finding two or more velocities which will produce the same effect as a given velocity is called the *resolution of a velocity*. When only two components are required, the process consists in finding the sides of a parallelogram whose diagonal represents the given velocity. The direction of the two components is usually given.

For example, it is required to resolve a velocity of 40 mi. an hour eastward into two velocities at an angle of 60°, one of which shall be southeastward. Draw AB (Fig. 12) to represent 40 mi. an hour eastward. Draw AE at an angle of 45° with AB, and AF making an angle of 60° with AE. Complete the parallelogram $ACBD$. The required components are AC and AD, and their numerical value may be obtained by measurement; or, if one wishes to be more precise, by plane trigonometry.

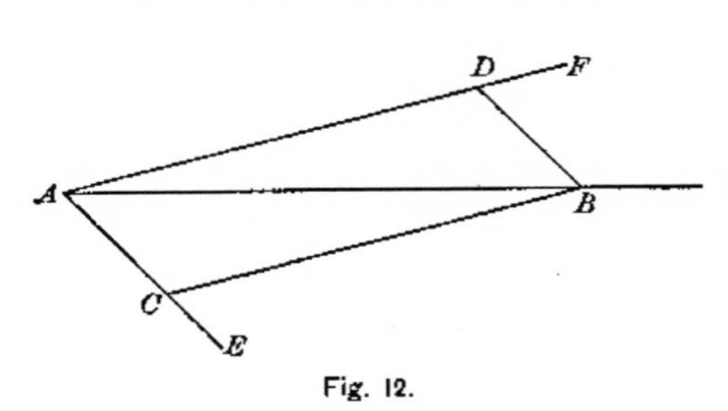

Fig. 12.

Problems.

NOTE. — Approximate answers only can be obtained by the graphic method.

See the Appendix for a method of constructing the angles used.

1. A body moving with a horizontal velocity of 20 m. per second, has a vertical velocity given to it of 30 m. per second. Find the resultant velocity.

2. What is the actual velocity of a vessel which is sailing 3 knots an hour eastward, while it is drifting with the current 2 knots an hour southeastward?

3. Two forces act simultaneously on a body at an angle of 60°, one tending to impart to the body a velocity of 20 m. a second, and the other a velocity of 30 m. a second. Find the actual velocity.

Ans. 43.6 m.

4. A ball moving with a velocity of 20 m. a second is acted on by a force at right angles to its direction of motion and capable of giving to it a velocity of 40 m. a second. Find the resultant velocity.

Ans. 44.721 m.

5. A ship is sailing at the rate of 5 mi. an hour, and a sailor climbs the mast 50 ft. high in 20 sec. Find his velocity relative to the earth.

6. A particle receives simultaneously two velocities, 100 ft. a second north, and 75 ft. a second northeast. Find the magnitude of the resultant velocity.

7. When a train is moving with a speed of 20 mi. an hour past a station platform, the conductor throws a parcel out with a horizontal velocity of 20 ft. a second in a direction at right angles to the motion of the train. What will be the velocity of the parcel at the beginning of its flight?

8. Three forces act on a body, the angle between the first two being 60°, and between the second and third 45°. The first force causes a velocity of 20 m. a second, the second 30 m. a second, and the third 40 m. a second. Find the actual velocity.

9. A body is acted on simultaneously by three forces capable of giving to it the velocities 50, 70, and 100 m. a second respectively. The

angle between the directions of the first two is 60°, and between the second and third 90°. Find the magnitude of the resultant velocity.
Ans. 110.6 m.

10. A body has four component velocities, namely: 60 ft. a second east, 70 ft. a second north, 80 ft. a second west, and 90 ft. a second south. Find the resultant velocity.

11. A train is moving southeastward with a velocity of 30 mi. an hour. How fast is it moving eastward, and how fast southward?
Ans. 21.213 mi. an hour.

12. A road is inclined to the horizon at an angle of 30°. What velocity does a carriage have vertically and horizontally when it moves up the slope at the rate of 8 mi. an hour?
Ans. 4 mi. vertically; 6.928 mi. horizontally.

13. The sail of a ship is set in such a manner that the wind strikes it at an angle of 30°. If the wind has a velocity of 15 mi. an hour, what is the value of its component velocity perpendicular to the sail?
Ans. 7.5 mi.

14. Replace a velocity of 24 m. a second by two components, making angles of 45° and 60°, respectively, with it in direction.
Ans. 21.5 m.; 17.6 m.

15. A velocity of 36 cm. a second is to be replaced by two component velocities, whose directions differ by 60°. One of these velocities is to be 24 cm. What is the other?

16. Resolve a vertical velocity of 25 cm. a second into two components at right angles, one of them to be inclined to the horizon at an angle of 60°.

17. A steamer, crossing a river flowing south at the rate of 5 mi. an hour, found that by steering at an angle of 30° N. of E. it landed at a point exactly east of the starting-point. Find the actual velocity of the steamer.

18. A stone thrown from a train has a certain velocity given to it at right angles to the direction of motion of the train, so that relative to the ground it has a velocity of 60 ft. a second in a direction making an angle of 60° with the direction of the train. Find the velocity of the train.

19. A ball moving eastward with a velocity of 10 m. a second is acted on by a force, with the effect of causing the ball to move northeastward without change of velocity. In what direction did the force act, and what velocity did it produce?

Ans. 22.5° W. of N.; 7.65 m. a second.

20. A ship is steaming due west, with a velocity of 15 mi. an hour. To a man on deck the wind appears to be blowing from the northwest, but its real direction is from the north. Find the velocity of the wind.

II. NEWTON'S LAWS OF MOTION.

39. Momentum. — We have so far considered motion in the abstract, without reference to the amount of matter moving, and without reference to the force producing the motion or change of motion.

It is plain that in any case of actual motion, there must be a definite quantity of matter moving, and the effect of a force in producing this motion will include the mass moved, as well as the speed imparted to it. It will, therefore, accord with the ordinary use of language, to speak of the "quantity of motion," and to consider it as proportional, first, to the speed of the body, and second, to the mass or quantity of matter in it (§ 9). The name given to the "quantity of motion" is *momentum.* It is the product of the mass and the velocity of a moving body.

$$\textit{Momentum} = \textit{mass} \times \textit{velocity}, \text{ or } M = mv. \qquad (5)$$

In the centimetre-gramme-second (C.G.S.) system, the unit of momentum is the momentum of 1 gm. of matter moving at the rate of 1 cm. a second.

40. Impulse. — Suppose a ball of 10 gm. mass to be fired from a rifle, with a velocity of 50,000 cm. a second. Its momentum would be 500,000 units.

If a truck weighing 500 kgm. moves at the rate of 1 cm. a second, its momentum is also 500,000 units. But the ball has acquired its momentum in a fraction of a second, while a minute or more may have been consumed in communicating the same momentum to the truck. In some sense, the propulsion required to set the ball in motion is the same as that required to give the equivalent motion to the truck, because the momenta of the two are equal. This equality is expressed by saying that the *impulse*, or the effect of the force in producing the same quantity of motion, is the same in the two cases. Since the effect produced is doubled if the value of the force is doubled, or if the time during which it acts is doubled, it follows that *impulse* is the product of the *force* and the *time* during which it acts.

In estimating the effect of a force, the time element and the magnitude of the force are equally important. The term *impulse* takes both into account.

41. Laws of Motion. — The relations of motions and changes of motion to the forces producing them are expressed in Newton's *laws of motion* They are to be regarded as physical axioms, and are incapable of rigorous experimental proof. They rest on convictions drawn from observation and experiment, and the results derived from their application in the higher fields of mechanics and astronomy are found to be invariably true.

I. *Every body continues in its state of rest or of uniform motion in a straight line, except in so far as it may be compelled, by impressed force, to change that state.*

II. *Change of motion is proportional to the impressed force, and takes place in the direction in which the force acts.*

III. *To every action there is always an equal and contrary reaction; or the mutual actions of two bodies are always equal and in opposite directions.*

By "*change of motion*" we must understand *change of momentum*, and by "impressed force," *impulse*.

42. Discussion of the First Law. — Matter has no capacity in itself to change its condition of rest or of motion. Not only this, but a body offers resistance to any such change in proportion to the *mass* of matter contained in it. This two-sided property of matter is expressed by the term *inertia*, and the first law of motion is often called the *law of inertia*.

When no force acts on a body, it persists in its state of rest or of uniform motion relative to other bodies. If at rest and left wholly to itself, it will remain at rest; if at one moment it be at rest, and afterward be found in motion, then it has been acted on by some force; or if it be in motion and its motion change either in rate or direction, then a force must have acted on it.

From this law we derive a definition of force, for the law asserts that the sole cause of change of motion is *force*.

43. Discussion of the Second Law. — The first law asserts that a change of momentum is due to force. The second law shows, first, how force may be measured. Maxwell has restated it so as to read as follows: "*The change of momentum of a body is numerically equal to the impulse which produces it, and is in the same direction.*" By a proper choice of units, impulse may be placed equal to the change of momentum which it produces, instead of proportional to it, or

$$Ft = mv. \tag{6}$$

Hence $$F = \frac{mv}{t} = m\frac{v}{t}.$$

The velocity of the mass m before the force acted is here supposed to be zero, and v is the final velocity. Force is therefore measured by the rate of change of momentum. Further, $\frac{v}{t}$ is the rate of change of velocity, or the *acceleration* a. We may then write

$$F = ma. \qquad (7)$$

Force may then be measured by the product of the mass moved and the acceleration due to the force; and acceleration is equal to the force producing it when the mass is unity.

This law teaches, further, that the change of momentum is always in the direction in which the force acts. Hence, when two or more forces act together, each produces its change of momentum independently of the others. We may therefore employ the same methods for compounding forces that we have already described for compounding velocities.

44. Units of Force. — Two systems of measuring force are in common use, viz., the *gravitational* and the *absolute*. The latter is usually in the C.G.S. system. The gravitational unit of force is the *weight* of a standard mass, as the *pound* or the *kilogramme*. These gravitational units are not strictly constant, but vary with the place on the earth's surface (§ 54). They are convenient for the work of the engineer, but are not suitable for precise measurements.

The absolute unit of force in the C.G.S. system is the *dyne*. *It is the force which, acting on a gramme mass for one second, produces a change of velocity of one centimetre a second.* This unit is invariable in value. The earth's attraction for a gramme in New York is 980 dynes, since gravity will impart to a gramme at that place a velocity of 980 cm. in one second. A dyne is therefore $\frac{1}{980}$ of the gramme as a gravitational unit of force.

Problems.

1. An 18-ton truck is moving at the rate of 30 mi. an hour; what is its momentum?

2. Compare the momentum of a 15-lb. cannon ball moving at the rate of 300 ft. per second, with that of a 3-oz. bullet which has a velocity of 420 yd. per second.

3. Two balls have equal momenta. The first weighs 100 kgm., and moves with a velocity of 20 m. a second. The other moves with a velocity of 500 m. a second. What is its mass?

Ans. 4 kgm.

4. What is the momentum of a mass of 15 kgm. moving with a velocity of 15 m. per minute?

5. With what velocity must a body whose mass is 10 gm. move to give it the same momentum as a body whose mass is one kilogramme with a velocity of 200 m. per minute?

6. Which has the greater momentum — a mass of 100 gm. moving 3 km. per hour or a mass of 10 gm. moving 150 m. in 5 sec.?

Ans. Second is greater by $21,666\frac{2}{3}$.

7. What mass must a body have whose velocity is 25 m. per hour in order to have the same momentum as a body whose mass is 25 gm. and velocity 500 cm. per second?

8. A constant force of 25 dynes acting on a body for 30 sec. gives it a speed of 1 km. per minute. Find the mass of the body.

Ans. 0.45 gm.

9. A mass of 100 gm. acquires a velocity of 30 cm. per second in 10 sec. Find the force acting on it.

10. A constant force acting on a mass of 15 gm. for 4 sec. gives it a velocity of 20 cm. per second. Compute the force.

11. What is the force that in 10 min. produces a velocity of a kilometre a minute in 100 gm.?

12. A mass of 15 gm. lying on a smooth flat table is acted on by a force of 60 dynes; what velocity will it have at the end of 2 sec.?

Ans. 8 cm. per second.

13. A force of 30 dynes acts for 12 sec. upon a body resting on a smooth horizontal plane, and imparts to it a velocity of 120 cm. per second; what is the mass of the body?

14. What velocity will a force of 100 dynes impart to a mass of 100 gm. in 100 sec.?

15. How long must a force of 20 dynes act on a mass of 20 gm. to change its velocity from 5 to 25 cm. per second?

16. What pressure per second will a man whose weight is 75 kgm. exert on the floor of an elevator which is ascending with an acceleration of 3 m. per second?

17. A mass of 20 kgm. which has been going 40 m. per second is now retarded by a constant force equal to the weight of 4 kgm. In what time will it stop it?

18. A load of 40 kgm. is being lowered by a cord from a height. Find the tension in the cord when the speed is increasing at the rate of 3 m. per second.

19. If a mass of 5 gm. be pushed by a force of 5 dynes without friction for 5 min., how much will the momentum of the mass be altered?

20. A certain force acting on a mass of 10 gm. for 5 sec. produces in it a velocity of 100 cm. per second. What velocity would this force impart to a mass of 50 gm. in 10 sec.?

45. How a Force is Measured. — The simplest device for measuring a force is the *dynamometer*. It consists of a coiled spring, to the bottom of which is attached a pointer, adapted to move in front of a graduated scale (Fig. 13). According to Hooke's law (§ 15) the division of the scale is in equal parts; it may be made to read in pounds, grammes, or dynes. The common drawscale is a dynamometer graduated in pounds and fractions of a pound.

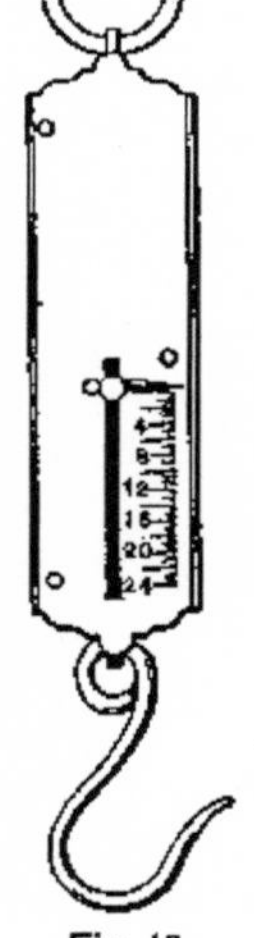
Fig. 13.

46. Graphic Representation of a Force. — A force is completely specified by its three elements: (*a*) *its point of application*; (*b*) *its direction*; and (*c*) *its magnitude.* These three particulars may be represented by a straight line drawn through the point where the force acts, in the direction in which the force changes the momentum of the body, and as many units in length as there are units of force. If a line 1 cm. long stands for a force of 1 dyne, a force of 10 dynes will be represented by a line 10 cm. long.

47. Composition of Forces acting at a Point. — The single force which will produce the same effect on a body as two or more forces acting together is called the *resultant.* A single force, *equal* and *opposite* to this resultant, is called the *equilibrant* of the forces. If the several forces and their equilibrant act together on a body, it will remain at rest. *Concurrent* forces are those acting at a point.

The following are the more important cases of the *composition of forces*: —

First. — *If two concurrent forces act on a body at an angle, the concurrent diagonal of the parallelogram constructed on the two lines representing the concurrent forces will represent their resultant.*

This is the principle of the *parallelogram of forces.* It may be verified as follows: —

Experiment. — Tie together three stout cords at *D* (Fig. 14) and fasten the free ends to the hooks of three drawscales, *A*, *B*, *C*, respectively. The drawscales may be graduated to read in grammes. Pass their rings over screws set in the edge of the blackboard or table at such distances apart that the drawscales will be stretched. Record the readings on the scales and mark the position of the knot *D* and the

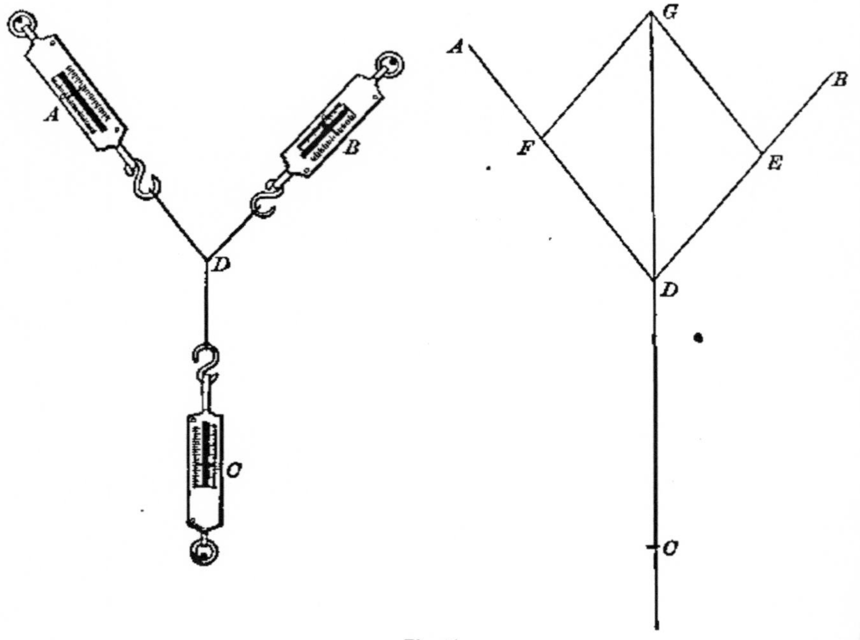

Fig. 14.

points of attachment of the cords to the hooks of the drawscales. Then remove the drawscales and draw the lines marking the positions of the cords.

Let the readings of *A*, *B*, and *C* be 100 gm., 90 gm., and 150 gm., respectively. Lay off on *DA*, *DB*, and *DC* the distances 100, 90, and 150, respectively, on some convenient scale. Complete the parallelogram *DFGE*, draw the diagonal *DG*, and find its value on the scale used in laying off the sides. If the work be carefully done, it will be found that *DG* and *DC* are in the same straight line and are of equal length. The single force *DG* is the resultant of the forces *DF* and *DE*, since it is equal to their equilibrant *DC*.

SECOND. — *If more than two concurrent forces act on a body, their resultant may be found by finding the resultant of any two of them, then of this resultant and a third force, and so on till all the forces have been included. The last resultant will be the one required.*

This principle may be verified by using four or more drawscales and proceeding as in the last experiment.

THIRD. — *The resultant of two parallel forces in the same direction is their sum, and its point of application divides the line joining the points of application of the two components into two parts inversely as the magnitudes of the forces.*

Experiment. — Suspend two drawscales A and B (Fig. 15) by cords from suitable supports in such a way that the two suspending cords shall be parallel. Let the graduated bar, from which the known weight W is suspended, be supported by the drawscales as shown. All the supporting cords will then be parallel. Read the two drawscales and the distances CE and ED. Change the position of the point of suspension E and read again. Then for each set of observations the weight W should equal the sum of the readings on the drawscales A and B. Moreover, in each case we should have

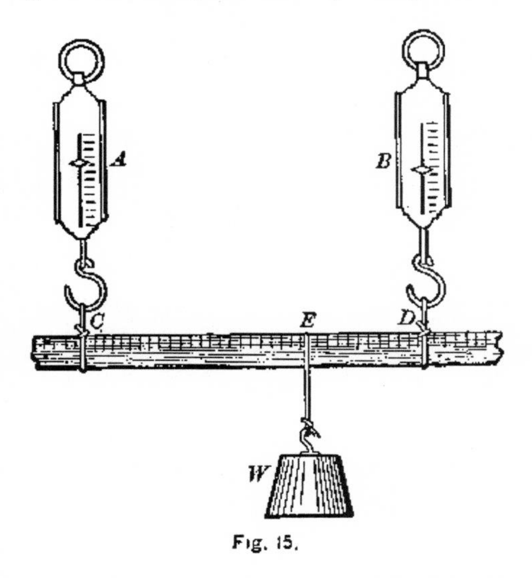

Fig. 15.

$$A : B :: ED : CE.$$

FOURTH. — *If two equal parallel forces act on a body at different points and in opposite directions, they produce a motion of rotation; they cannot be replaced by a single force.*

Such a pair of forces is known as a *couple*. A magnetic needle, suspended so as to swing freely in a horizontal plane, is acted on by a couple when it is displaced from a north-and-south position of equilibrium. One end of the needle is attracted toward the north and the other end toward the south with equal and parallel forces. The effect is to rotate the needle about its vertical axis of suspension till it returns to its north-and-south position.

48. Resolution of a Force.—A force, like a velocity, may be resolved into components in any given directions, by the same method as that given for velocities in Article 38. The most common case is to resolve a force into two components along directions at right angles to each other.

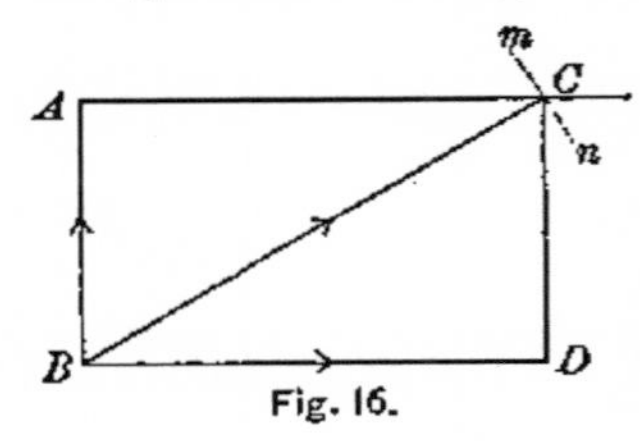

Fig. 16.

To illustrate: Let it be required to resolve a force equal to the weight of 32 lb. into two rectangular components, one of them to be 12 lb. The problem is to construct a rectangle with a diagonal of 32 and one side 12, and to find the other side.

Draw *AB* (Fig. 16) 12 units in length, and at *A* draw *AC* perpendicular to *AB*. With *B* as a centre, and with a radius of 32 units, draw the arc *mn* cutting *AC* at *C*. Complete the rectangle *ABDC*. Then *BA* and *BD* are the two components of *BC*, and *BD* is the one required. Its value may be found by a scale and a pair of dividers, or it may be calculated from the right triangle *BDC*.

$$BD = \sqrt{\overline{32}^2 - \overline{12}^2} = 29.66.$$

As a second illustration, let a body *W*, whose weight is 20 lb., be supported on the smooth inclined surface *AB*

(Fig. 17). What force parallel to the plane will be necessary to maintain it at rest? Draw *WH* perpendicular to *AB*, and let the vertical line *WF* represent the weight of 20 lb. Draw *WE* parallel to *AB*, and complete the rectangle *WHFE*. Then *WH* and *WE* are components of *WF*. *WH* represents the pressure on *AB*, and *WE* the force urging the weight down the inclined plane. To maintain *W* at rest, it will be necessary to apply to it a force *WK* equal and opposite to *WE*.

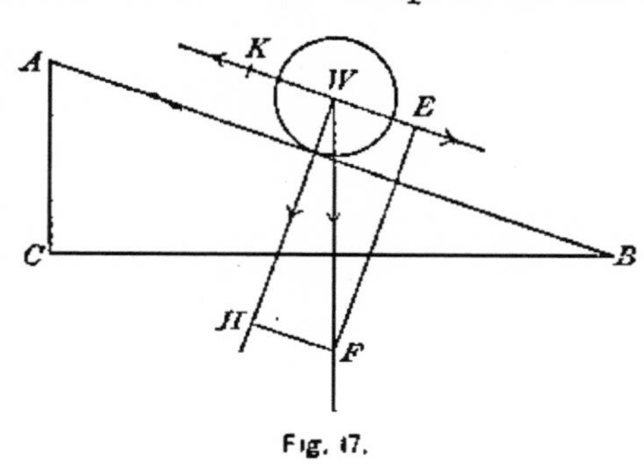

Fig. 17.

49. Discussion of the Third Law.—The meaning of this law of *reaction* is that force is always due to the *mutual action* between two bodies or parts of bodies. This mutual action is called a *stress*, and stress is always a two-sided phenomenon. It includes both *action* and *reaction*, and is called the one or the other according as the attention is directed to one aspect of it or the other; just as trade includes both sale and purchase, and is designated by the one term or the other according to the side of the trade which one considers.

The stress in a stretched elastic cord *pulls* the two things to which it is attached equally in opposite directions; the stress in a piece of compressed india rubber exerts an equal *push* both ways. The former is called a *tension* and the latter a *pressure*. Now, every force is one of a pair of equal and opposite forces composing a stress. Imagine a rope supporting a weight. The tension in the rope is a stress tending to part it by pulling adjacent portions in opposite directions. The same is true if two men are pulling at the two ends of the rope.

The same conditions of action and reaction exist when the medium is invisible. A stone attracts the earth with the same force that the earth exerts on the stone. The action between a magnet and a piece of iron is a mutual one, the magnet attracting the iron and the iron the magnet with precisely the same force.

Suspend two elastic ivory or celluloid balls by long threads so that the balls just touch. Draw one aside in the plane of the threads and let it fall so as to strike the other one. It will be brought to rest, while the second ball takes the motion. The impact is a stress. The *action* of the first ball on the second sets the latter in motion, while the *reaction* of the second on the first brings the first to rest.

The third law of motion asserts that the two phases of a stress, the action and the reaction, are always equal to each other and in opposite directions.

Problems.

1. Two forces, 60 and 90 gm., act at an angle of 90°. Find their resultant.

2. Two forces, 5 and 8 gm., act at an angle of 60°. Find the resultant.

3. The angle between two forces is 45°. Find the resultant if one force is 18 and the other 24 dynes.

4. A boat is pulled by a force of 30 kgm. toward the south, and by a force of 20 kgm. toward the northwest. What other force is acting upon it if it remains at rest?

5. Forces of 20, 30, and 40 dynes act on a particle ; the angle between the directions of the first two is 45°, and that between the directions of the last two is 60°. Find the magnitude of the resultant,

6. The angles between three forces of 25, 30, and 40 kgm., respectively, are 120°. Find the resultant.

7. Resolve a force of 25 dynes into two components at right angles, one of the forces to be 10 dynes. Find the other.

Ans. 22.91 dynes.

8. Two forces act at an angle of 60°. Their resultant is 40 kgm., and one of the forces is 25 kgm. Find the other force.

Ans. 21.14 kgm.

9. A weight of 25 kgm. is suspended by two strings, inclined to the vertical at 30° and 60°. Find the tension of each string.

Ans. 21.65 kgm.; 12.5 kgm.

10. Find the point of application of the resultant of two forces, 7 and 13 gm., acting in parallel lines 40 cm. apart and in the same direction.

11. A bar 8 m. long is supported in a horizontal position by props placed at the extremities. Where must a weight of 100 kgm. be hung so that the pressure on one of the props shall be 20 kgm.?

Ans. 1.6 m. from end.

12. Find the point of application of the resultant of two forces acting along parallel lines 1 m. apart, so that one may be four times the other and their resultant 120 kgm.

13. A uniform ladder 6 m. long, weighing 30 kgm., is supported horizontally by two men at distances of 1 and 1.5 m. respectively from its ends. Find the weight borne by each man.

Ans. 12.86 kgm.; 17.14 kgm.

14. Two parallel forces of 20 and 30 dynes have the point of application of their resultant distant 40 cm. from the larger force. Find the distance between the points of application of the forces.

Ans. 100 cm.

15. Resolve a force of 50 dynes into two parallel components distant 10 and 15 cm. respectively from the given force.

Ans. 30 and 20 dynes.

16. Two men carry a weight of 100 kgm. on a pole 3 m. long. Where must the weight be placed so that for every 2 kgm. carried by the first man the second man shall carry 3 kgm.?

17. Three forces acting in the same direction have their points of application in a straight line and distant apart 20 and 30 cm. If the value of these forces be 10, 15, and 20 dynes, find the magnitude and point of application of the resultant.

Ans. 45 dynes, and $21\frac{1}{9}$ cm. from 20 dynes.

18. Weights of 3, 5, 7, and 9 kgm. are suspended from a light rigid rod 3 m. long, at points equally distant from each other. Find where a force must be applied to the rod to support it.

III. GRAVITATION.

50. The Fall of Bodies. — The early philosophers thought that light bodies fall more slowly than heavy ones. Galileo was the first to find out the truth experimentally by dropping various bodies together from the top of the leaning tower of Pisa. He found that they fell to the ground in nearly the same time, whatever their size or weight. The lighter bodies fell slightly slower than the heavier ones, and the difference he rightly ascribed to the resistance of the air.

Fig. 18.

The "guinea and feather tube," devised since the invention of the air-pump, shows that all bodies fall toward the earth with the same acceleration.

Experiment. — Place a coin and a feather, or a pith ball and a shot, in a long tube (Fig. 18) closed at one end and fitted with a stopcock at the other. Hold the tube in a vertical position and suddenly invert it; the coin or the shot falls to the bottom first. Now exhaust the air as thoroughly as possible. When the tube is then suddenly inverted the two objects fall to the lower end of the tube in the same time.

The inference from this experiment is that all bodies would fall from rest through the same height in the same time if the resistance of the air were wholly removed. In air, the body which has the larger surface in proportion to its mass falls the more slowly because it meets with more resistance in falling.

The resistance offered by the air to bodies falling through it is illustrated by its effect on a small stream of water flowing over a high precipice. It breaks up the stream into a fine spray as it descends. In a vacuum water falls like a solid. The *water hammer* (Fig. 19) is an instrument devised to illustrate this fact. It consists of a heavy glass tube half filled with water, the air having been expelled by boiling the water for some time just before sealing the upper end. When it is suddenly inverted, the water falls like a solid, with a metallic ring.

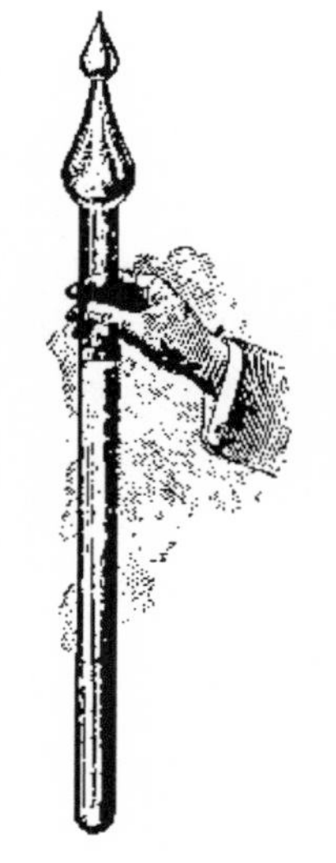

Fig. 19.

51. Weight. — Since all bodies fall in a vacuum with the same acceleration, and so traverse the same distance in the same time when starting from rest, the forces acting on them, due to the earth's attraction and called *gravity*, are proportional to their masses. This force of gravity is called *weight*. The proportionality of mass and weight was first demonstrated by Sir Isaac Newton. We have already had the relation $F = ma$ in equation (7). Plainly, if the acceleration a is the same for all masses, the force F

is proportional to the mass m. In the case of gravity the particular force is the *weight*, denoted by W, and the particular acceleration is the *acceleration of gravity*, denoted by g. Making these substitutions in equation (7), we have

$$W = mg. \tag{8}$$

52. Direction of Gravity. — The path described by a falling body is a *vertical line*. A line or plane perpendicular to it is said to be *horizontal*. The direction of a vertical line at any point may be determined by suspending a weight by a cord passing through the point. The weight and cord are called a *plumb-line*. The direction of the plumb-line is perpendicular to the surface of still water. Vertical lines converge toward the earth's centre; but vertical lines drawn through neighboring points may be considered parallel without sensible error. Vertical lines 100 ft. apart make an angle with each other of approximately one second of arc. At the poles of the earth and at the equator, the direction of gravity is that of the plumb-line; elsewhere there is a slight variation on account of the rotation of the earth on its axis (§ 66).

53. Law of Universal Gravitation. — Toward the end of the seventeenth century Sir Isaac Newton discovered the law of *universal gravitation*. He derived this great generalization from a study of the results obtained by two eminent astronomers, Copernicus and Kepler. The law may be expressed as follows: —

Every particle of matter in the physical universe attracts every other particle with a force whose direction is that of the line joining the two particles, and whose magnitude is directly as the product of the two masses, and inversely as the square of the distance between them.

This law expressed in symbols is

$$F = G\frac{mm'}{d^2}, \tag{9}$$

where m and m' are the masses of the particles, d is the distance between them, and G is a proportionality factor or constant of gravitation to be determined by experiment. When applied to bodies like the earth, whose dimensions are large compared with that of any body on its surface, Newton proved that the distance involved in the law of gravitation is the distance from the body to the earth's centre; for any spherical mass like the earth or the sun attracts another body exactly as if all the matter it contains were concentrated at its centre.

54. Law of Weight. — Since the earth is flattened at the poles, it follows from the law of gravitation that the intensity of gravity, and therefore the weight of a body, increase, in going from the equator toward the poles. If the earth were a stationary sphere, the value of g would be the same all over its surface. It would then vary only on ascending above the surface or descending below it; as the inverse square of the distance on ascending, and simply as the direct distance on descending, assuming the density (§ 140) uniform. The value of g at the equator in C.G.S. units is 978.1 and at the poles 983.1. At New York it is a little over 980 cm. per second per second, or 32.15 ft. per second per second.

55. Centre of Gravity. — A solid body is composed of particles all acted on by gravity with a force equal to the product of their mass and the intensity of gravity g (§ 51). For bodies of ordinary size these forces are all parallel.

Hence the weight of a book or of a quoit disk, for example, is the resultant of an infinite number of parallel forces of gravity. The point of application of all these parallel forces, *however the body be turned about*, is called the *centre of gravity* of the body. So long as the forces of gravity on the parts are strictly parallel, the centre of gravity coincides with what is known as the *centre of mass* and the *centre of inertia*. It is the point in a body about which the mass is evenly disposed. In a uniform sphere, it is evidently its centre. In a uniform ring, it is not in the body of the ring at all, but at its centre.

56. Equilibrium. — A body is in equilibrium when the resultant of all the forces acting on it is zero. If the resultant force is zero, the acceleration is evidently zero. Equilibrium does not mean that the velocity is zero, but that the acceleration is zero. Rest means zero velocity; equilibrium, zero acceleration.

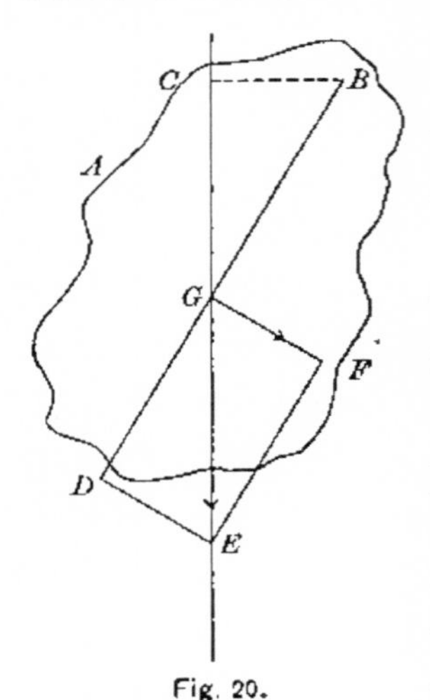

Fig. 20.

In the case of a body constrained to turn about a horizontal axis, it can be in equilibrium only when the vertical line through its centre of gravity passes through the axis. Let A be a body, whose centre of gravity is at G, and let the axis supporting it pass through B (Fig. 20). Represent its weight by GE through the centre of gravity. When the line GE does not pass through B, it can be resolved into rectangular components GD and GF, the former producing pressure on the axis, and the latter causing the centre of gravity to move to the right. As the body swings, shortening the line BC, the point G

approaches the vertical through *B*, and *GF* approaches zero. When *G* is on the vertical through *B*, the resultant force and the acceleration become zero, and the body is in equilibrium. The only forces then are the weight acting downward through *G*, and the equal and opposite reaction of the axis upward and along the same vertical line.

When a body rests in equilibrium on a plane, the vertical line through its centre of gravity falls within its base of support. If this vertical line falls outside the base, the body will overturn, for the supporting force, consisting of the reaction of the plane, does not act opposite to the weight, but parallel to it, forming a couple. If a body, like a chair, is supported on legs, the base is the polygon formed by lines connecting the points of support.

57. Three Kinds of Equilibrium. — **Experiment.** — Fill a round-bottomed Florence flask one-quarter full of shot, crowding paper into the remaining space to keep the shot in place (Fig. 21). Tip the flask over; after a few oscillations it will return to the upright position. If the experiment be repeated with a similar flask empty, it will be impossible to find any other position of equilibrium for the flask than on its side. We may then roll it about, and it will remain in any position in which the top and bottom rest on the supporting plane.

Fig. 21.

This experiment illustrates the three kinds of equilibrium of position. (1) The centre of gravity of the loaded flask is nearer the supporting plane than that of the empty one. (2) In overturning the loaded flask, its centre of gravity is raised, and at the same time the vertical line through it is thrown outside the point of support, so that the reaction of the plane upward, and the weight of the flask and contents downward, form a couple which returns the flask to its upright position. In overturning the empty flask, its centre of gravity is lowered, and the vertical line through it falls between the two points of contact with the plane. (3) When the empty flask is rolled about on its side, its centre of gravity is neither raised nor lowered. We have thus three kinds of equilibrium of position: *stable*, in which the centre of gravity rises when the body is moved; *unstable*, in which the centre of gravity falls when the body is moved; and *neutral*, when the height of the centre of gravity is not changed by the movement of the body.

58. Illustrations.—The rocking-horse and the rocking-chair are familiar examples of stable equilibrium. The half of a split ball, or any segment of a sphere, will rock in stable equilibrium on its rounded side. An egg lying on its side has neutral equilibrium for rolling, and stable equilibrium for rocking; it is unstable in every direction when balanced on either end. In the case of a body pivoted at a point, it is stable when the centre of gravity is below the point, and unstable when it is above. With the balls in the position shown in Fig. 22, the movable system is in stable equilibrium; but if the balls are raised above the level of the pivot, the equilibrium

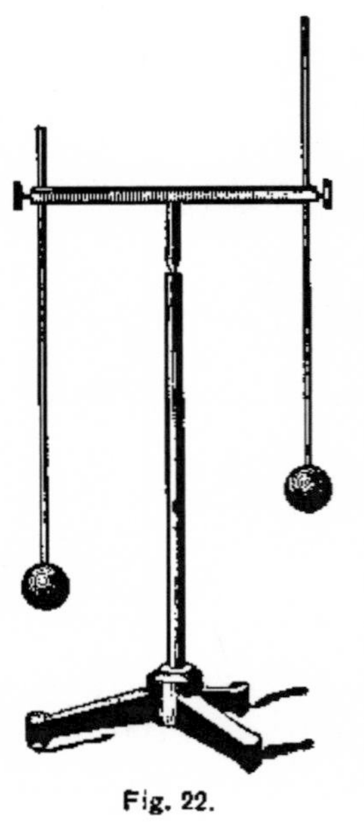

Fig. 22.

becomes unstable. The nearer the centre of gravity of a beam balance is to the point of support, the smaller is its stability, and the greater its sensitiveness; but its centre of gravity must be slightly below its support, or it will be unstable and topple over (§ 95).

59. Stability. — The most useful measure of *stability* is the work required to upset a body (§ 78); that is, it is the product of its weight and the difference between the

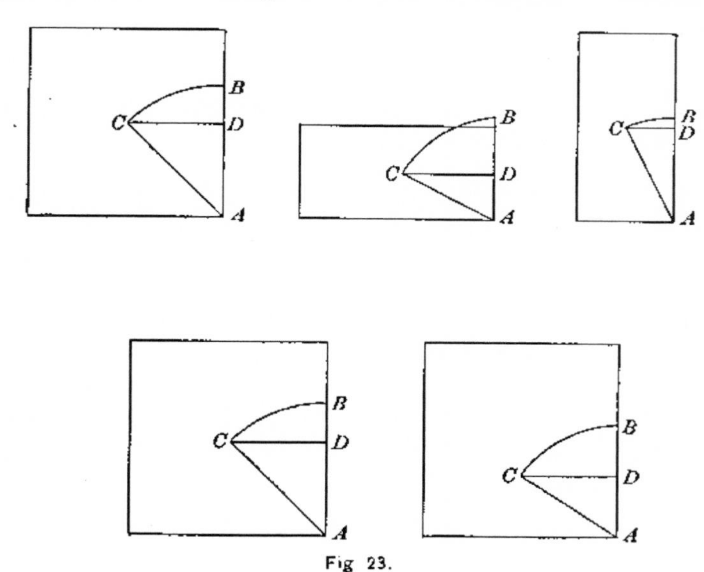

Fig 23.

distances AC and AD in Fig. 23. An inspection of these diagrams shows that the stability is increased by lowering the centre of gravity, and by enlarging the base. C denotes the centre of gravity or centre of mass, A the point about which the body is turned, and BD the height through which the weight is lifted, to bring the body to the position of unstable equilibrium. If the masses are equal, BD is a measure of the stability.

A brick has less stability when standing on end on a

table, than when lying on edge; and it has less stability on edge than when lying on its broad side. In the first position, the centre of gravity is the highest, and the base the smallest; in the last position, the centre of gravity is the lowest, and the base the largest.

Experiment. — Make a disk or short cylinder of wood, and load it with lead on one side, near the circumference (Fig. 24). Find by trial the position G of the centre of gravity on one end of the cyl-

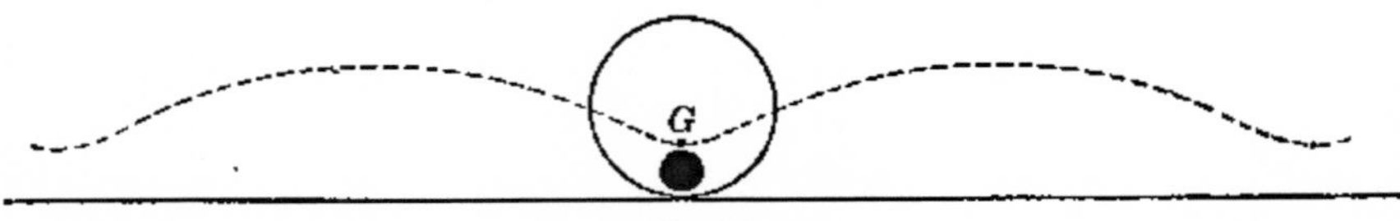

Fig. 24.

inder, and mark it. This cylinder may be placed on a slightly inclined plane, with the centre of gravity near the highest point, in such a position that the cylinder will roll up the plane into a position of stable equilibrium. Roll the cylinder along a horizontal plane, and watch the path described by its centre of gravity. It will be a curve, with crests and hollows, similar to the one in the figure; every hollow corresponds to a position of stable equilibrium, with the centre of gravity in its lowest position, and every crest is a position of unstable equilibrium. If the path were a straight line, the equilibrium would be neutral.

Problems.

1. Explain why a ball rolls down hill, but a cubical-shaped block will not.

2. Why does a person lean forward in climbing a hill?

3. Why is a pyramidal-shaped structure very stable?

4. Why can an old man walk better with the aid of a cane?

5. If the attraction of the apple for the earth is equal that of the earth for the apple, why does not the earth meet the falling apple halfway?

6. How far below the surface of the earth will a 10-lb. ball weigh only 4 lb., considering the earth's radius as 4000 mi.?

7. A body at the earth's surface weighs 900 lb. What would it weigh 8000 mi. above the surface, the earth's radius being taken as 4000 mi.?

8. How far below the surface of the earth must an avoirdupois pound weight be placed to weigh an ounce, the earth's radius being 4000 mi.?

9. What would a body weighing 550 lb. on the surface of the earth weigh 3000 mi. below the surface, the radius being 4000 mi.?

Ans. 137.5 lb.

10. If a body weighs 20 kgm. with a spring balance at the equator, where $g = 978.1$ cm., what will it weigh at the pole, where $g = 983.11$ cm.?

11. A mass weighed in a spring balance at the foot of a mountain appears to be 1000 lb. Carried to the top, the weight seems to be half a pound less. What is the height of the mountain, the earth's radius being taken as 3950 mi.?

12. If a body falls 16 ft. in a second at the surface of the earth (radius = 4000 mi.), how far would it fall in one second at the distance of 240,000 mi. from the earth's centre?

13. The mass of the sun is 332,000 times that of the earth, and the radius of the sun is 433,250 mi. Find how far a body will fall in one second on the sun, if the earth's radius is 3950 mi. and a body falls 16 ft. in one second on its surface.

14. If a body weighs 100 lb. on the earth, how much will it weigh on the sun, the sun's mass being 332,000 times that of the earth and its radius 109.5 times that of the earth?

IV. LAWS OF FALLING BODIES.

60. Uniform Acceleration applied to Falling Bodies. — Since the acceleration g, due to gravity, is constant for small distances above the earth's surface, the formulæ already obtained for uniformly accelerated motion may be directly applied to falling bodies. The relations between velocity, time, space, and acceleration are expressed by the equations, $v = at$ and $s = \frac{1}{2} at^2$. Substituting g for a, we have

$$v = gt, \tag{10}$$

and

$$s = \frac{1}{2} gt^2. \tag{11}$$

If in equation (11) t is made one second, then $s = \frac{1}{2} g$; or, the space described in the first second, when the body falls from rest, is half the value of the acceleration of gravity. A body falls 490 cm. the first second; the velocity attained in one second, and the acceleration, are 980 cm.

To find the space passed over in any one second, find the space described in t seconds and in $(t - 1)$ seconds, and subtract the latter from the former. Denoting the distance sought by s',

$$s' = \frac{1}{2} gt^2 - \frac{1}{2} g(t - 1)^2 = \frac{1}{2} g(2t - 1). \tag{12}$$

The distance passed over in any second is equal to half the product of g and one less than double the number of the second. By combining equations (10) and (11) we have

$$v^2 = 2 gs. \tag{13}$$

61. Laws. — The laws embodied in the preceding formulæ may be expressed as follows: —

I. *The velocity attained by a falling body is proportional to the time of falling.*

II. *The space described is proportional to the square of the time.*

III. *The acceleration is twice the space through which a heavy body falls in the first second.*

62. Experimental Proof. — The laws of falling bodies were first verified experimentally by Galileo. His method consisted in rolling a ball down a smooth inclined plane, thus reducing the acceleration by making a part only of the force of gravity effective in producing motion (§ 48). It then becomes comparatively easy to measure the distances described in successive seconds.

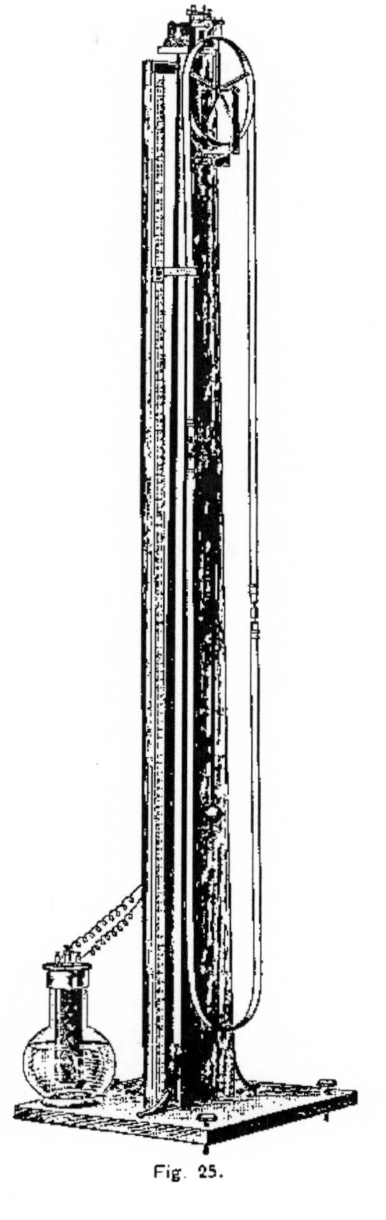
Fig. 25.

In the Hawkes-Atwood machine (Fig. 25) a different method is employed. Two unequal weights are suspended by a flexible ribbon of tissue paper over a very light wheel. The wheel, shown on a larger scale in Fig. 26, has a wood rim and aluminum spokes. It must be perfectly balanced, and must run as nearly without friction as possible on agate bearings. At the top of the supporting column is an electromagnet, carrying a locking device and a small inked camel's-hair brush as a marker. The current through this magnet is controlled by a seconds pendulum, which closes the electric circuit by sweeping through a drop of mercury on a small metallic shelf at the bottom of its swing. At the first swing of the pendulum a mark is made on the ribbon, the wheel is unlocked, and it starts to rotate because of the excess weight on one side. The paper ribbon is made endless, so that the excess weight shall remain constant. An electric brake enables the operator to stop the wheel at any desired point.

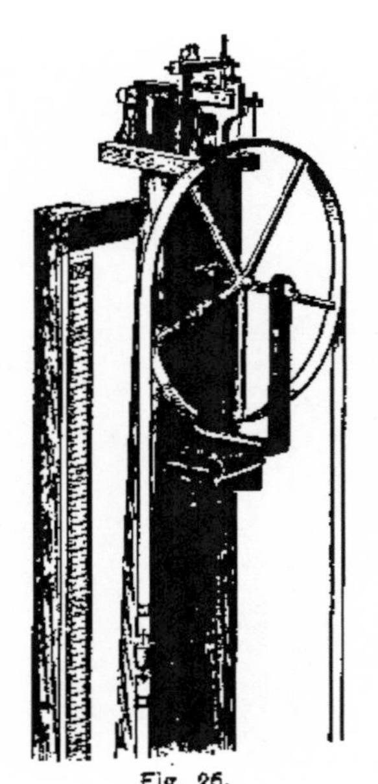

Fig. 26.

If the two suspended masses are a and b grammes, then the force effective in producing motion of the system is the force of gravity on $(a - b)$ grammes, and this force must overcome the inertia of a mass of $(a + b)$ grammes, neglecting the wheel and paper. Hence the effective acceleration is reduced from g to $\frac{a - b}{a + b}g$.

The distances described in 1, 2, 3, 4, etc., seconds may be found by measuring the distances from the first mark on the ribbon to the others in succession. They will be found to be as the numbers 1, 4, 9, 16, etc. The distances described in successive seconds, the first, second, third, etc., are the distances between successive marks on the recording ribbon. They are as the odd numbers 1, 3, 5, 7, etc. It is not necessary that the pendulum beat seconds. Any other interval will serve as well for the verification of the laws of falling bodies. The distances on the ribbon may be measured without removing it from the wheel, by using the slider and scale on the column.

In the table, the second column gives the actual distances s read from the ribbon; s' stands for the successive differences of s in the second column; v, the quantities under

s' diminished by 2.91; and a, the successive differences of v, or the acceleration. From the formula

$$s' = \tfrac{1}{2}\,g(2t - 1) = gt - \tfrac{1}{2}\,g,$$

equation (12), it will be evident that column four denotes the accumulated velocity gt at the end of the successive time intervals.

t	s	s'	v	a	s-Ratios	s'-Ratios	a-Ratios
1	2.91	2.91	5.81	5.81	1.000	1.000	1.000
2	11.63	8.72	11.64	5.83	3.997	2.997	1.003
3	26.18	14.55	17.47	5.83	8.997	5.000	1.003
4	46.56	20.38	23.29	5.82	16.000	7.003	1.002
5	72.76	26.20	29.11	5.82	25.004	9.004	1.002
6	104.78	32.02	34.92	5.81	36.007	11.003	1.000
7	142.61	37.83			49.007	13.000	

63. Projection Upward. — When a heavy body is thrown vertically upward, the acceleration is negative, and its velocity is diminished each second by g units (980 cm. or 32.15 ft.). Hence, the time of ascent to the highest point will be the time taken to bring the body to rest. If the velocity of projection upward is v, then we have, from equation (10), neglecting atmospheric resistance,

$$t = \frac{v}{g}. \qquad (14)$$

If the velocity lost is g units a second, the time required to lose v units of velocity will be the quotient of v divided by g. If, for example, the velocity of projection upward is 1470 cm. a second, the time of ascent, if there were no resistance of the air, would be $\frac{1470}{980}$, or 1.5 sec. The time of ascent is, therefore, the time of descent again to the starting-point; and the body will return to the starting-point with a velocity equal to its velocity of projection in the opposite direction.

V. CURVILINEAR MOTION.

64. Uniform Circular Motion. — Hitherto we have dealt only with the change in the *magnitude* of the velocity of a body. But force may also change its *direction*. The motion will then be *curvilinear*. Let us consider the simplest case in which a force acts to produce only *curvature* in the path of a body without affecting the speed or rate of motion.

Imagine a body moving round and round in a circle with constant speed. A heavy ball attached to a string and whirled around the hand has such a motion. The velocity along the circle is constant, but its direction is constantly changing, and there is a constant pull on the string. If this pull should cease, the body would instantly move on in the direction of the tangent line, by the first law of motion. The constant force which incessantly deflects the body from a rectilinear path, and compels it to move in a circle, is called the *centripetal force*. It is always applied at right angles to the motion of the body, and therefore cannot change its velocity along the circle.

65. Centripetal and Centrifugal Force. — The tension in the string which restrains the whirling stone is of course a stress. One aspect of it is the *action* of the hand or other central body on the revolving ball; the other is the *reaction* of the ball on the hand or central body. The former is the *centripetal force*, or force toward the centre; the latter is the *centrifugal force*, or force away from the centre. The two are necessarily equal. Centrifugal force is the resistance which the body offers, on account of its inertia, to deflection from a straight path.

66. Centripetal Acceleration and Centripetal Force.—Since the rate of deflection of the revolving body from a straight line in uniform circular motion is constant, the acceleration is constant; and it is directed toward the centre of the circle at every point because there is no change in speed along the circle. If the acceleration were not toward the centre, it could be resolved into two components, one toward the centre and the other along the tangent to the circle; the latter would change the velocity in the circle. But the velocity is uniform, and therefore there is no tangential component, or the acceleration is wholly toward the centre. Uniform circular motion, then, is compounded of a uniform motion along the tangent to the circle and a uniformly accelerated motion along the radius. The problem is to find the value of this acceleration toward the centre and the force producing it.

Let ABC (Fig. 27) be the circle in which the body revolves, and AB the minute portion of the circular path described in a very small interval of time t, with the constant velocity v. Denote the length of the arc AB by s. Then, since the motion along the arc is uniform, $s = vt$. Complete the figure as shown. AB is the diagonal of a very small parallelogram with sides AD and AE. The latter is the distance through which the revolving body is deflected toward the centre while traversing the *very small arc* AB. Since the acceleration is constant, $AE = \frac{1}{2}at^2$,

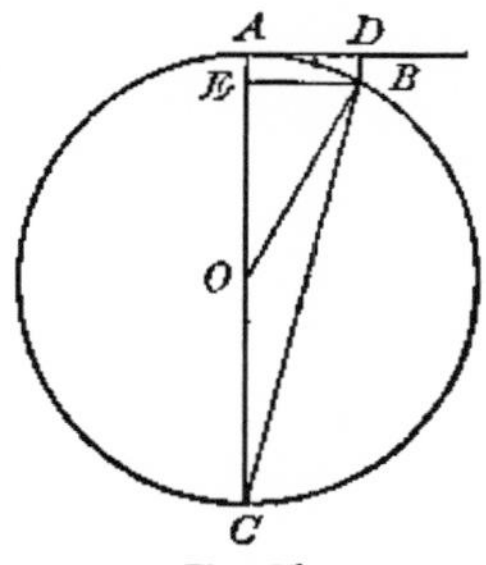

Fig. 27.

where a is the centripetal acceleration required. The two triangles ABE and ABC are right-angled and similar. Hence the proportion,

$$AE : AB :: AB : AC, \text{ or } \overline{AB}^2 = AE \cdot AC.$$

Calling the radius of the circle r and substituting for AB, AE, and AC their values,

$$v^2t^2 = \tfrac{1}{2}\,at^2 \times 2\,r = at^2r.$$

Then
$$a = \frac{v^2}{r}, \tag{15}$$

or the centripetal *acceleration* is equal to the square of the velocity along the circle divided by the radius.

Since $F = ma$, equation (7), we have here for the centripetal *force*,

$$F = \frac{mv^2}{r}. \tag{16}$$

If m is in grammes and v and r in centimetres, F is expressed in dynes. In the gravitational system of units, the weight w must be divided by g to get the mass (equation 8). The result will then be in pounds or grammes, according to the units employed. For example, if a body weighing 10 lb. move in a circle of 5-ft. radius with a speed of 20 ft. a second, then $F = \dfrac{10 \times 20^2}{5 \times 32.2} = 24.85$ lb.

67. Illustrations of Centrifugal Force. — The occasional bursting of grindstones and balance wheels when run at high speed illustrates the increase of centrifugal force with velocity. When the centripetal force becomes insufficient to hold the body to the centre, it flies off along a tangent line. A stone is thrown by whirling in a sling and finally releasing one of the strings. The overturning of a carriage in rounding a corner at too high speed is an instance of centrifugal force. Drying machines are now made on the centrifugal principle. Instead of drying clothes in a laundry by first squeezing them in a wringer and then letting the moisture evaporate in the air, the modern way is to place the wet clothes in a large cylinder with openings in the sides, and then to whirl it rapidly. Centrifugal force causes the water to pass out through the holes in the cylinder.

In the process of refining sugar, the sugar crystals are separated from the syrup by centrifugal machines. Honey is extracted from the comb in a similar way. New milk is an emulsion of fat and a heavier liquid. If allowed to stand, the little fat globules rise slowly to the top and form the cream. But if the fresh milk be whirled in a dairy separator, the cream and milk will form distinct layers, and they may then be collected in separate chambers. In a Watt's steam engine governor, the balls open out by centrifugal force when the speed increases, and this motion is made to control the supply of steam.

The figure of the earth is an oblate spheroid, flattened at the poles. This flattening was doubtless caused by the centrifugal force of rotation when the earth was still in a liquid or at least a plastic condition, before cooling down to its present state.

68. The Simple Pendulum. — A small lead ball, suspended by a long silk string without sensible mass, represents a *simple pendulum.* The length of the pendulum may be taken as the distance from the point of support to the centre of the ball. When at rest the string hangs vertically like a plumb-line; but if the ball be pulled to one side and let go, it will swing to and fro, or oscillate, about its position of rest, each swing being "damped" a little by the air. Its excursions on either side become gradually smaller; but if the time of the swing be carefully noted, it will be found to remain unchanged, if the arc described be small. This feature of pendular motion first attracted the attention of Galileo, who observed it in the oscillations of a "lamp" or bronze chandelier, suspended by a long rope from the roof of the cathedral in Pisa. This "lamp" may still be seen in the same place. Galileo noticed the even time of the oscillations as the path of the swinging chandelier became shorter and shorter. Such a motion, which continues to repeat itself in equal time-intervals, is said to be *periodic*.

69. Pendular Motion Explained. — *A* (Fig. 28) is the point of suspension, and *AN* is the length of the simple pendulum. Let the ball be drawn aside to the position *B*, and released. Let *BG* represent the *weight* of the ball. Resolve this force into *BD* in the direction of the string, and *BC* at right angles to it, or tangent to the arc *BNE*. Only the latter component is effective in producing motion of the ball toward *N*. As the ball approaches *N*, the component *BC* becomes smaller and vanishes at *N*. (The student should convince himself of this by drawing a figure with *B* nearer *N*.) In falling from *B* to *N*, the ball moves in the arc of a circle under the influence of a force which is greatest at *B* and vanishes at *N*. The motion is,

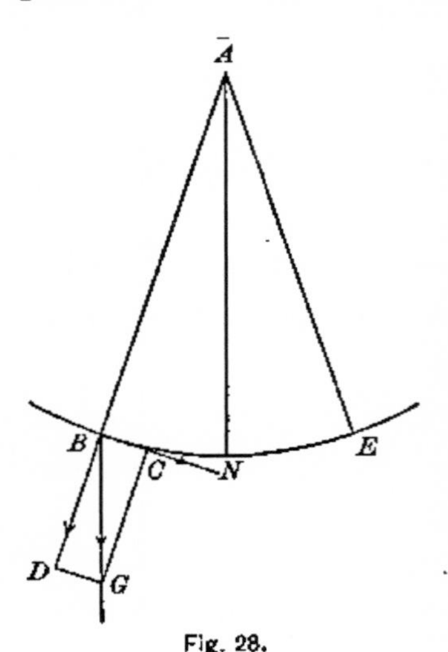

Fig. 28.

therefore, accelerated all the way from B to N, but the acceleration is not uniform. The velocity increases continuously from B to N, but at a decreasing rate. The ball passes through N with its greatest velocity, and continues on toward E. From N to E the component of the force of gravity along the tangent, which is always directed toward N, is opposite to the motion, or the acceleration is negative. Hence the pendulum is brought to rest again at E. It then retraces its path, and continues to oscillate with a periodic or pendular motion.

70. Definition of Terms. — A *single vibration* is the motion from N to either B or E, and back again to N; a *complete* or *double vibration* is the motion from N to B, across to E, and then back again to N, the motion at the end of the complete vibration being in the same direction as at its beginning. The *period* of a complete oscillation is the time consumed in making a complete vibration, or the time-interval between two successive passages of the pendulum through N *in the same direction.* The *period* of a *single* vibration is half that of a double vibration. The *amplitude* is the arc BN.

71. Laws of the Pendulum. — When the amplitude does not much exceed three degrees, the period of the vibration depends solely on the length of the pendulum and the acceleration of gravity. For a single vibration the formula is

$$t = \pi\sqrt{\frac{l}{g}}. \qquad (17)$$

In this formula, l is the length of the pendulum and π is the ratio of the circumference of a circle to the diameter, 3.1416. This constant is evidently the "proportionality factor" of the equation. The following are the laws of a simple pendulum: —

I. *The period of vibration is independent of the amplitude, if the latter is small.*

II. *The period of vibration is proportional to the square root of the length.*

III. *The period of vibration is inversely proportional to the square root of the acceleration of gravity.*

72. Experimental Illustrations. — The laws of the pendulum may be illustrated in the following manner: —

Experiment. — Suspend three small lead balls by fine silk threads as shown in Fig. 29. Make the lengths of the pendulums, to the centre of the ball in each case, 1 m., $\frac{1}{4}$ m., and $\frac{1}{9}$ m., respectively. Find the period of a single vibration for each pendulum by counting the number made in, say, 20 sec. These periods will be 1 sec., $\frac{1}{2}$ sec., and $\frac{1}{3}$ sec. nearly, showing that they are directly proportional to the square root of the lengths.

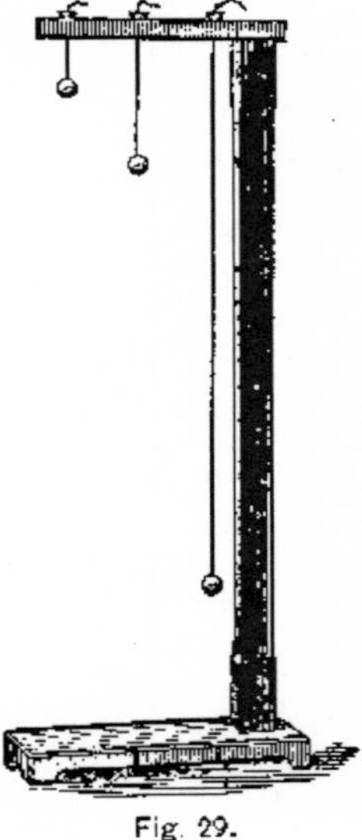
Fig. 29.

Experiment. — Again, make a pendulum by suspending an iron ball over a strong permanent horseshoe magnet, so that the ball will just fail to touch the poles of the magnet as it swings over them. Determine the period of vibration, first, with the magnet in position, and second, after removing it. The period with the magnet in place under the ball will be perceptibly shorter than without it. The attraction of the magnet is equivalent to an increase in the intensity of gravity. The relative intensity of gravity may therefore be measured by observing the period of vibration of the same pendulum at different places. A pendulum will oscillate more slowly on the top of a high mountain than at sea level, and more slowly at the equator than at the poles.

73. A Seconds Pendulum is one making a single vibration in one second. Its length for the latitude of New York may be computed by making t equal to unity, and g, 980.19 in the formula $t=\pi\sqrt{\frac{l}{g}}$, and solving for l. (To be done by the pupil.) Since g increases from the equator toward either pole, it follows that the length l of a pendulum to beat seconds increases in the same ratio as g.

74. The Compound Pendulum. — Any body suspended so as to oscillate about a horizontal axis is a *compound or physical pendulum.* Every actual pendulum is really compound. A simple pendulum is an ideal one about whose length there is no ambiguity. The equivalent length of a compound pendulum, to be used in the formula for the period, is not obvious.

75. Centres of Suspension and Oscillation. — Let AB (Fig. 30) be a bar suspended so as to have freedom of motion about a horizontal line through C. Then C is the *centre of suspension.* Let the centre of the mass be at G. This compound pendulum has a period of vibration equal to that of some ideal simple pendulum, oscillating about the same axis, and with the whole mass collected at a point. Suppose that point to be at D. Then the distance CD is the length l of the simple pendulum which will vibrate in the same time as the physical bar; and the point D on the line CG produced, is called the *centre of oscillation.* The length l of the compound pendulum is therefore the distance between the centres of suspension and oscillation.

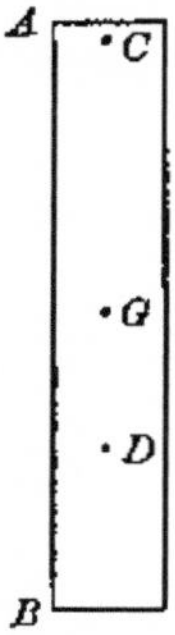

Fig. 30.

The centre of oscillation is also called the *centre of percussion*, because, if the suspended body be struck at this point in a direction at right angles to the axis of suspension, it will be set swinging without any jar. A baseball club, or a cricket bat, has a centre of percussion, and it should strike the ball at this point to avoid breaking the bat and jarring the hands.

Experiment. — Hold a thin wood strip a metre long by the thumb and forefinger near one end. Strike the flat side with a soft mallet at different points. One point may be found where the blow will not throw the wood strip into shivers, but will only set it swinging like a pendulum.

76. The Two Centres Interchangeable. — Huyghens, a celebrated Dutch physicist, discovered that the centres of suspension and oscillation are interchangeable, that is, that the period of vibration is the same whether the pendulum swings about the one as an axis or the other. This discovery led to the invention by Captain Kater of a pendulum with two parallel axes of suspension, and with adjustable weights which can be moved till the pendulum will swing in the same time about the two axes. The distance between them is then the length l.

77. Utility of the Pendulum. — The discovery of Galileo suggested a most obvious use of the pendulum as a time-keeper. The common clock is an instrument in which the to-and-fro motion of the pendulum regulates the rotary motion of the hands. The train of wheels is kept in motion by a weight or a spring, and the regulation is effected through an escapement (Fig. 31). One tooth of the escapement wheel escapes from the pallet with every double vibration, but the movement of the wheel round its axis is about equally divided between the two halves of the double vibration. The pendulum controls the escapement, and it receives in turn small impulses which keep it swinging against friction and opposition of the air. The pendulum itself is supported by a thin flexible spring at the top.

The vibration period is affected by any change in the length of the pendulum. To secure a uniform rate this length must be kept invariable; and corrections must be made for changes in length due to changes of temperature. The common clock, with an uncompensated pendulum, loses time in hot weather and gains in cold. A correction may be made by raising or lowering the bob by means of a running nut shown at the bottom of the figure.

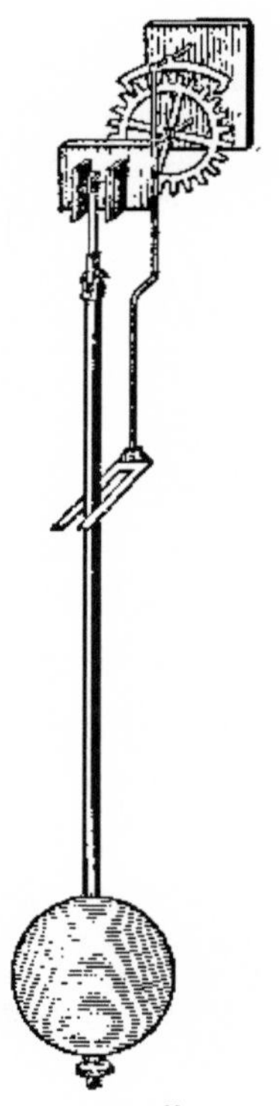

Fig. 31.

Astronomical clocks and clocks for precise physical measurements have compensated pendulums, which adjust themselves automatically when the temperature changes. The mercurial pendulum, commonly used for this purpose, has in it a mass of mercury, which expands upward while the pendulum rod expands downward. Compensation is thus effected. In one form the mercury in glass tubes forms the pendulum bob; in another the bob is lens-shaped, and the mercury partly fills a steel tube carrying the bob, similar to the one shown in the figure. For perfect compensation, the height of the mercury in the tube is dependent upon the latitude of the place and the height above sea level.

Problems.

1. If a tower were 200 ft. high, with what velocity would a stone dropped from the top strike the ground?

2. A rifle ball is shot vertically upward with a velocity of 1500 ft. per second. In what time will it reach the ground?

3. A stone thrown over a tree reaches the ground in 3 sec. What is the height of the tree?

4. How far must a stone fall in order to acquire a velocity of 100 m. per second?

5. A stone dropped from a bridge is seen to strike the water in 3 sec. What is the height of the bridge and the velocity of the stone on reaching the water?

6. The stick of a rocket reaches the ground 5 sec. after it leaves the ground. With what velocity did it start out and how high did it ascend?

7. If a falling body is found to pass over 75 m. during a certain second, how far had it fallen at the time of observation?

Ans. 250.7 m.

8. What velocity must be given to a stone that it may just pass over a tree 50 m. high?

9. A ball is let fall from the top of a high tower, and 2 sec. later a second ball is dropped. How far will they be apart at the end of 3 sec. from the starting of the second ball?

10. A stone is let fall from the top of a wall, and it is observed to pass over the last half of the height in half a second. Find the height of the wall.

11. A toy car whose mass is $\frac{1}{2}$ lb. runs at the rate of 5 mi. an hour on a level circular railway 20 ft. in circumference. Calculate the horizontal pressure on the rails in pounds.

12. A mass of 20 lb. is revolving uniformly, once in 5 sec., in a circle whose radius is 3 ft. Find the centrifugal force.

Ans. 2.94 lb.

13. A body of mass 2 kgm. is attached to the end of a string a metre long, and is whirled around at a uniform rate, making 20 revolutions in a minute. What is the tension in the string?

Ans. 877,302.272 dynes.

14. A mass of 75 gm. is connected to a fixed point by a string 2 m. long, and is whirled around in a circle twice in 3 sec. Find the tension in the string, in grammes.

15. A locomotive whose mass is 25 tons passes round a curve in the shape of a part of a circle whose radius is half a mile, with a velocity of 30 mi. an hour. What is the centrifugal force?

Ans. 1140.5 lb.

16. If a string can just stand a force of 1000 dynes without breaking, what is the greatest length of it which can be used, to which a 5-gm. mass attached and whirled around a fixed point at the rate of two revolutions in 5 sec. just breaks it?

17. In what time will a pendulum 1 m. long make a vibration if $g = 980.29$ cm.?

18. The acceleration due to gravity at Paris is 9.81 m. per second. What is the length of the seconds pendulum at that place?

Ans. 99.39 cm.

19. What must be the value of g in order that a pendulum 1 m. long shall vibrate seconds?

20. What must be the length of a pendulum that shall make 70 vibrations per minute, when $g = 32.16$ ft.?

21. A pendulum 10 ft. long makes 10 complete vibrations in 35 sec. What is the value of g at that place?

22. The force of gravity on Jupiter is 2.65 times greater than on the earth. Find the length of the Jovian seconds pendulum.

Ans. 2.631 m.

23. If $g = 980.29$ cm., what will be the length of a pendulum that makes 25 vibrations in 30 sec.?

24. A pendulum, the length of which is 1.95 m., makes 61,680 oscillations in a day at a certain place. Find the length of the seconds pendulum at that place.

25. If g were increased by 10 per cent, by what per cent would the length of the seconds pendulum be affected and in what direction?

Ans. Increased by 10 per cent.

26. A certain pendulum designed for a seconds pendulum loses 40 sec. per week. Find its length, if $g = 981$ cm.

VI. WORK AND ENERGY.

78. Work. — Whenever an agent exerting a force produces any effect on a body, and the point of application moves in the direction of the force, then the agent exerting the force is said to *do work*. For example, steam exerts pressure on the piston in the cylinder of an engine, causes it to move, and does work. Gravity does work on the weight of a pile driver, causing it to descend; a horse does work in pulling a wagon up an inclined roadway; the electric current, by means of a motor, does work when it drives a pump and forces water up into the tank of a water tower.

Unless the point of application moves in the direction of the force, no work is done, however great the force may be; the pillars supporting a pediment over a portico do no work, though manifestly they support a great weight and exert force.

Work is the production of an effect in bodies by means of a force whose point of application moves in the direction of its own line of action, and it is measured by the product of the force and the distance moved, or

$$W = Fs. \tag{18}$$

79. Units of Work. — The three units of work in common use are: —

1. The *foot-pound*, or the work done by a force of one pound working through a distance of one foot. This unit is the one still used by English-speaking engineers. It is open to the objection that it is variable, on account of the variation of the weight of a pound with the latitude (§ 54).

2. The *kilogramme-metre*, or the work done by a force of one kilogramme working through a distance of one metre. This is the gravitational unit of work in the metric system; it is open to the same objection as the foot-pound.

3. The *erg*, or the work done by a force of one dyne working through a distance of one centimetre. The erg is the absolute unit in the metric system and is invariable.

Gravity gives to a gramme in a second a velocity of about 980 cm. a second. It is therefore equal to 980 dynes. If, then, a gramme be lifted vertically one centimetre, the work done against gravity is 980 ergs; or, one erg of work is done in lifting $\frac{1}{980}$ gm. 1 cm. high.

A silver dollar weighs about 26.73 gm., and the height of an ordinary table is about 76.3 cm. The work done in lifting a silver dollar from the floor to the top of a table is then the continued product of 26.73, 76.3, and 980, or 2,000,000 ergs nearly.

The erg is, therefore, an excessively small unit, and it is more convenient to use a multiple for practical measurements. The multiple commonly employed is the *joule*, which is equal to 10^7, or 10,000,000, ergs. Expressed in this larger unit the work done in lifting the silver dollar is 2 joules.

80. Time not an Element in Work. — It necessarily takes time to do work, but the *amount of work done* has nothing whatever to do with the time taken to do it. To lift the silver dollar from the floor to the table top requires the expenditure of 20 million ergs of work, whatever the time consumed in lifting it. If a man weighing 150 lb. walks up the nine hundred steps leading to the highest attainable level in the Washington monument, 500 ft. high, he does work against gravity equal to 75,000 foot-pounds, irrespective of the time taken in the ascent.

81. Power. — It is obviously necessary to take into account the time an agent takes to do a certain quantity of work. Then the work done in a given time, divided by the time, is called *power* or *activity*. *Power is the time-rate of doing work.*

In the English gravitational system, the unit of power is the horse-power (H.P.); it is the rate of doing work equal to 33,000 foot-pounds a minute, or 550 foot-pounds a second.

In the C.G.S. system the unit of power is the *watt*. It equals work done at the rate of one joule (10^7 ergs) a second. One horse-power is equivalent to 746 watts. A kilowatt (K.W.) is 1000 watts. It is therefore very nearly $\frac{4}{3}$ horse-power. To convert kilowatts into horse-power, add one-third; to convert horse-power into kilowatts, subtract one-fourth. For example, 60 K.W. equals 80 H.P., and 100 H.P. equals 75 K.W.

82. Energy. — In general, a body upon which work has been done is found to have an increased power of doing work itself. It is then said to possess more *energy* than before. The increase of energy acquired by the body is the most essential part of the effect produced when work is done on it. When work is done on a quantity of water by lifting it to a high level, its energy is increased because it is capable of doing work by flowing down again through a water motor. The winding of a clock lifts its weight against gravity, and the clock thereby acquires enough energy to keep it running against resistance for an entire week. The rate of putting energy into it is very large compared to the rate at which the clock doles it out.

Energy is then the capacity of doing work.

83. Potential Energy. — The energy or capacity to do work, possessed by a lifted weight or by a coiled spring, or in general the energy which a body has by virtue of its position relative to some other body, or the relative position of its parts, is called *potential energy*. In bending a bow work is done in distorting it, or placing it under stress; and it then possesses potential energy due to its altered shape, for it can do work on an arrow and give it rapid motion. Potential energy is often called *energy of position* or *energy of stress.*

The potential energy of the weight of a pile-driver is the work done against gravity in lifting it. The measure of this energy is then

$$E = Wh = mgh, \tag{19}$$

where m is its mass and h the vertical height through which it has been lifted.

84. Kinetic Energy. — A moving body has the capacity of giving motion to another body. It then possesses energy. The energy which a body has by virtue of its motion is called *kinetic energy.* Work is done on a cannon ball by the agency of the gases due to the explosion of the powder. The ball acquires high speed and something more than that; for it then has the capacity of overcoming resistance. This moving mass may imbed itself in earthworks, demolish fortifications, or pierce the nickel-steel armor of a battleship. The energy which the ball acquires from the explosion is *kinetic energy*, or energy of motion. Motion is the essential fact in a body possessing kinetic energy.

85. Measure of Kinetic Energy. — Not only is energy a measurable quantity, but it is measured in terms of the same units as those used in measuring work.

Let a body of mass m, moving with a velocity v, be acted on by a constant force F in a direction opposing the motion; and let it be brought to rest after it has passed over the distance s. Then the work done by the moving body against the force F, before it is brought to rest, is Fs (equation 18). But $F=ma$ (equation 7), and $s=\frac{v^2}{2a}$ (§ 34), since the body loses velocity v in a space s.

Therefore $$E=ma\times\frac{v^2}{2a}=\tfrac{1}{2}mv^2. \tag{20}$$

The measure of kinetic energy in terms of the mass and velocity of the moving body is therefore half the product of the mass and the square of the velocity. If m is expressed in grammes and v in centimetres per second, the kinetic energy is in ergs. In the English gravitational system

$$E=\frac{Wv^2}{2g}=\frac{Wv^2}{64.3}, \tag{21}$$

where, if W is expressed in pounds, and v in feet per second, E is in foot-pounds.

In the latitude of New York ergs can be reduced to kilogramme-metres by dividing by 98,000,000 (§ 79).

86. Transformations of Energy. — If a ball be thrown vertically upward, it gradually loses its motion and its kinetic energy, but it gains energy of position. When it reaches the highest point its energy is all potential. It then descends, and again acquires energy of motion at the expense of energy of position. The one form of energy is therefore convertible into the other.

The pendulum illustrates the same principle. While it is moving from the lowest point of its path toward either extremity, its kinetic energy is converted into potential energy; and the reverse transformation sets in when the pendulum reverses its motion. All physical processes involve energy changes, and such changes are in ceaseless progress. A machine is only an instrument or device for the transformation of energy and the turning of it to useful account. A watch when wound has a small store of potential energy which it expends very slowly in the work of turning the train of wheels against friction and the resistance of the air, and producing the sound of ticking. In a week a watch distributes the energy of winding it seven times among over 3,000,000 ticks.

Potential energy is the highly available or useful form. It always tends to revert to the kinetic form, but in such a way that only a portion of the kinetic energy is available to effect useful changes in nature or art. The remainder goes into useless heat. The energy of the solar system is therefore becoming all the time less and less available. Strictly, the capacity which a body possesses of doing work does not depend on the total quantity of energy which it may possess, but only on that portion which is *available*, or is capable of being transferred to other bodies. We have to deal chiefly with the variations of energy in a body, and not with its total value. In subsequent sections we shall have occasion to consider many other forms of energy than those already mentioned, and we shall find that they are all mutually convertible the one into the other.

87. Conservation of Energy. — The question arises, when work has been done on a body and energy communicated to it, has the energy been made out of nothing, or has it been transformed? The answer of science is that the latter is the truth. Innumerable facts and experiments show that it is as impossible to create *energy* as to create *matter*.

Whenever energy appears as the result of work done on a body or system, it is always at the expense of some other form of energy which existed previously.

The agent, or body which does work, always loses energy; the body which has the work done on it gains the same amount. On the whole, there is neither gain nor loss of energy, but only transference from one body to the other.

The law of *Conservation of Energy* means that no energy is created or destroyed by the action of forces that we know anything about.

88. Matter and Energy. — "All that we know about matter relates to the series of phenomena in which energy is transferred from one portion of matter to another till in some part of the series our bodies are affected, and we become conscious of a sensation. We are acquainted with matter only as that which may have energy communicated to it from other matter. Energy, on the other hand, we know only as that which in all natural phenomena is continually passing from one portion of matter to another. It cannot exist except in connection with matter."[1]

Problems.

1. A man whose weight is 150 lb. walks up Mt. Washington, height 6288 ft. How many foot-pounds of work does he do?

Ans. 943,200 ft.-lb.

2. A horse exerting a force of 100 lb. pulls a load of 1000 lb. up a hill 100 ft. high. How much work is done? How long is the hill?

Ans. 100,000 ft.-lb.; 1000 ft.

3. A stone weighing 75 kgm. is carried to the top of a structure 25 m. high. Calculate the amount of work done.

4. Which does the most work, a man who carries a ton of coal to a height of 40 ft. in 2 hr., or a man who carries two tons of coal to a height of 10 ft. in 4 hr.?

Ans. 80,000 ft.-lb. and 40,000 ft.-lb. respectively.

---

[1] Maxwell's *Matter and Motion*, p. 163.

5. What is the work done by a force of 50 dynes acting on a body whose mass is 1 kgm., moving it through 75 cm.?

6. A man can pump 25 gal. of water per minute to a height of 15 ft.; how many foot-pounds of work does he do in a day of 10 hr.?

Ans. 1,800,000 ft.-lb.

7. A mass of 10 kgm. is lifted vertically 20 m. How many ergs of work are done?

8. How many foot-pounds of work are done in carrying 10 bbl. of water to the top of a hill 50 ft. high?

9. A weight of 2 tons drops on the head of a pile from a height of 32 ft., and drives the pile through a distance of 1 ft. What is the average resistance of the ground, the weight of the pile being neglected?

10. A machine is constructed in such a way that a weight of 3 tons, by falling 3 ft., is able to lift a weight of 168 lb. to a height of 100 ft. Find the work done by the falling body, and what part of the work is used up in overcoming the friction of the machine.

Ans. 18,000 ft.-lb. done by falling body.
16,800 ft.-lb. effective work.
1,200 ft.-lb. used up in friction.

11. James Watt found that an English dray horse could travel at the rate of 2.5 mi. an hour, and at the same time raise a weight of 150 lb. by means of a rope led over a pulley. How much work was done per minute?

12. At what rate is an engine working which raises 2000 tons of coal per day from a pit 250 ft. deep?

13. A tank of 1000 gal. at an elevation of 50 ft. is filled with water in 2 hr. by a steam pump. Calculate its horse-power.

14. In a town of 15,000 inhabitants each inhabitant uses water at the rate of 50 gal. per day. If the water has to be pumped to an elevation of 200 ft., of what horse-power is the pump?

15. The diameter of the piston of an engine is 2 in., the length of stroke is 15 in., the number of complete strokes is 300 per minute, and the mean pressure on the piston is 80 lb. per square inch. Compute the horse-power.

16. A man whose weight is 160 lb. carries 60 lb. on his back, and climbs a ladder to a height of 40 ft. in $1\frac{1}{2}$ min. Find his power (1) in foot-pounds per minute; (2) as a fraction of horse-power; (3) in watts.

17. What is the activity of an agent, expressed in watts, which can raise 1000 kgm. 10 m. in 5 sec.?

18. What is the horse-power of an engine which draws a train at a uniform rate of 40 mi. an hour, against a resistance equal to a 1000-lb. weight?

19. The falls of Niagara are about 160 ft. high, and deliver about 700,000 tons of water per minute. What is the horse-power of the falls?

20. A steamer is going at the rate of 20 mi. an hour; the horse-power of her engines is 5000. What is the resistance to motion through the water?

21. Compute the energy of a ball of 15 gm. mass moving with a velocity of 150 cm. per second.

22. Find the ratio of the energy of an ounce ball fired with a velocity of 1600 ft. per second, to that of the recoil of the gun, the mass of the gun being 10 lb.

23. What is the energy of a mass of 5 kgm. moving with a velocity of 50 m. per second?

24. Compute the kinetic energy that a stone would have on reaching the ground, if its mass was 15 kgm., and it fell from a chimney 50 m. high.

25. A cannon ball weighing 70 kgm. is moving at the rate of 600 m. per second. How much energy does it possess?

26. A train, of mass 150 metric tons, moving with a velocity of 50 km. an hour, is brought to rest in 100 m. What is the force applied to stop the train?

27. A bullet of 15 gm. mass, moving with a velocity of 300 m. per second, strikes a piece of wood and penetrates 8 cm. What is the resistance to penetration?

28. A train whose mass, together with that of the engine, is 250 metric tons, is going at the rate of 30 km. per hour. It is fitted with a Westinghouse brake, capable of furnishing a frictional force of 10 metric tons weight. Find how far from a station the steam must be shut off and the brake applied in order that the train may stop at the proper place.

29. A 20-gm. bullet leaves the mouth of a rifle with a velocity of 500 m. per second. If the barrel be 1 m. long, calculate the mean pressure of the powder, neglecting all friction.

30. A ball weighing 15 gm. is projected vertically upward with a velocity of 500 m. per second, and rises 1000 m. What work, in kilogramme-metres, is done against the resistance of the air in the ascent?

VII. MACHINES.

89. A Machine is a device designed to transform or transfer energy, and to do useful work. An electric lighting and power plant illustrates both features of a useful machine or collection of machines. The heat energy of the steam actuates the moving parts of the engine, and is thence transferred mechanically to the armature of the dynamo, where it is converted into the energy of an electric current.

Simple machines, or *mechanical powers*, are restricted to devices for merely transferring energy. They are six in number, the *lever*, *pulley*, *wheel and axle*, *inclined plane*, *wedge*, and *screw*. Since it will appear from what follows that the pulley and the wheel and axle are only modified levers, and the screw and wedge are modified inclined planes, it is therefore possible to reduce the six simple machines to two, the lever and the inclined plane. All complex machines are mechanically only combinations of two or more simple machines.

90. Mechanical Advantage. — The effective force exerted by the agent losing energy, and the force exerted by the body receiving energy in a simple machine, may be denoted by two terms introduced by Rankine, namely, *effort* and *resistance*. The problem in simple machines consists in finding the ratio of the resistance to the effort, and this ratio is known as the *mechanical advantage*. In elementary discussions it is customary to neglect friction and to assume that the parts of a machine are rigid and without weight.

91. General Law of Machines. — Every machine must conform to the principle of the conservation of energy, or *the work done by the effort must equal the work done in overcoming the resistance*, except that some energy may be dissipated as heat or may not appear in a mechanical form. A machine can never produce an increase in the quantity of energy.

Denote the effort by F and the resistance by R, and let d and D denote the distances through which they act, respectively. Then we have, from the law of conservation,

$$Fd = RD, \qquad (22)$$

or the effort multiplied by the distance through which it acts is equal to the resistance multiplied by the distance through which it is moved.

92. Efficiency. — If a machine could be made that would waste no energy, that is, one in which the resistance is all *useful* and not *wasteful*, the machine would be perfect and its efficiency would be unity. But in practice there is always some wasteful resistance due to friction (§ 107), rigidity of cords, etc. The work done is therefore always partly *useful* and partly *wasteful*. The efficiency of a machine is the ratio of the useful work done by it to the total work done on it. Efficiency is always, therefore, a proper fraction, and it is expressed as a percentage. An efficiency of 90 per cent means that the energy recovered is 90 per cent of the energy put into the machine. A machine which will do either useful or useless work continuously, without a supply of energy from without, is thus clearly impossible.

93. A Lever is a rigid bar, straight or curved, turning about a fixed axis called the *fulcrum*. The perpendicular distances between the fulcrum and the lines of action of the effort and the resistance are called the *arms* of the lever. A *straight* lever has the arms in the same straight line.

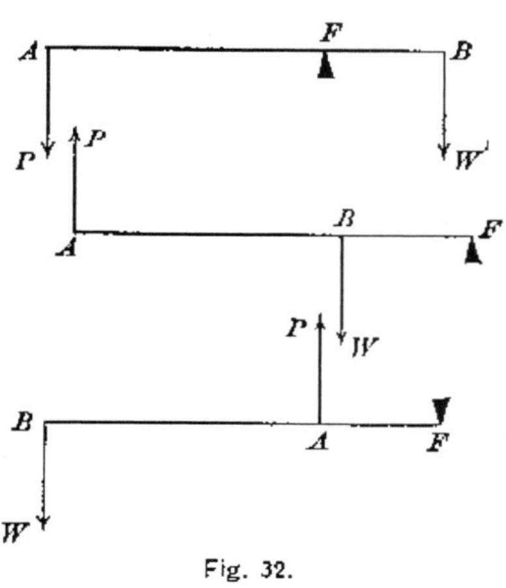

Fig. 32.

If the fulcrum is between the points of application of the effort and the resistance, the lever is of the *first class* (Fig. 32); if the resistance is between the effort and the fulcrum, the lever is of the *second class;* if the effort is between the resistance and the fulcrum, the lever is of the *third class*.

94. Mechanical Advantage of the Lever. — Let AB (Fig. 33) be a straight lever whose arms are FA and FB. Let P be the effort and W the resistance, or the weight lifted. Tilt the lever into the position aFb. If there is no friction, $P \times aC$ is the work done by the effort against

$W \times bD$ of the resistance. Putting these equal to each other, $P \times aC = W \times bD$, or $\frac{W}{P} = \frac{aC}{bD}$. But the ratio of the lines aC and bD is the same as that of the arcs aA and bB; and the ratio of these is equal to that of their radii, FA and FB; therefore $\frac{W}{P} = \frac{FA}{FB}$. Hence, *the mechanical advantage of the lever equals the inverse ratio of its arms.*

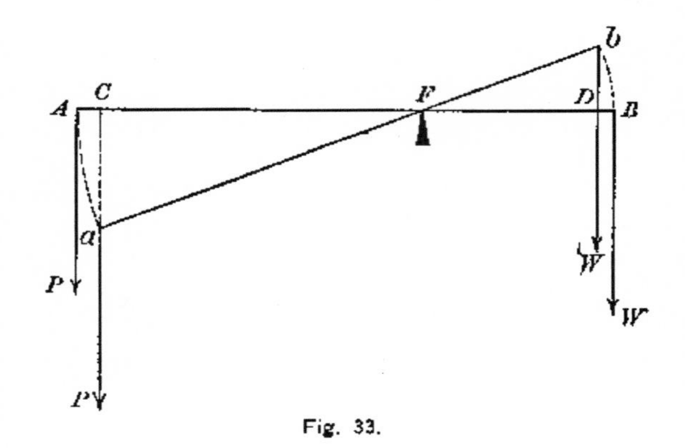

Fig. 33.

95. Illustrations. — The *common balance* (Fig. 34) is a lever of the first class with equal arms. P is therefore equal to W. For accuracy the two arms of the beam must be strictly of the same length; and for high sensibility, the friction must be small, the beam light, and the centre of gravity only slightly below the "knife-edge" forming the fulcrum. If the arms are not of equal length, the true weight of a body may be found by weighing it first in one scale pan and then in the other, and taking the square root of the product of these apparent weights.

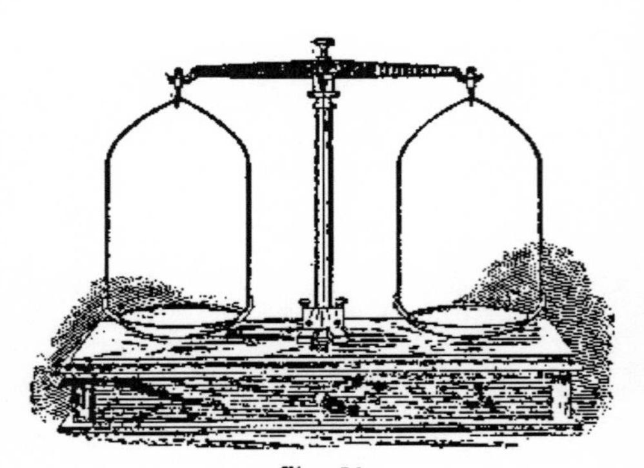
Fig. 34.

The *steelyard* (Fig. 35) is a lever of the first class with unequal arms. Scissors are double levers of the first class. A crowbar, used to lift a weight with one end on the ground, is a lever of the second class. Nutcrackers are

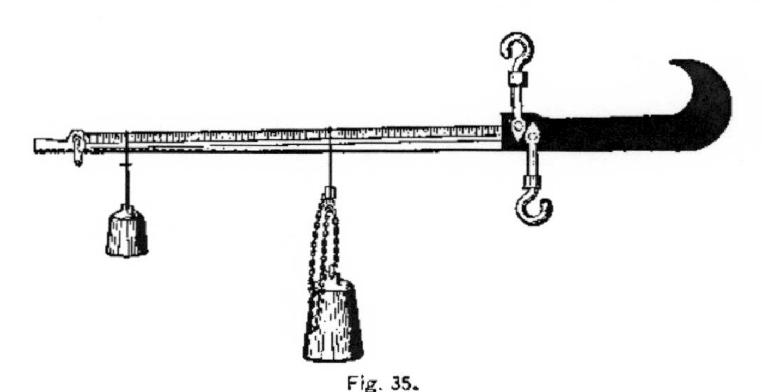

Fig. 35.

double levers of the same class. When a weight is held in the palm of the hand, the forearm acts as a lever of the third class; for the fulcrum is at the elbow and the effort is applied through the tension of the muscles which are attached between the elbow and the hand.

96. The Wheel and Axle (Fig. 36) consists of a cylinder and a wheel of larger diameter turning together on the same axle. In the figure the axle passes through C, the radius of the cylinder is BC and that of the wheel is AC. If the weights P and W are suspended by ropes wrapped around the circumferences of the two wheels, the arrangement is a continuous lever of the first class with the fulcrum at C.

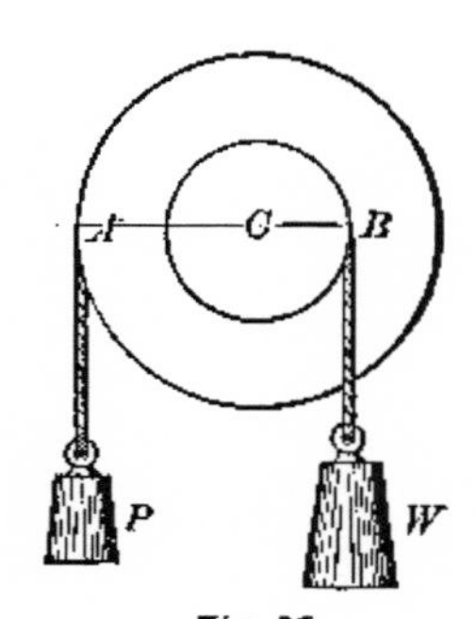

Fig. 36.

Therefore, $$\frac{W}{P}=\frac{AC}{BC}=\frac{R}{r},$$

where R and r are the radii of the wheel and the axle respectively. The weight P may represent the effort applied at the circumference of the wheel, and the weight W the resistance at the circumference of the axle. Hence, *the mechanical advantage is the ratio of the radius of the wheel to the radius of the axle.*

97. Applications.—The *derrick* is a form of wheel and axle much used for raising heavy weights. The essential parts are shown in Fig. 37. This may be looked upon as a double wheel and axle. The axle of the first system works upon the wheel of the second by means of the spur gears. The cranks or handles of the first system answer the same purpose as a wheel. The mechanical advantage in this case is the ratio of the product of the radii of the wheels to the product of the radii of the axles.

Fig. 37.

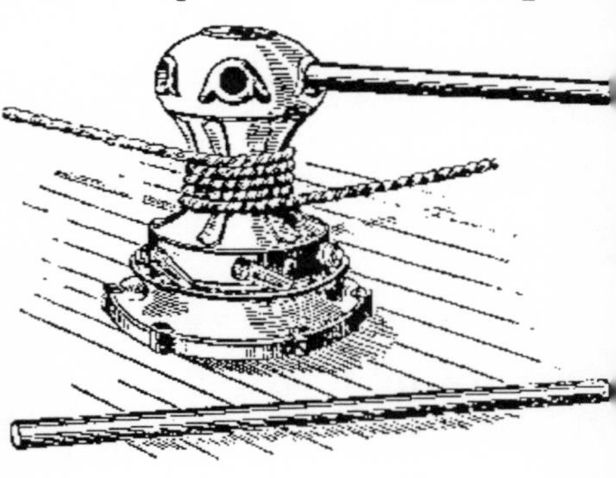

Fig. 38.

In the *capstan* (Fig. 38) handspikes inserted in the holes at the top are used instead of a wheel; while the rope, on which the work is done, is wrapped around the body of the capstan as an axle.

98. The Pulley is a wheel, called the *sheaf*, free to turn about an axle in a framework called a *block*. The effort and the resistance or weight are attached to a rope which lies in a groove cut in the circumference of the wheel. A simple fixed pulley is used to change the direction of the motion produced by a force, and the effort and resistance are equal to each other; for, neglecting friction and the stiffness of the rope, the tension throughout the rope is the same.

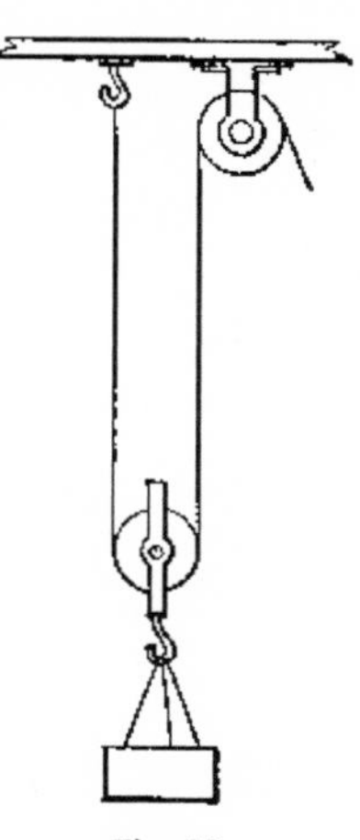
Fig. 39.

In the movable pulley (Fig. 39) a weight may apparently be supported by a force half as great as itself; but the other half of the force is supplied by the fixed hook to which the cord is attached. If the weight is lifted it rises only half as fast as the free end of the cord travels. The mechanical advantage of a simple fixed pulley is thus one, and of a simple movable pulley two.

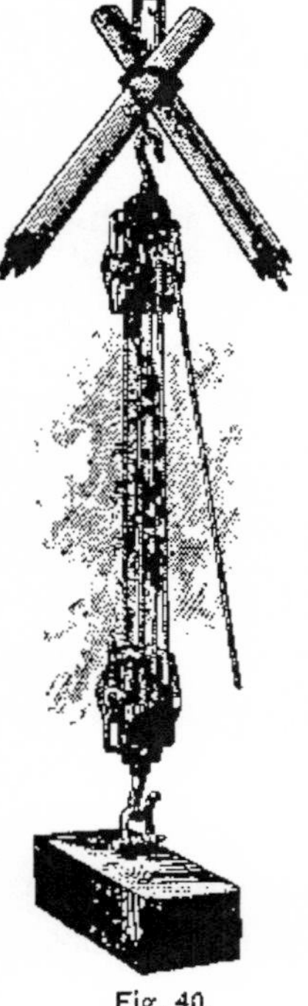
Fig. 40.

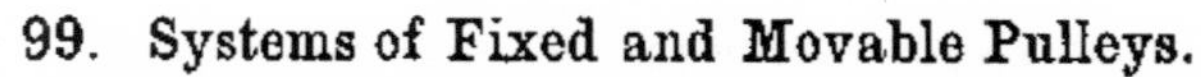

99. Systems of Fixed and Movable Pulleys.—The most useful combination of pulleys consists of two blocks, each with several sheafs which usually turn on the same axle. One of these blocks is attached to a fixed point, while to the other is attached the resistance or weight (Fig. 40). This is the common "block and tackle."

100. Mechanical Advantage of the Pulley.—The principle involved in determining the mechanical advantage of the pulley is the transmission of the same tension to all parts of the cord or rope. The only purposes served by the wheels of the pulley are to diminish friction and to change the direction of the effort.

When the cord passes in succession around each pulley, as in Fig. 40, it is obvious that the weight is sustained by the several parts of the cord, the tension in each part being P, the effort applied at the free end. If there are n parts to the cord, the total tension supporting the weight W will be nP; that is, $W = nP$ and $\frac{W}{P} = n$.

The mechanical advantage of a system of pulleys, when a single cord is used, is therefore equal to the number of times the cord passes from one block to the other.

101. The Inclined Plane. — Suppose a body rests on an inclined plane without friction. The weight of the body acts vertically downward, while the reaction of the inclined plane is perpendicular to its surface; so that to maintain the body in equilibrium on the incline, a third force must be applied. Two principal cases occur; first, when the force applied is parallel to the *face* of the plane; second, when it is parallel to the *base* of the plane.

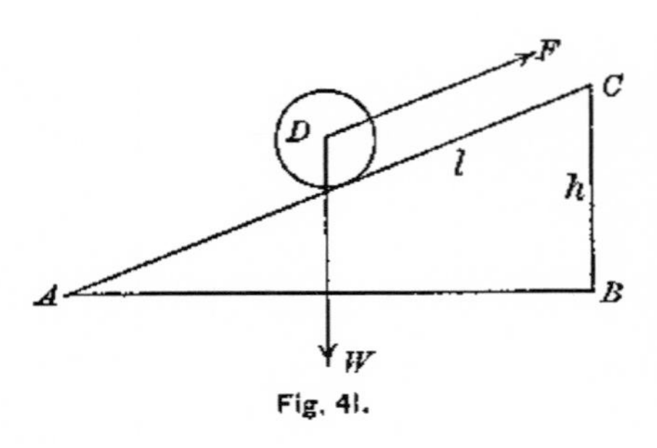

Fig. 41.

102. Mechanical Advantage of the Inclined Plane. — *Case I:* When the force is applied parallel to the face of the plane. The most convenient method of obtaining the relation between the force F (Fig. 41) and the weight W is to apply the principle of work. Suppose D to move under the influence of the force F from A to C. Then the work done by F is $F \times AC$. The work done on the body D, of weight W, is at the same time $W \times BC$, since W is lifted through a vertical distance, BC. Therefore, $F \times AC = W \times BC$, or

$$\frac{W}{F} = \frac{AC}{BC} = \frac{l}{h},$$

or *the mechanical advantage, when the force is applied parallel to the face of the plane, is the ratio of the length of the plane to its height.*

Case II: When the force is applied parallel to the base of the plane. If, in this case, we estimate the work done by the force F in moving the body up the plane from A to C (Fig. 42), we must take the distance moved in the direction of the force. Now the displacement of the body in the direction of the force in going from A to C is not AC, but the base of the inclined plane AB. Therefore, the work done is $F \times AB$. The work done on the weight W is the same as in the first case. Hence, $F \times AB = W \times BC$, or

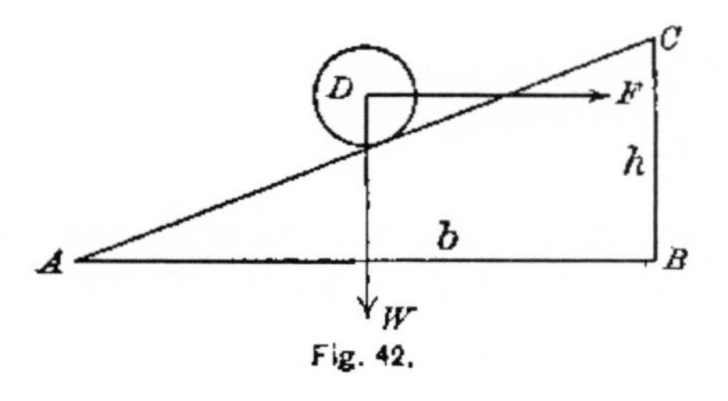

Fig. 42.

$$\frac{W}{F} = \frac{AB}{BC} = \frac{b}{h}.$$

The mechanical advantage, when the force is applied parallel to the base of the plane, is the ratio of the base of the plane to its height.

103. The Wedge (Fig. 43) is a double inclined plane, with the effort applied parallel to the base, so as to enlarge an opening or lift a weight. The effort is generally ap-

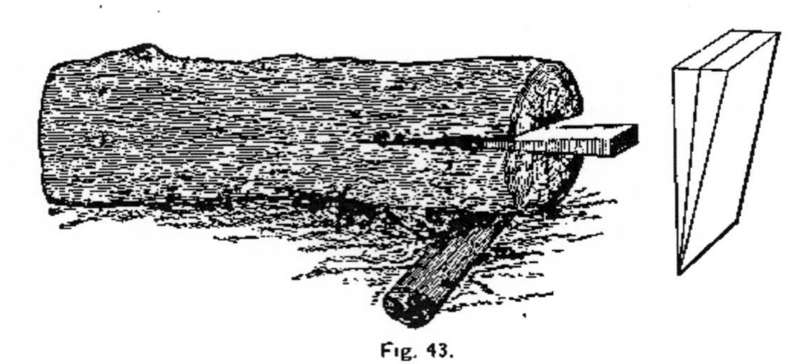

Fig. 43.

plied by a blow with a heavy body. Although the principle of the wedge is the same as that of an inclined plane, yet no exact statement of the mechanical advantage is possible, since the resistance has no definite relation to the faces of the plane, and friction cannot be neglected.

104. The Screw is a cylinder, on the outer surface of which is a uniform spiral projection called the *thread*. The faces of this thread are inclined planes, as may be seen by wrapping a long triangular strip of paper around a rod like a pencil (Fig. 44). The base of the triangle is perpendicular to the axis of the cylinder, and the hypotenuse traces a spiral like the thread of a screw. The screw works in a block called the *nut*, on the inner surface of which is a groove. This groove is the exact counterpart of the thread (Fig. 45). The effort

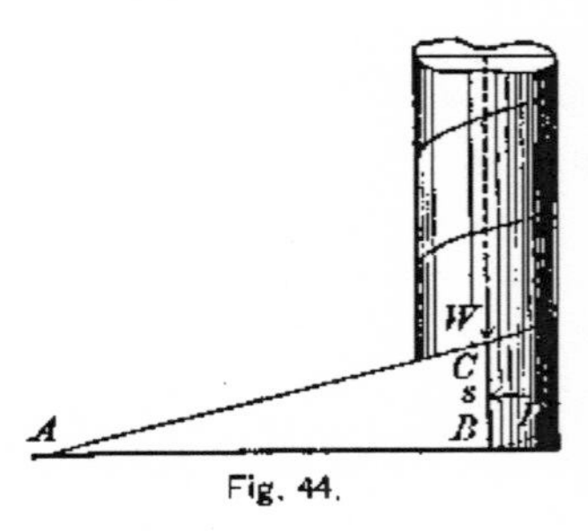

Fig. 44.

is applied at the end of a lever, fitted either to the screw or to the nut. When either makes a complete turn, the screw or nut moves through a distance equal to that between two contiguous threads, measured parallel to the axis of the screw cylinder. This distance (s in Fig. 44) is called the *pitch* of the screw.

Fig. 45.

105. Mechanical Advantage of the Screw. — Since a screw is usually combined with the lever, the simplest method of finding the mechanical advantage is to apply the principle of work expressed in the general law of machines (§ 91). If the pitch be denoted by s, and the lever arm by l, then when P makes a complete revolution, the work done is $P \times 2\pi l$, or $\frac{W}{P} = \frac{2\pi l}{s}$; that is, *the mechanical advantage of the screw equals the ratio of the distance traversed by the effort in one revolution to the pitch of the screw.*

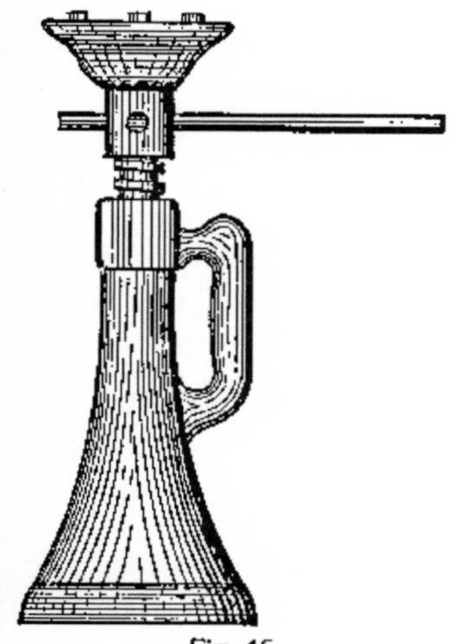

Fig. 46.

106. Applications. — Familiar uses of the screw are illustrated by the lifting jack (Fig. 46), copying and book presses, cotton and hay presses, the screw propeller of ships, and air fans. In most cases it depends for its efficiency on friction, as in holding together the parts of machinery and woodwork.

It is also used for the purpose of accurately measuring small dimensions, as in the *wire micrometer*

(Fig. 47). An accurate screw, *C*, has its head *D* divided into a number of equal parts, so as to register any portion of a complete revolution. If, for example, the pitch of the screw is $\frac{1}{2}$ mm., and the head is divided into 25 equal parts, then for each revolution the end of the screw advances $\frac{1}{2}$ mm.; and if the head of the screw be turned through one of the 25 divisions, the end of the screw will advance $\frac{1}{25}$ of $\frac{1}{2}$, or $\frac{1}{50}$ mm.

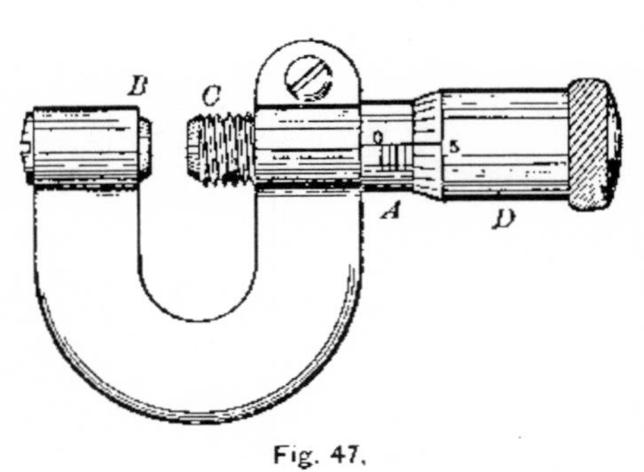

Fig. 47.

107. Friction. — The *resistance* which is brought into play when we attempt to slide one body over another is called *friction*. Friction is called into action only when a force is applied to make one surface move over another, and it always resists this motion. It arises from inequalities in the surfaces, and is diminished by the use of a lubricant. The amount of friction depends on the pressure between the surfaces in contact, and it has a limiting value, which is its value just before motion takes place.

108. Limiting Angle. — A body may rest on an inclined plane without sliding, the force to maintain equilibrium being supplied by friction. For every pair of substances there is an angle of elevation, which is the greatest at which a plane made of one substance can be inclined to the horizontal before a body made of the other will start to slide on it. This angle is called the *limiting angle of friction*, or the *angle of repose*.

109. Laws of Friction. — The following laws of limiting friction have been established by experiment: —

I. *The angle of friction remains the same for all weights so long as the nature of the surfaces remains the same.*

II. *The angle of friction is independent of the size or shape of the surfaces in contact.*

The meaning of these laws is that the friction per unit area is proportional to the normal pressure per unit area.

110. Friction of Motion. — Friction continues after the sliding motion has begun, and opposes the motion; but its magnitude in general is less than the friction of rest at the moment the slipping begins. At speeds less than 100 feet a minute the friction on journals diminishes with increase of speed; at higher speeds it increases as the square root of the speed. This is independent of the load, but is dependent on the extent of the bearing surface.

111. Loss of Energy Due to Friction. — Friction acts in every case as a resistance opposing motion. Whenever any displacement actually takes place, work must be done against frictional resistance. The energy equivalent to this work is converted into heat, which is gradually diffused among neighboring bodies, and this energy is then no longer available. Friction, therefore, decreases the efficiency of machinery and wastes energy.

112. Uses of Friction. — Friction, like every other affliction, has its uses. Screws and nails hold entirely by friction. We are able to walk because the friction between the shoe and the pavement is less than the limiting friction. Shoes with nails in them are dangerous on cast iron plates, because the limiting friction is small between smooth iron surfaces. The friction between leather and wood or iron permits of driving machinery by leather belts. They should not slip on the pulleys.

The friction between the driving wheels of a locomotive and the rails acts forward to prevent the wheels from slipping on the track. The effort that an engine can make to "pull a train" is limited to the friction of its driving wheels on the rails.

Problems.

1. On a lever of the second kind, a weight of 20 kgm. is suspended at a distance of 20 cm. from the fulcrum; the effort is 5 kgm. Find the length of the lever.

2. If the weights 10 and 16 kgm. balance on the ends of a lever 5 m. long, find the position of the fulcrum.

Ans. 1.923 m. from greater weight.

3. A lever of the first kind is 10 m. long, with the fulcrum 1.5 m. from one end. What weight will an effort of 20 kgm. balance, the lever being a uniform bar weighing 1 kgm. to the metre?

Ans. $136\frac{2}{3}$ kgm.

4. A uniform bar 4 ft. long weighs 10 lb., and weights of 30 lb. and 40 lb. are appended to the extremities. Where must the fulcrum be placed to produce equilibrium?

Ans. 1.75 ft. from the greater weight.

5. A false balance is one whose arms are unequal. If a kilogramme of tea, weighed from the long arm of such a balance, is sold for a dollar, how much do you really pay per pound, the arms being in the ratio of 19 to 20?

6. The circumference of the axle is 50 cm., to which is attached a weight of 100 kgm. Find what force applied to a crank 1 m. long will be necessary to balance it.

7. If the capstan of a ship is 10 in. in diameter, and the circle described by the handspikes is 10 ft. in diameter, what force is exerted in raising an anchor that weighs 2000 lb.?

8. The axle of a wheel and axle is 10 cm. in diameter, and has a cord wound around it which bears a weight of 80 kgm.; this is kept in equilibrium by another weight at the end of a cord passing similarly around the wheel, which is 100 cm. in diameter. Find the latter weight.

9. A windlass is turned by a crank 16 in. long. This is attached to a wheel with 10 teeth, which works a wheel with 40 teeth. This wheel is attached to the barrel, which is 8 in. in diameter; the efficiency of the machine is 70 per cent. Find the force necessary to raise a weight of half a ton.

SUGGESTION. — The circumferences are proportional to the number of teeth.

10. Suppose that we have seven weightless pulleys, three movable and four fixed, connected by a single cord, and that a weight of 200 lb. is raised. Find the force applied.

11. With a system of six pulleys and one cord, what force will be necessary to balance 250 kgm.?

12. With six pulleys arranged in two blocks, so that a single rope is used, it is found that a force of 60 lb. is necessary to raise 250 lb. What is the efficiency of the machine?

13. What is the least number of pulleys in a movable block that, with a force of 25 kgm., will support a weight of 100 kgm.?

14. An inclined plane is 300 ft. long, and its perpendicular height is 10 ft. What power acting parallel to the plane can support 2240 lb. on the plane? Parallel to the base?

15. What length must be given to a plane, in order that a force of 25 kgm. can support a weight of 500 kgm., the height of the plane being 5 m.?

16. If a force of 8 lb. applied parallel to an inclined plane supports a weight of 20 lb., what is the pressure on the plane?

Ans. 18.33 lb.

17. The arm of a jackscrew is 2 ft. long, and the screw rises 2 in. when it is turned around 10 times. What force must be applied to produce a thrust of a half-ton weight?

18. The distance between two consecutive threads of a screw is $\frac{1}{4}$ in., and the length of the lever is 5 ft. What weight will be sustained by a force of 10 lb.?

19. The thread of a screw makes 12 turns in a foot of its length; the effort is applied at the end of an arm 2 ft. long; it is found that when the effort is 30 lb., it can just raise 1200 lb. What portion of the effort is used in overcoming friction, and how many foot-pounds of work are done by the effort when the weight is raised 2 ft.?

Ans. 22.043 lb.; 9047.8 ft.-lb.

20. The radii of a wheel and axle are 5 ft. and 6 in. respectively. Some of the work is lost in friction. If a force of 125 lb. is required to overcome a resistance of half a ton, what is the efficiency of the machine?

CHAPTER III.

MECHANICS OF FLUIDS.

I. MOLECULAR PHENOMENA IN LIQUIDS.

113. Characteristics of a Fluid. — A solid has rigidity or elasticity of form (§ 14), but a fluid cannot resist a stress unless it is supported on all sides. It offers no sensible resistance to forces tending only to change its shape, and it has therefore only elasticity of volume. Every fluid, however, offers some resistance to change of shape on account of internal friction. This property is called *viscosity*. Viscosity varies through wide limits, being large, for example, in tar and very small in hydrogen gas. A *perfect fluid* would be one entirely without rigidity and viscosity.

114. Liquids and Gases. — Fluids are divided into liquids and gases by means of two distinguishing properties: —

First, liquids, such as water and mercury, are but slightly compressible, while *gases*, such as air and hydrogen, are highly compressible. A *liquid* offers great resistance to forces tending to diminish its volume, while a *gas* offers relatively small resistance to reduction of its volume. Both have perfect elasticity of volume, but the measure of these elasticities differs widely. Water is reduced 0.00005 of its volume by a pressure of one atmosphere (§ 146), while air is reduced 0.5 of its volume by the same pressure.

Second, *gases* are distinguished from *liquids* by the fact that any mass of a gas introduced into a closed vessel always completely fills it, whatever its volume. A liquid has a bulk of its own, but a gas has not. This second characteristic may be regarded as a corollary of the first, since a gas expands indefinitely as the pressure on it decreases.

115. Cohesion in Liquids. — If a clean glass rod be dipped into water and then withdrawn, a drop will adhere to the end of the rod. If enough water runs down the vertical rod to enlarge the drop sufficiently, its weight will tear it away from the rod, and it will fall as a little sphere of water.

If by means of a pipette a large globule of oil be placed in a mixture of alcohol and water, the mixture having the same mass per unit volume as the oil, the globule of oil will assume a spherical form, because the influence of gravity on it is eliminated, and it will float anywhere in the mixture.

In both cases the spherical form is due to the *cohesion between the molecules of the liquid. Cohesion in a liquid is the attraction existing among its molecules.*

116. Surface Conditions of a Liquid. — **Experiment.** — Place a sewing needle on the surface of clean water. If carefully done, the needle will float. On close examination it will be seen that the surface of the water round the needle is depressed, the latter resting in a little hollow large enough to hold perhaps four such needles. If the needle is forced below the surface, it will at once sink.

Experiment. — Float two wooden toothpicks on water, placing them parallel and separated by a few millimetres. Let a drop of alcohol fall on the water between them, and they will suddenly fly apart.

The needle indents the surface of the water as if the surface were a tense membrane or skin, and tough enough to support the needle. The second experiment indicates that this membrane is stretched, the effect of the alcohol being to weaken it between the toothpicks, thus permitting the parts to separate through the superior tension of the portions outside.

Experiment. — Spread a thin film of water over a very clean glass plate, and touch it with a drop of colored alcohol on a glass rod. The alcohol makes a weak spot in the film. It breaks, and the tension around it draws the water away, leaving a dry area about the alcohol.

117. Surface Tension. — The surface of a liquid is physically different from the interior. The molecules composing the surface are not under the same conditions of equilibrium as those within the liquid. The latter are attracted equally in all directions by the surrounding molecules, while those composing the surface layer are attracted downward and laterally, but not upward. The result is that this surface layer is compressed and tends to contract. The contraction means that the surface acts like a stretched membrane, and, hence, when it is curved it exerts a pressure toward the centre of curvature.

By reason of this surface tension the surface always contracts so that it shall be as small as possible. Liquids in small masses always tend, therefore, to become spherical, since the surface is then the smallest that will enclose the given volume. Tears, dewdrops, and drops of rain are for this reason spherical. Surface tension rounds the end of a glass rod or a stick of sealing-wax when softened in a flame. It also breaks up a small stream of molten lead, and molds the detached masses into spheres, which cool as they descend, and form shot. This is only another way of saying that the spherical form is due to cohesion.

118. Further Illustrations of Surface Tension.—**Experiment.**—Make a ring 8 or 10 cm. in diameter, of stout iron wire, with a supporting handle. Tie to this a loop of thread, so that the loop may hang near the middle of the ring (Fig. 48). If now the ring is dipped into a soap solution, and a film is formed across it, the loop of thread will spring out into a circle when the film inside the loop is carefully broken by thrusting a hot wire through it. The tension in the film pulls the thread outward in all directions equally. If the ring be tilted, the circle will float about on the film.

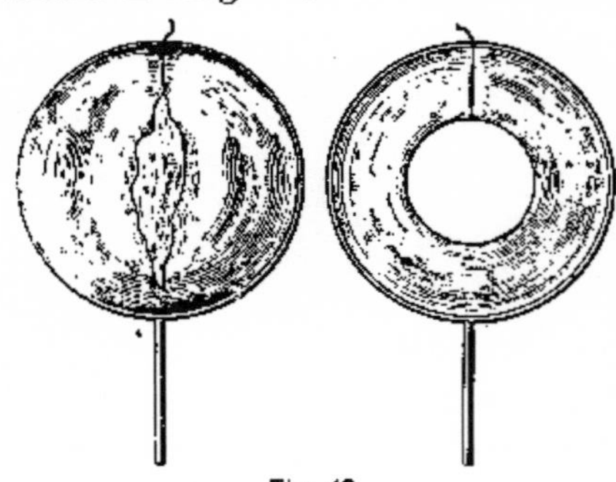

Fig. 48.

Experiment. — Blow a soap bubble on the wide end of a thistle tube. Hold the other end of the tube close to the flame of a candle. The bubble, which has both an outer and an inner surface film, will contract and drive a current of air through the tube with perhaps sufficient force to extinguish the flame.

Experiment. — Place some fragments of camphor gum on *clean water.* The camphor dissolves unequally at different points and produces an unequal weakening of the film. This causes the particles of camphor to move about in the most erratic manner.

An interesting modification of the above experiment is to make a miniature tin or wooden boat, having a notch cut in the stern, in which rests a bit of camphor gum (Fig. 49). The camphor weakens the tension astern, and the tension at the bow draws the boat onward.

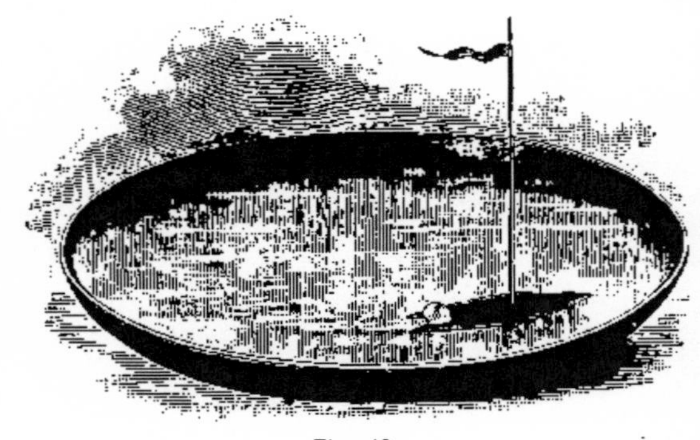

Fig. 49.

Caution. — In all experiments on the surface tension of water, the greatest care should be taken to keep the surface chemically clean. The slightest trace of oil, or a touch with a greasy finger, will often cause a failure in the experiment.

119. Capillary Phenomena. — The first phenomenon of surface tension to be observed and studied was the rise of liquids in capillary tubes, or those with a fine bore.

Experiment. — Support a clean strip of glass in water. The water will be seen to reach up above its level on the glass, making the surface near it concave upward. If mercury be used instead of water, the surface is depressed and is convex.

Experiment. — Support vertically two clean plates of glass inclined at a small angle (Fig. 50), with their lower edges in water. The height to which the water rises at different points is inversely as the distance between the plates at the points; and therefore the water line is a curve known in mathematics as a rectangular hyperbola.

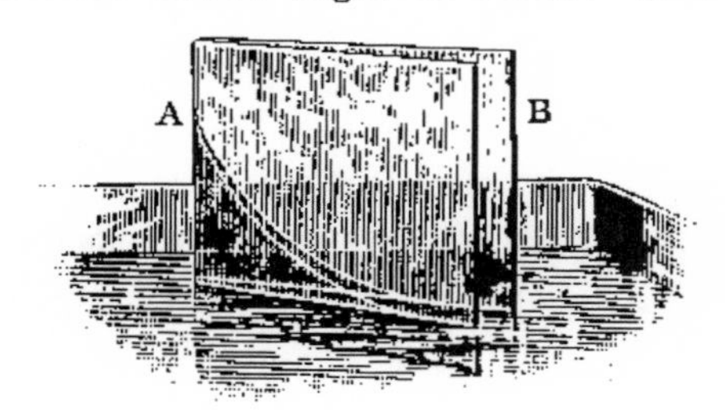

Fig. 50.

Experiment. — Support vertically several clean glass tubes (Fig. 51) of small internal diameter in a vessel of pure water. The water will be seen to rise in these tubes, highest in the one of smallest diameter and least in the one of the greatest. If mercury be used, instead of water, it will be depressed, the most in the smallest tube. On examining the surface of the liquid within the tubes, it is found to be concave upward when the liquid rises and convex when it is depressed.

Fig. 51.

120. Laws of Capillary Action. — The following laws have been established by experiment: —

I. *Liquids ascend in tubes when they wet them, that is, when the surface is concave; and they are depressed when they do not wet them, that is, when the surface is convex.*

II. *The elevation or the depression is inversely as the diameter of the tube.*

III. *The elevation or the depression decreases as the temperature increases.*

121. Familiar Illustrations of capillary action are numerous. Blotting paper absorbs ink because it is porous, and oil rises in a wick by capillarity. A sponge absorbs water for the same reason. The spread of water through a lump of sugar is explained in a similar manner. Small objects float together on water or cling to the sides of the vessel because of capillary action. Water rises around a fine wire dipping into it and interferes with its free rotation.

122. Explanation of Capillary Action. — It has already been shown that the attraction of glass for water is greater than the attraction of water for itself (§ 17). When a solid thus attracts a liquid, the liquid wets it and rises with a concave surface upward. The surface tension in a curved film produces a normal pressure toward its centre, as shown in the case of the soap-bubble. When, therefore, a liquid rises in a glass tube (Fig. 52) this normal force produced by surface tension in the concave film at *A* is upward. Where the surface is level at *C* and *D* there is no curvature, and the surface tension has no effect on the level of the liquid. The liquid therefore rises to

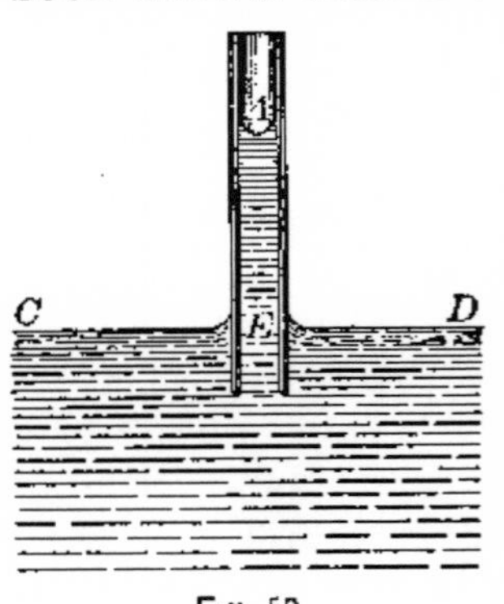

Fig. 52.

such a level in the tube that the pressure of the liquid column AE downward just equals the resultant normal force of the film upward. When the liquid does not wet the tube, the normal pressure of the film is downward, and for equilibrium this must be counterbalanced by pressure of the liquid on the outside.

II. PRESSURE IN FLUIDS.

123. Laws of Fluids. — There are three fundamental principles of pressure in fluids which may be called the *laws of fluids*: —

I. *Fluid pressure is normal to any surface on which it acts.*

II. *Fluid pressure at a point in a fluid at rest is of the same intensity in all directions.*

III. *Fluid pressure, neglecting the weight of the fluid, is the same at all points throughout the mass of the fluid.*

Fluid pressure is measured by the force exerted per unit area.

The pressure at a point is estimated by supposing the pressure equal to it to be exerted uniformly over a unit of area. The total force on unit area is called the pressure at the point.

124. Pascal's Principle. — The first of these laws is a consequence of the mobility of a fluid. It exhibits no friction at rest, and therefore yields under any force not normal to its surface. If at rest, therefore, the resultant force on it at any point is normal to the surface.

The other two laws are included in Pascal's principle of the equal transmission of pressure in all directions. A solid transmits pressure only in the direction in which the force acts; but a fluid transmits pressure in every direction. Hence Pascal's law: —

Pressure applied to any area of an enclosed fluid is transmitted in all directions and without diminution to every part of the fluid and of the interior of the containing vessel.

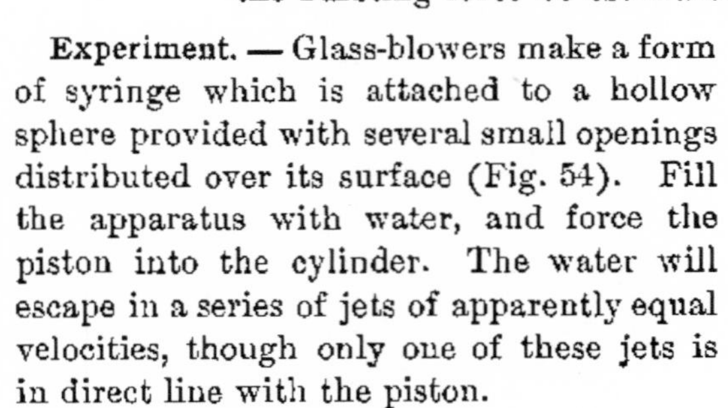
Fig. 53.

This is the fundamental law of the mechanics of fluids, and it applies to both liquids and gases. It was first enunciated by Pascal in 1653.

125. Illustrations. — **Experiment.** — Fit accurately to the mouth of a thin-walled pint bottle a close-grained cork (Fig. 53). Fill the bottle full of water, and then force in the cork by pressure, using a lever if necessary. The bottle will probably break. Explain. How could the bursting force be estimated?

Experiment. — Glass-blowers make a form of syringe which is attached to a hollow sphere provided with several small openings distributed over its surface (Fig. 54). Fill the apparatus with water, and force the piston into the cylinder. The water will escape in a series of jets of apparently equal velocities, though only one of these jets is in direct line with the piston.

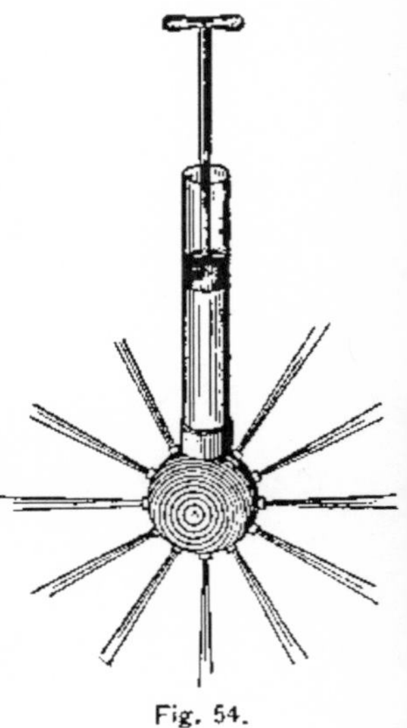
Fig. 54.

Experiment. — Fit a glass tube to a toy-balloon. Blow air into the tube; the balloon will swell out equally in all directions, showing equality of pressure.

126. The Hydraulic Press. — An important application of Pascal's principle is the *hydraulic press*, a machine employed for exerting great pressure, as in baling hay and cotton, making lead pipe, and lifting heavy masses in Bessemer steel mills, locomotive works, and on warships. It was invented by Bramah in 1795, and is shown in section in Fig. 55. Two heavy metal cylinders are connected by a strong tube *K*. A cast-iron piston passes water-tight through the collar *n* of the large cylinder; while in the smaller cylinder, the piston *p* is worked up and down as a force pump (§ 160), and pumps water from a reservoir at the bottom and forces it through the tube *K* into the cylinder *B*. When the plunger *p* of the

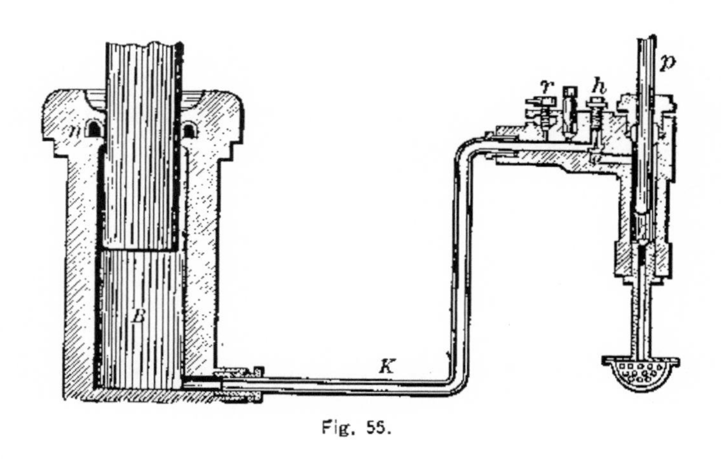

Fig. 55.

pump is forced down, the liquid in the machine transmits the pressure to the base of the large piston or ram, which is forced up with its load. If the cross-sectional area of the plunger of the pump is *a* and the downward force on it is *P*, then the pressure on the water is $\frac{P}{a}$. This pressure is transmitted to the cylinder *B*, and a pressure $\frac{P}{a}$ acts on each unit of surface of the base of the ram. If the area of the base is *A*, the total upward force, *W*, exerted on it is $\frac{AP}{a}$. Hence, the mechanical advantage is

$$\frac{W}{P}=\frac{A}{a}=\frac{D^2}{d^2},$$

if D is the diameter of the ram and d that of the plunger. If, for example, the diameter of the larger piston is 25 cm. and that of the smaller one 5 cm., then the force P applied is multiplied $\frac{25^2}{5^2} = 25$ times.

This machine conforms to the principle of work, for it is evident that the small piston moves as many times farther than the large one as the force exerted by the large one is greater than the effort applied to the small one.

127. Pressure due to Gravity. — The weight of each layer of a liquid is transmitted to every layer at a lower level.

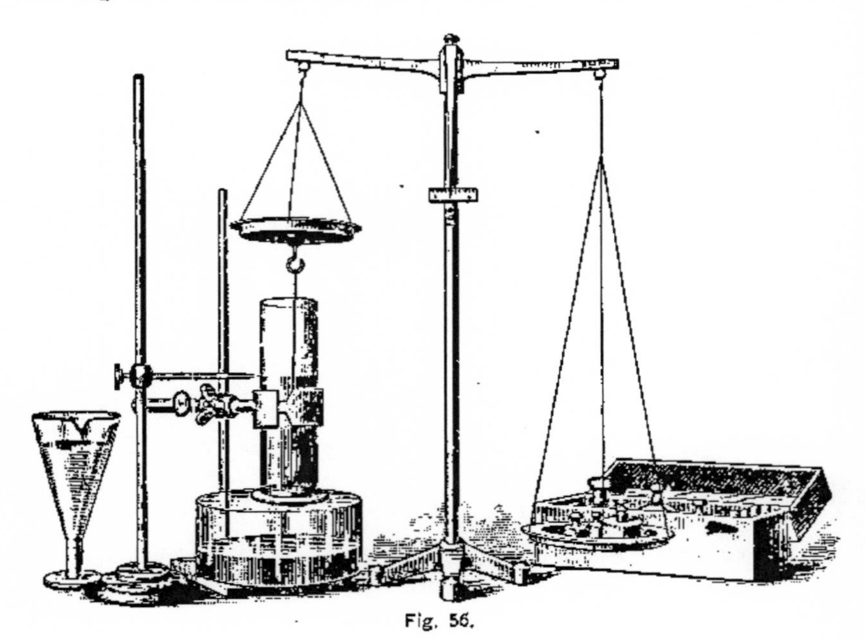

Fig. 56.

Experiment. — Grind to a true plane one end of a cylindrical lamp-chimney till a metal disk closes it water-tight. Suspend the disk by a cord from one end of a scale beam and counterpoise it (Fig. 56). Now place, say, 200 gm. on the scale pan and pour water into the cylinder until its pressure detaches the disk, and mark the depth. Repeat the experiment with 400 gm. on the pan. It will be found that the depth of the water when the disk is detached is twice as great as before.

Hence, *the downward pressure of a liquid is proportional to the depth.*

Experiment. — With the apparatus of the last experiment it is found that with a given weight on the scale pan a fixed depth of water is necessary to detach the disk. With the same weight in the pan, use a heavier liquid, such as a saturated solution of common salt. Determine its density (§ 144). Then the depth necessary to detach the disk will be decreased in the same ratio as the density is increased.

Hence, *the downward pressure of a liquid is proportional to its density.*

128. Pressure at a Point. — **Experiment.** — Bend three glass tubes into the J-forms shown in Fig. 57. Each one has a long and a short arm. The short arms are of equal length and are bent so as to open in different directions — upward, downward, and sidewise. Place the same depth of mercury in each tube. Now lower them into a tall jar filled with water. When the openings of the short arms are all as nearly as possible at the same point, the change of mercury level is seen to be the same in each.

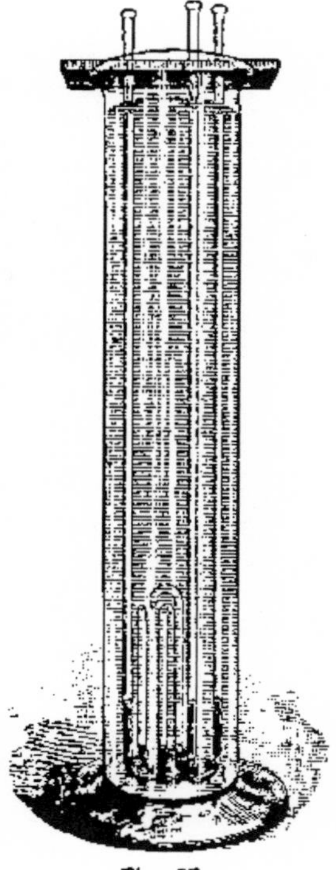

Fig. 57.

Hence, *the pressure at a point in a liquid is the same in all directions.*

It is immaterial whether this pressure is due to the weight of the liquid or is applied from without. The equality of pressure in all directions is a consequence of the equal transmission of pressure in all directions. The absence of currents in a vessel containing a liquid of uniform temperature demonstrates this principle, since any unbalanced pressure would produce motion in the liquid.

129. Pressure Independent of Shape of Vessel.—Experiment.— Using successively Pascal's vases, vessels differing in shape but having equal bases (Fig. 58), it will be found that with a given weight in the scale pan, the disk will be detached when the depth of water is the same in each case, notwithstanding the difference in amount of water used.

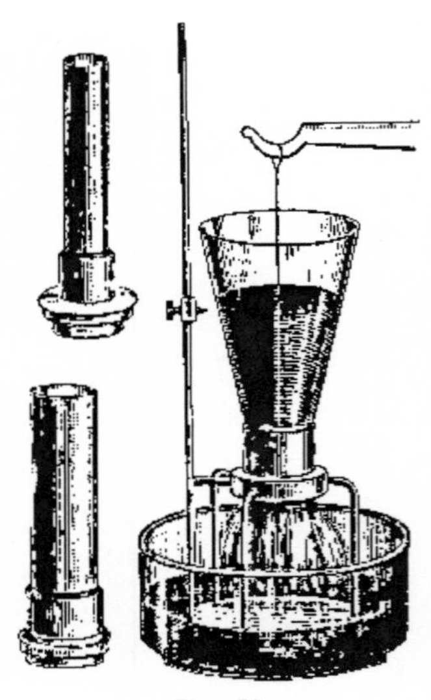

Fig. 58.

Therefore, *the pressure on the bottom of a vessel, or the downward pressure, is independent of the shape of the vessel.* The apparent contradiction of unequal masses of water producing equal pressures is often known as the *hydrostatic paradox.*

130. Total Pressure on any Surface. — Since liquid pressure depends on the depth and the density of the liquid, we may calculate the pressure on any horizontal area as follows:—

Let A denote the area pressed upon, H its depth, and d the weight of the unit of volume. Then the whole pressure on this area will be

$$P = AHd. \qquad (23)$$

In the metric system, d for water is 1 gm. per cubic centimetre. (For other liquids, consult Table of Densities in the Appendix.) In the English system, d is 62.4 lb. per cubic foot for water. For any other liquid, multiply by the specific gravity (§ 141) of the liquid.

Hence, *the total pressure of a liquid on any horizontal surface is equal to the weight of a column of the liquid whose base is the area pressed upon, and whose height is the depth of this area below the surface of the liquid.*

The pressure on any immersed surface, whatever its inclination, is found by computing the pressures on all the elementary areas and adding them together. The result is expressed as follows: —

The total pressure of a liquid on any immersed surface is equal to the weight of a column of the liquid whose base is the area pressed upon, and whose height is the distance of the centre of figure, or the centre of gravity, of this area below the surface of the liquid.

131. Surface of a Liquid at Rest. — The free surface of a liquid under the influence of gravity alone, is horizontal. If it were not horizontal, then the weight BW (Fig. 59) of a particle of the liquid surface would have a component BC parallel to the surface of the liquid. Since the air pressure on the surface is everywhere the same, there is no hydrostatic pressure to resist this force; and as there is no friction of rest in a liquid, the particle B would move. When the surface is level, BC vanishes and there is no motion. Even very viscous liquids assume a horizontal surface in course of time.

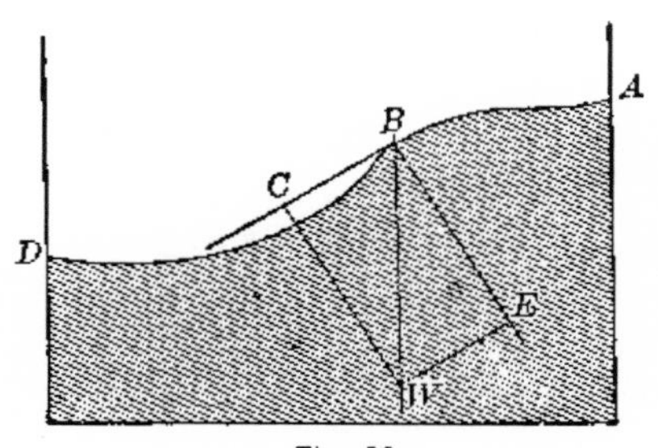

Fig. 59.

The sea or any large expanse of water is a part of the

spheroidal surface of the earth. It is therefore curved. A small liquid surface is practically plane, but not absolutely so.

132. Liquid in Communicating Tubes. — In Fig. 60, tubes of various shapes open into a connecting horizontal arm attached to the jar on foot. If a colored liquid be poured into the jar, it will rise to the same level in all the tubes. There is then equilibrium, because the pressures on the opposite sides of any imaginary cross-section of the liquid in the connecting tube are equal, since they are all due to liquid columns of the same height. Liquids are therefore said to "find their own level." The water supply of towns depends on this principle. When the water is pumped into a reservoir on a height, it is thence distributed everywhere by its own weight. Springs, fountains, and water and spirit levels illustrate the same principle.

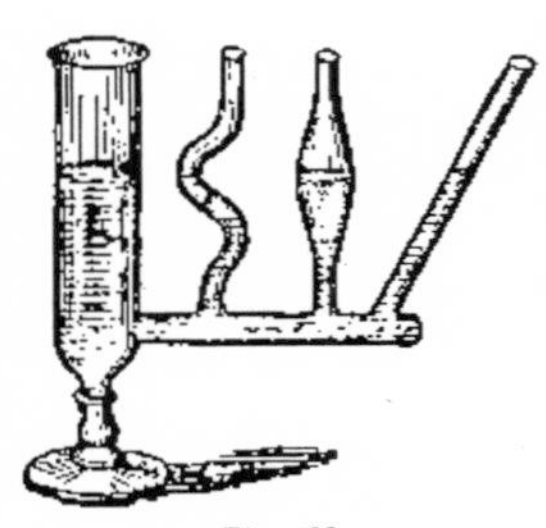

Fig. 60.

133. Superposed Liquids in Equilibrium. — If several liquids of different densities (§ 140) be placed in a vessel, with the heaviest at the bottom, and the others in order to the lightest at the top, they will show horizontal surfaces of separation, especially if no two in contact mix readily. Figure 61 shows mercury, water, oil, and alcohol with well-defined surfaces of separation between them. The contrast will be the more striking if the water and the alcohol are colored with an aniline dye. If a small iron ball be dropped into the

Fig. 61.

jar, it will settle down as far as the mercury and float there. In a tidal river the salt water flows up underneath the fresh water, which floats on top.

Experiment. — Place an egg in fresh water in a tall jar. It will sink to the bottom. Some strong brine may then be poured down a glass tube to the bottom of the jar. It will remain underneath the fresh water, and the egg will be seen suspended at the surface of separation at the top of the brine.

Problems.

1. Why are the embankments of canals made thicker at the bottom than at the top?

2. Which would require the stronger dam, and why — a mill-pond of large area and shallow or one of small area and deep?

3. The diameters of the cylinders of a hydraulic press are respectively 4 in. and 1 in.; the operating lever of the pump is 5 ft. long, with the piston-rod attached 10 in. from the fulcrum end. What resistance can be overcome by a force of 100 lb.?

4. The area of the small piston of a hydraulic press is 1.5 sq. in., and that of the larger one is 200 sq. in.; the arms of the lever by which the force pump is worked are to each other as 21 to 2. What force applied at the extremity of the lever will produce a pressure of 75,000 lb.?

5. A hydraulic lift is constructed to carry 8000 lb. If the pressure gauge attached to the pipe supplying the water registers 50 lb. per square inch, what must be the diameter of the large piston?

Ans. 14.27 in.

6. The water in a swimming tank is 5 ft. deep; the sides are vertical and 50 ft. long. Compute the pressure on the bottom; also on one side.

7. What is the amount of pressure exerted against a vertical milldam whose length is 250 ft., the water being 8 ft. deep?

8. A cylindrical jar 4 in. in diameter and 24 in. high is full of mercury. Compute the total lateral pressure.

9. Calculate in dynes per square centimetre the pressure on the bottom of a vessel 1.75 m. deep, full of mercury.

10. The spout of a common pump (§ 159) is 16 ft. from the piston; the diameter of the pump barrel is 2 in. Find the tension of the piston rod when the pump is full of water.

11. A rectangular board 3 ft. long and 2 ft. broad is immersed in water with its length parallel to the surface. Its upper edge is 3 ft. below the surface, and its lower edge $4\frac{1}{2}$ ft. below the surface. Find the pressure on the upper surface of the board.

12. To what depth may an empty closed vessel just capable of sustaining a pressure of 20 kgm. to the square centimetre be sunk in water before it breaks?

13. An inch board 12 ft. long and 12 in. wide is submerged in water. Its position is horizontal, face upward, and 6 ft. below the surface of the water. What is the upward pressure?

Ans. 4555.2 lb.

14. What head of water is equivalent to 14.7 lb. pressure per square inch?

Suggestion. — "Head" means "height."

15. Calculate the pressure that will be exerted on the face of a piston 1 ft. in diameter by a head of water of 400 ft.

16. When the pressure gauge was applied to a water pipe in a building it was found to register 75 lb. per square inch. How high was the top of the water in the supply reservoir above that point?

Ans. 173.077 ft.

17. The pressure in the water pipe in the basement of a building is 60 lb. per square inch, and at the top of the building it is 20 lb. Find the height of the building.

18. A closed cubical box, each edge of which is 2 ft., is filled with water. A vertical pipe 10 ft. high also filled with water is inserted in its upper surface. Compute the total water pressure on the six sides of the box.

III. DENSITY AND SPECIFIC GRAVITY.

134. Buoyancy. — A marble will sink in water, but will float on mercury. A piece of oak floats in water, but a piece of the dense wood known as "lignum vitæ" sinks. An egg will sink in fresh water and float in brine. When a swimmer wades up to his neck in sea water, he is nearly lifted off his feet by the water, which buoys him up.

Experiment. — Suspend a weight by a string from the hook of a spring balance and note the reading. Now submerge the weight in water. The index reading will be less. If salt water be used, the apparent loss of weight will be greater; if kerosene, it will be less.

Experiment. — Balance two half-kilogramme weights by a silk string over an easily running pulley, and bring a beaker of water under one of them. On lifting the beaker, the weight will not enter the water and remain immersed, but will rise to the surface.

These experiments show that the resultant pressure of a liquid on a body immersed in it is a vertical force upward, and it counterbalances a part or the whole of the body's weight. The upward pressure of a liquid is known as its *buoyancy*.

135. The Principle of Archimedes. — The law of buoyancy was discovered by Archimedes about 240 B.C. while attempting to determine the composition of King Hiero's crown.[1] It is as follows: —

A body immersed in a liquid is buoyed up by a force equal to the weight of the liquid displaced by it.

---

[1] *Encyclopædia Britannica*, Vol. II, Art. "Archimedes."

Let a cube be immersed in water (Fig. 62). The pressures on the vertical sides *a* and *b* are equal and in opposite directions. The same is true of the other pair of vertical faces. There is therefore no resultant horizontal pressure. On *d* there is a downward pressure equal to the weight of the column of water having the face *d* as a base and of a height *dn*. On *c* there is an upward pressure which is equal to the weight of a column of water whose base is the area *c*, and whose height is *cn*. The upward pressure therefore exceeds the downward pressure by the weight of the prism of water whose base is the face *c* of the cube, and whose height is the difference between *dn* and *cn*, or *cd*; and this is the weight of the volume of water displaced by the cube.

Fig. 62.

136. Experimental Proof. — Experiment. — A metallic cylinder 3.5 cm. long and 1.9 cm. in diameter has a volume of 10 cm^3. nearly. Suspend it by a fine thread from one of the arms of a balance (Fig. 63), and counterpoise. Then place a vessel of water under it so that the cylinder is completely submerged. The equilibrium will be destroyed, and may be restored by placing a 10-gramme weight in the pan above the cylinder. Since the cylinder displaces 10 cm^3. of water weighing 10 gm., and loses 10 gm. in weight when submerged, it follows that the body is buoyed up by a force equal to the weight of the water displaced. The resultant fluid pressure on a body of fixed volume immersed in a given liquid is the same whatever its substance. It depends on its volume only.

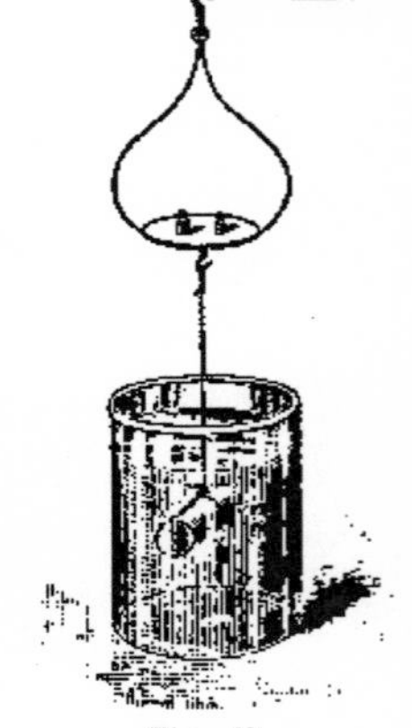
Fig. 63.

137. The Cartesian Diver. — Descartes illustrated the principle of Archimedes by means of a grotesque figure, since called a *Cartesian diver*, or a bottle imp. It is made of glass, is hollow, and the tail has a small opening at the end. The figure is partly filled with water so that it just floats in a jar of water (Fig. 64). When pressure is applied to the sheet of rubber tied over the top of the jar, it is transmitted to the water, more water enters the imp through the tail, and the air in it is compressed. It then displaces less water and sinks. When the pressure is withdrawn, the air in the diver expands and forces water out again. The displacement is then increased and the figure rises. The water in the diver may be so nicely adjusted that the little figure will sink in cold water, but will rise again when the water has reached the temperature of the room and the air in the figure has expanded. A good substitute for the diver is a slender inverted test-tube or a homœopathic vial containing a little air.

Fig. 64.

138. Equilibrium of Floating Bodies. — When a body is immersed in a fluid, it may displace a weight of fluid *less* than, *equal* to, or *greater* than its own weight. In the first case, the upward pressure will be less than the weight of the body and the body will sink. In the second case, the upward pressure will equal the weight of the body and the body will be in equilibrium, remaining in the liquid wherever placed. In the third case, the upward pressure will exceed the weight of the body, and the body will rise till these forces become equal. In liquids the buoyancy is practically independent of the depth so long as the body is wholly immersed, but will decrease as soon as it begins to emerge from the liquid. Hence,

When a body floats on a liquid it sinks to such a depth that the weight of the liquid displaced equals its own weight.

The weight of a body acts vertically downward, and the resultant pressure of the liquid acts vertically upward through the centre of gravity of the displaced liquid, which is called its *centre of buoyancy*. These two forces must be equal, and in the same vertical line for equilibrium.

139. Equilibrium of Floating Bodies Demonstrated.—**Experiment.**—Make a wooden bar 20 cm. long and exactly 1.5 cm. square. Bore a hole in one end and fill with enough shot to give the bar a vertical position when floating, with nearly its whole length in water. Fill the pores of the wood with hot paraffin. Graduate the bar in millimetres along one edge. Find the weight of the loaded bar in grammes, and then observe the length of the bar immersed when it floats in a tall jar of water. Calculate the immersed volume in cubic centimetres. This will also be the volume of water displaced; and since one cubic centimetre of water weighs 1 gm., we have the measure of the buoyancy. It will be found very nearly equal to the weight of the bar and shot. Hence, *a floating body displaces its own weight of the sustaining liquid.*

140. The Density of a body is the number of units of mass of it contained in a unit of volume. In the metric system it is the number of grammes per cubic centimetre. If m denotes mass, v volume, and d density, then

$$d = \frac{m}{v}, \quad v = \frac{m}{d}, \text{ and } m = vd. \tag{24}$$

141. The Specific Gravity of a body is the ratio of the mass of any volume of it to the mass of the same volume of pure water at 4° C. Specific gravity is, therefore, only the *relative density* as compared with water. It is also evident that the specific gravity of solids and liquids is numerically equal to the density when expressed in grammes per cubic centimetres, since the density of water is then unity.

Let m be the mass of a body and m' the mass of an equal volume of the standard, as water. Then the specific gravity $s = \frac{m}{m'}$ and $m = m's$. If the mass m' of water is expressed in pounds and its volume v in cubic feet, then $m' = v \times 62.4$, and $m = v \times 62.4 \times s$.

Since the density of water in the C.G.S. system is sensibly unity, there is no occasion to use the term specific gravity unless the mass and volume are given in some other system of measurement.

142. Density of a Solid.—To find the density of a body it is necessary to know its mass and volume. Its mass is ascertained by a balance. The most accurate and convenient method of obtaining the volume is furnished by Archimedes' principle. The buoyant effort of a liquid equals the difference between the weight of the body in air and its weight when immersed in the liquid. This difference is the weight of a volume of the liquid equal to that of the body. Hence, if this difference be divided by the density of the liquid, the quotient will be the volume of the liquid and also that of the body. The mass divided by this volume will be the density.

Water is the liquid generally used, and in the metric system its density is sensible unity. If the solid is soluble in water, then a liquid of known density, in which the solid is not soluble, must be used.

In case the solid is lighter than the liquid, a sinker

sufficiently heavy to sink the body must be employed. By subtracting the buoyant effort on the sinker from the buoyant effort on both, the weight of the liquid displaced by the given body is obtained. Then by proceeding as in the first case, the density of the body can be computed.

143. Examples. — First, *for a body heavier than water.*

Weight of body in air . . .	10.5 gm.
Weight of body in water . .	6.3 gm.
Weight of water displaced . .	4.2 gm.

Since the density of water is 1 gm. per cubic centimetre, the volume of the water displaced is 4.2 cm^3. This is also the volume of the body. Therefore, $10.5 \div 4.2 = 2.5$ gm. per cubic centimetre is the density.

Second, *for a body soluble in water.* Suppose it is insoluble in alcohol, the density of which is 0.8 gm. per cubic centimetre.

Weight of body in air . . .	4.8 gm.
Weight of body in alcohol . .	3.2 gm.
Weight of alcohol displaced . .	1.6 gm.

The volume of alcohol displaced is $1.6 \div 0.8 = 2$ cm^3. This is also the volume of the body. Therefore, the density of the body is $4.8 \div 2 = 2.4$ gm. per cubic centimetre.

Third, *for a body lighter than water.*

Weight of body in air . . .	4.8 gm.
Weight of sinker in water . .	10.2 gm.
Weight of body and sinker in water	8.4 gm.

The combined weight of the body in air and the sinker in water is then $4.8 + 10.2 = 15$ gm. But when the body is attached to the sinker, their apparent combined weight is only 8.4 gm. Therefore the buoyant effort on the body is $15 - 8.4 = 6.6$ gm., and this is the weight of the water displaced by the body, and hence its volume is 6.6 cm^3. The density is then $4.8 \div 6.6 = 0.73$ gm. per cubic centimetre.

144. Density of Liquids. — *First, by the specific gravity bottle.* With liquids, as with solids, the chief feature of the problem is to ascertain the volume. The simplest method is by the use of the *specific gravity bottle.* This bottle is usually made to hold a definite amount of distilled water at a specified temperature, as 25, 50, 100, or 1000 gm. at 15° C. (Fig. 65). To use it, find the weight of the bottle when empty and when filled with the given liquid. The difference of these weights will be the mass of the liquid, which, divided by the volume of the bottle, will be the density.

Fig. 65.

To check the volume of the bottle, weigh it filled with ice-cold water and subtract its weight when empty and dry. The difference will be its volume in cubic centimetres. The stopper is a ground capillary tube for convenience in filling completely.

Second, by a glass sinker. Weigh a glass sinker in air and then in the liquid. The difference will be the mass of the liquid displaced by the sinker. (Why?) Then weigh the sinker in water; the loss divided by the density of water will be the volume of water displaced by the sinker, and hence the volume of the liquid whose mass has been found. Divide the mass of the liquid displaced by the volume displaced and the quotient will be the density.

Third, the hydrometer method. The common *hydrometer* is usually made of glass and consists of a cylindrical stem and a bulb weighted with mercury or shot to make it float vertically. Within the hollow glass stem is a scale graduated in some arbitrary manner or by trial, the zero being

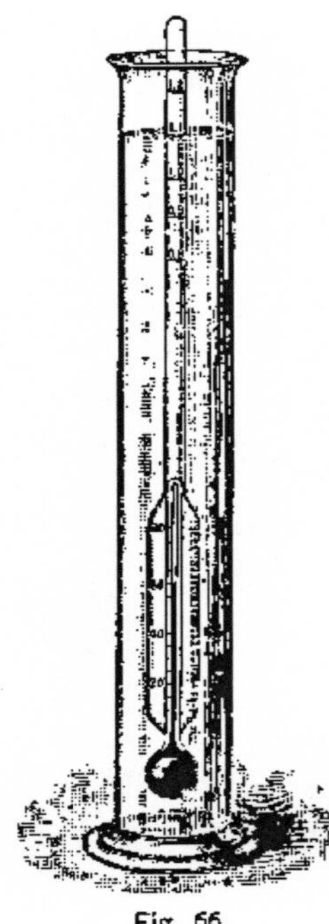
Fig. 66.

the point to which it sinks in distilled water at either 4° C. or at 60° F. The mark to which the instrument sinks in the liquid under test determines the density, either directly or by referring to an accompanying table. These instruments are often provided with a thermometer in the stem (Fig. 66), to give the temperature of the liquid at the time of taking the density.

Problems.

1. Why can stones be moved under water so much more easily than out of water?

2. A body weighs 62 gm. in air and 42 gm. in water. Find its density.

3. A solid weighs 100 gm. in air and 64 gm. in a liquid whose density is 1.2. What is its density? What is its specific gravity?

4. A bottle when filled with water weighs 64.485 gm., and when filled with methylated spirit 53.462 gm. The bottle weighs 15.063 gm. What is the density of the liquid?

5. A solid weighs 120 gm. in air, 90 gm. in water, and 78 gm. in a solution of zinc sulphate. What is the density of the solution?

Ans. 1.4 gm.

6. Find the density of a body from the following data: —

Weight of body in air	0.5	gm.
Weight of sinker in water	3.5	"
Weight of both in water	3.375	"

Ans. 0.8 gm.

7. If a cubic centimetre of iron weighs 6.23 gm. under water, what does it weigh in air?

8. A bar of aluminum ($d = 2.6$) weighs 54.8 gm. in air. When placed in water what will be the buoyant force?

9. A body weighs 24 gm. in air, and 20 gm. in water. What will it weigh in alcohol ($d = 0.8$)?

10. A tube 120 cm. long holds 600 gm. of mercury ($d = 13.6$). Calculate its internal diameter.

11. Calculate the diameter of an iron ball ($d = 7.8$) whose weight is 300 gm.

12. If the density of sea water is 1.025, what fraction of an iceberg ($d = 0.917$) floats above water?

13. An ounce of silver (sp. gr. = 10.15) is suspended in water. Find the tension of the supporting string.

14. Compute the weight of a lead ball (sp. gr. = 11.3) 1 in. in diameter.

15. A body weighs 10 oz. in air, and 7 oz. in water. Find its specific gravity.

16. What must be the density of a body, one-tenth of which floats out of water?

17. A piece of copper wire 2 m. long weighs 860 gm. in air, and 760 gm. in water. Compute the diameter of the wire.

Ans. 0.798 cm.

18. A cubic foot of cork weighs 250 gm. and floats on water. What pressure will be required to hold it completely under water?

Ans. 28.054 kgm.

19. A body ($d = 0.5$) weighing 50 gm. in air is attached to a lead ball which weighs 75 gm. under water. What weight is required in the opposite scale pan for equilibrium?

20. If the density of a mixture of glycerine and water be 1.094 gm., find the relative weights of glycerine and water in the mixture, the density of pure glycerine being 1.126 gm.

21. How much silver is contained in a gold crown which weighs 985 gm. in air and 918 gm. in water, taking gold as nineteen times and silver eight times as heavy as water?

22. Find the specific gravity of a body which weighs 120 grains in air, 215 grains in water with an attached sinker, the sinker alone weighing 275 grains in water.

23. Find the volume of a pound of sulphuric acid (sp. gr. = 1.84).
Ans. 8.341 liquid ounces.

SUGGESTION. — 1 gal. of water weighs 8.34 lb. and contains 128 liquid ounces.

24. An alloy (brass) of zinc ($d = 7.2$) and copper ($d = 8.95$) has a mass of 467 gm. Its volume is 60 cm³. Find the volume of each component.

IV. PRESSURE OF THE ATMOSPHERE.

145. Air has Weight. — **Experiment.** — Cement to a thin glass globe a brass cap provided with a stopcock for attachment to an air-pump (§ 151). (A thin metal globe may be used instead.) Suspend it from the scale pan of a balance and counterpoise. Then exhaust the air from the globe and hang on the balance again. It will be lighter than before. Open the stopcock; the air will rush in and the equilibrium will be restored. Compress the air in it and weigh again. It will now be heavier than when the stopcock was open. The mass of 1 litre (cubic decimetre) of air at 0° C. and 760 mm. pressure of mercury is 1.296 gm.

146. The Torricellian Experiment. — In the middle of the seventeenth century Galileo was called upon to explain why certain pumps erected by the Duke of Tuscany would not cause the water to rise more than about 30 feet. He suspected that the pressure of the air sustained a column of water of this height, but died without demonstrating it. Torricelli, a pupil of Galileo, first measured the pressure of the atmosphere in 1643 by the following method: —

Experiment. — Select a stout glass tube 80 or 90 cm. long and closed at one end. Fill with mercury, close the open end with the finger, and invert it in a cup of mercury (Fig. 67). When the finger is removed the mercury will settle in the tube a few centimetres, leaving a vacuum, called a *Torricellian vacuum*, above it. This column of mercury *AB* in the tube is supported by the pressure of the atmosphere on the mercury in the larger vessel at the bottom.

The demonstration was completed by Pascal, who found that the height of the mercury was less on the top of a high tower in Paris than on the ground; and that it fell nearly eight centimetres when the apparatus was carried to the top of the Puy-de-Dôme, about 1000 metres high, showing that the atmospheric pressure was less at that height.

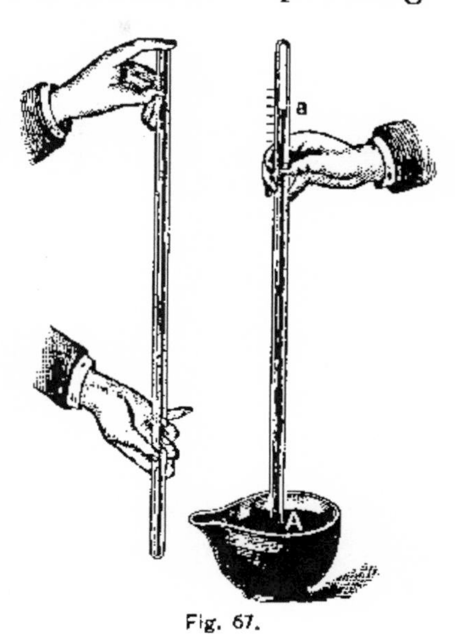

Fig. 67.

The height of the mercurial column supported by the atmosphere varies considerably from time to time. A standard value of 76 cm. has therefore been adopted to represent the mean pressure of the atmosphere at sea level. The height of the column is independent of the cross-section of the tube. Suppose an internal cross-sectional area of 1 cm^2. The volume of mercury supported by the atmospheric pressure on 1 cm^2. will then be 76 cm^3. The density of mercury at 0° C. is 13.596. Hence 76 cm^3. weighs 76 times 13.596, or 1033.3 gm. At sea level, then, the atmosphere exerts an average pressure of 1033.3 gm. per cm^2. This is equivalent to 14.7 lbs. per square inch. Either is called a pressure of one *atmosphere*.

147. Illustrations of Air Pressure. — **Experiment.** — Fill a tumbler full of water, cover it with a sheet of paper, and invert (Fig. 68) without letting the water escape. The air exerts a pressure on the paper more than sufficient to support the weight of the water.

Experiment. — Select two test-tubes, one wider than the other. The smaller one should fit the larger one rather loosely. Fill the larger one with water, insert the smaller one and quickly invert them. As the water escapes, the air-pressure will force the smaller tube upward into the larger one against gravity and hold it there.

Fig. 68.

Experiment. — Fasten a string to a round piece of leather. Wet the leather so as to make it pliable, and press it down evenly on a smooth flat stone. The stone, if not too heavy, can be lifted by the string, the pressure of the air keeping the leather pressed down on it.

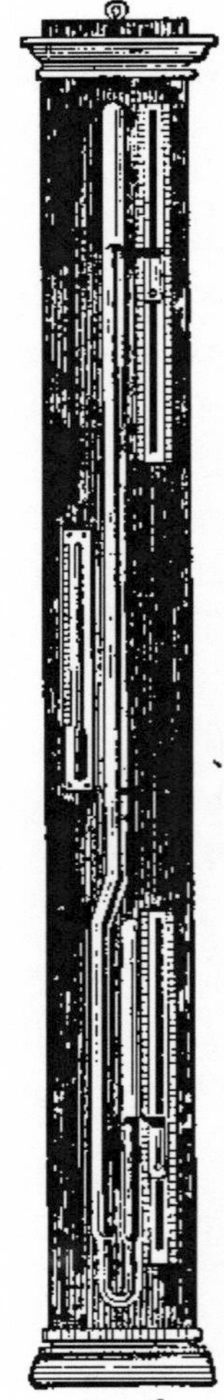

Fig. 69.

148. The Mercurial Barometer. — The mercurial *barometer*, for measuring atmospheric pressure, in its simplest form consists of a Torricellian tube about 86 cm. (nearly 34 in.) long, attached to a supporting board. A scale, whose zero is at the surface of the mercury in the cistern, is fastened by the side of the tube, to give the height of the mercury column. Torricelli suggested that a J-shaped tube be used, the short open arm taking the place of a cistern. The form shown in Fig. 69 was designed by Gay-Lussac. The short arm has a small pin-hole near the top for the admission of air. The height of the mercury column is given by the difference of the readings of two pointers on the scales on the right; for example, if the upper pointer reads 78.45 cm., and the lower one 4.23 cm., the pressure is 74.22 cm. of mercury. Readings must be taken with the tube in a vertical position. (Why?) When accuracy is required, corrections must be made for temperature, capillarity, and gravity.

A good barometer must contain clean mercury, and the mercury must be boiled in the glass tube to expel air and moisture.

149. Barometric Variations. — Since the mercury in the tube of the barometer is sustained by the pressure of the column of air resting on the mercury outside, any change in this pressure will produce a change in the barometric reading. Changes of this kind are going on continually at every place. Certain very slight changes are found to be periodic, but the greater changes follow no known laws. These irregular movements point to corresponding fluctuations in the pressure of the air, and consequently herald important atmospheric movements.

150. Uses of the Barometer. — The barometer is a faithful indicator of all changes in atmospheric pressure, and constant use is made of it by the Weather Bureau in forecasting changes of weather. Experience has shown that barometric changes are generally indicative of changes in the state of the weather, according to the following rules: —

I. *The rising of the barometer indicates the approach of fair weather.*

II. *The rapid fall of the barometer denotes the near approach of a storm.*

III. *A high, unchanging barometer, indicates continued settled weather.*

Since the pressure of the atmosphere diminishes with the elevation above the surface of the earth, the difference in the altitude of two stations may be computed from barometric readings taken at the two places simultaneously. The various rules proposed to express the relation between the height of the barometer and the elevation above sea level are more or less arbitrary, but they are used for determining the heights of mountains or other places with considerable accuracy. A simple rule for places near the sea level is to allow 0.1 inch for every 90 feet of ascent.

Problems.

1. Compute the height in inches of a mercurial barometer when the atmospheric pressure is 14.7 lbs. per square inch.

2. Calculate in dynes per square centimetre the atmospheric pressure when the barometer reads 28.5 in.

3. Calculate by what number a barometer-reading in inches must be multiplied in order to give the atmospheric pressure in pounds per square inch.

4. If the average barometer reading at the level of the sea is 30 in., and a cubic centimetre of air weighs 0.00129 gm., what would be the height of the atmosphere on the supposition that its density is uniform throughout?

5. What would be the height of a water barometer when a mercurial one reads 28.5 in.?

6. The diameter of a pair of Magdeburg hemispheres is 10 cm. If the barometer reads 74 cm. and the pressure gauge on the pump shows a pressure of 10 cm. within the hemispheres, what force in kilogrammes will be necessary to separate them?

7. If the surface of the body of a person of average size is 18 sq. ft., what atmospheric pressure does he sustain when the barometer reading is 29 in.?

8. If a barometer, whose reading is 743 mm., is placed with its cistern 50 cm. under water, what will be the reading?

9. Calculate the air pressure on a sphere 10 cm. in diameter, the barometer reading being 75 cm.

V. INSTRUMENTS DEPENDING ON PRESSURE OF THE AIR.

151. The Air-pump, as the name denotes, is a device for removing air or any gas from a vessel and depends for its action on the fact that gases are indefinitely expansible. The first pump was devised by Otto von Guericke about 1650.

Figure 70 represents the general appearance of one of the best forms made at the present time. Figure 71 shows the essential parts in section. A piston P, with a valve S in it, works in a cylindrical barrel, communicating with the outer air by a valve V at its upper end, and with the receiver on the pump table by a tube. The valve S' is carried by a rod which passes through the piston, fitting tightly enough to be lifted by the piston when the up stroke begins; but its ascent is almost immediately arrested by a stop near the upper end of the rod, and the piston slides on this rod during the remainder of the up stroke. This allows the air from the receiver to flow into the space below the piston. In the top plate of the cylinder is a lever, one end of which covers the valve rod. When the piston reaches the top of the cylinder it strikes this lever, and the lower valve, S', is thus closed. In the down stroke of the piston the valve S opens automatically, and the

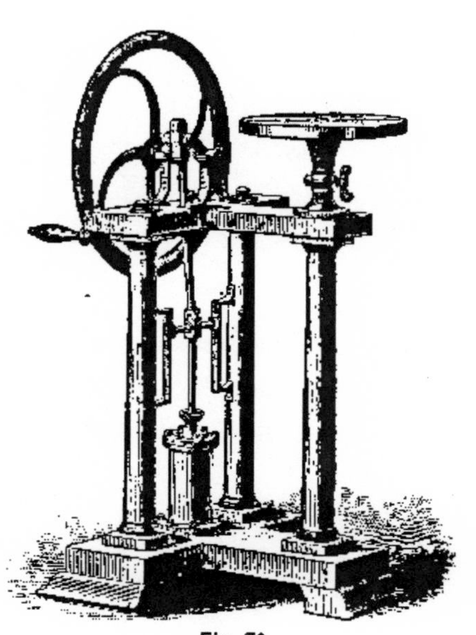

Fig. 70.

enclosed air passes through it into the upper part of the cylinder. The ascent of the piston again closes it; and as soon as the air is sufficiently compressed, it opens the valve *V* and escapes. Each complete double stroke of the piston removes a cylinder full of air; but as the air grows rarer with each double stroke, the mass removed each time is less. On account of the tendency of a gas to fill the containing vessel, irrespective of quantity, the removal of all the air from the receiver is not possible, although we might continually approach a vacuum were it not for the unavoidable mechanical defects of the pump, such as leakage of valves, untraversed space, etc.

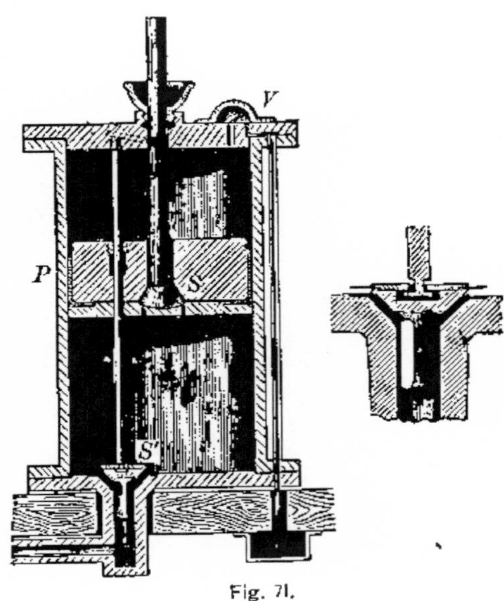

Fig. 71.

152. Experiments with the Air-pump. — **1. Football.** Fill a small rubber football half full of air, and place it under a bell-jar on the air-pump table. Exhaust the air from the bell-jar and notice that the ball expands till it is free from all wrinkles. What property of air is illustrated?

2. The Bladder Glass. Over one end of a glass cylinder tie a piece of bladder or paste a piece of paper (Fig. 72). Place it on the air-pump table and exhaust the air. The membrane or paper will break with a loud report. Why?

Fig. 72.

Repeat the experiment with sheet-rubber tied over the glass (Fig. 73). The rubber will be pressed into the glass. If the glass be turned on its side the effect is the same. Explain.

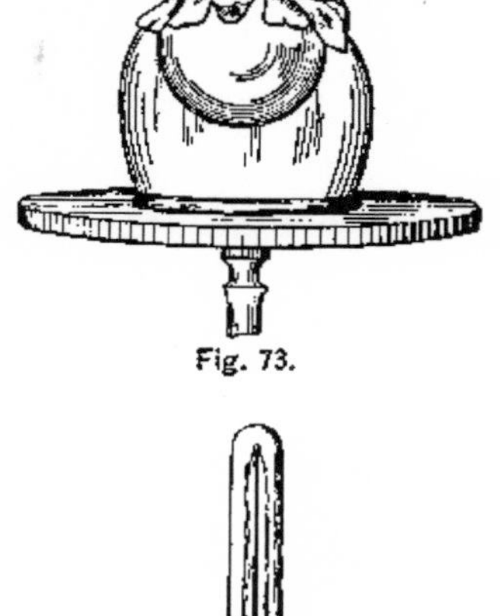

Fig. 73.

3. **The Bacchus Experiment.** Select two bottles; fit to one of them a perforated stopper. Connect the two by a bent tube reaching nearly to the bottom of each (Fig. 74).

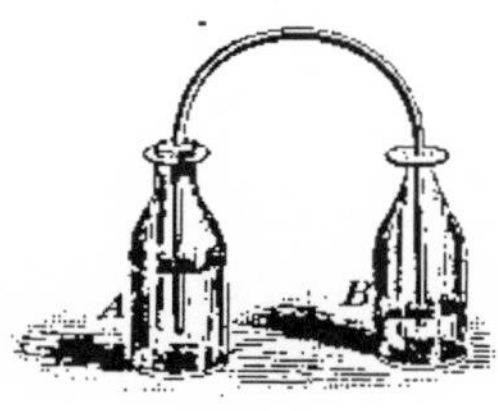

Fig. 74.

Fill the stoppered one nearly full of water and place them under a bell-jar on the air-pump table. Exhaust the air. Explain why the water flows out of the stoppered bottle and then flows back on admitting air into the bell-jar.

4. **The Vacuum Fountain.** A tall glass vessel is provided with a stopcock and jet-tube. (A bottle fitted with a rubber stopper can be used.) Having exhausted the air, place the mouth of the jet-tube in water and open the stopcock (Fig. 75). Why does the water rush into the vessel? Is it possible to determine how much air was not removed?

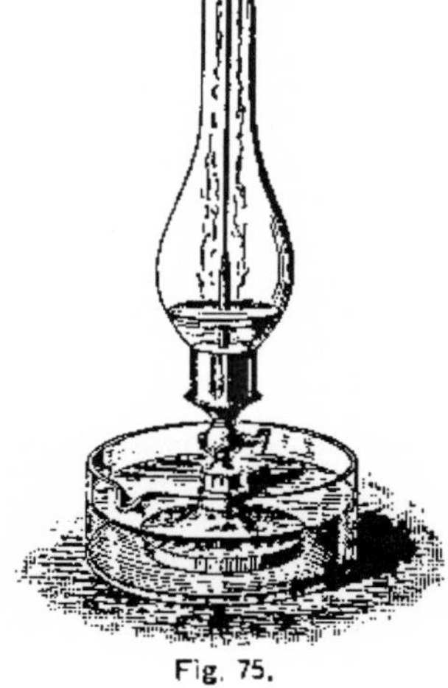

Fig. 75.

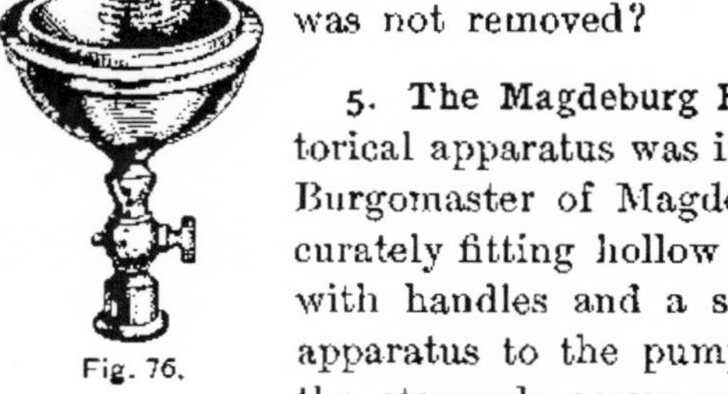

Fig. 76.

5. **The Magdeburg Hemispheres.** This famous historical apparatus was invented by Otto von Guericke, Burgomaster of Magdeburg. It consists of two accurately fitting hollow metallic hemispheres, provided with handles and a stopcock (Fig. 76). Attach the apparatus to the pump and exhaust the air. Close the stopcock, screw on the handles, and try to pull the hemispheres apart. How could it be shown that they are held

together by atmospheric pressure? In computing the pressure that holds them together, which is the surface to be considered, the spherical surface or the cross-sectional area? Why?

153. Buoyancy of the Air. — The principle of Archimedes applies to gases as well as to liquids. The resultant pressure of the atmosphere on bodies in the air is an upward force equal to the weight of air displaced. A body therefore weighs more in a vacuum than in the air, unless the volume of air displaced by it is the same as that displaced by the weights.

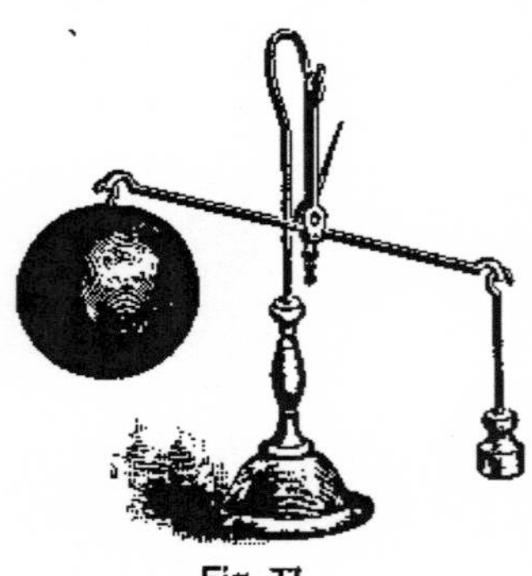

Fig. 77.

The *baroscope* is an instrument designed to exhibit the upward pressure of the air. A thin hollow globe is slightly overbalanced by a lead or brass weight on a small pair of druggist's scales (Fig. 77). (A cork sphere may be used in place of the hollow sphere.) When the baroscope is placed under a large receiver and the air is exhausted, the hollow sphere or the cork sinks, showing that it is really heavier than the counterpoise, but in the air it is buoyed up more because its volume is greater. Why would this experiment fail if the globe were not air-tight?

154. Balloons.—The upward pressure of the air is utilized in balloons. A balloon must be filled with a gas lighter than air, so that the weight of the gas and of the balloon with its car and contents shall be less than that of the air displaced. A balloon is not quite filled with gas at first, but as it rises it expands as the pressure of the air decreases. Its buoyancy then decreases but little as it rises into a rarer atmosphere.

With hydrogen, the ascensional force is about one kilogramme per cubic metre of gas; with common illuminating gas it is about half as great, but the latter is much less expensive.

On September 30 and October 9, 1900, two long-distance balloon races were made from Paris in an easterly direction. One of the contestants, Count de la Vaulx, the winner in both races, reached Russian territory in both, having travelled, the first time, a distance of 766 mi. in 21 hr. and 34 min.; and the second time, a distance of 1193 mi. in 35 hr. and 45 min. The maximum altitude reached was 5700 m., or 18,700 ft.

The aeronauts testify that when the sun shone on the balloon and heated it, the expansion of the gas enlarged the balloon and increased its buoyancy, so that it shot up to higher altitudes. It became necessary, in consequence, to let out some gas to cause the balloon to descend again. In the night, when the temperature fell, the buoyancy, on the other hand, decreased. Ballast was then thrown out to lighten the balloon and prevent its descent. These alternate losses of gas and ballast at length exhausted the capacity of the balloon to keep afloat, and it finally descended to the ground.

155. The Condensing Pump. — If the discharge pipe of an air-pump were connected to a suitable vessel, air would be forced into the vessel during the action of the pump. Such a device would be a *condensing pump*. Since the valves of the ordinary air-pump will not stand high pressure, a pump designed as in Fig. 78 is more suitable for compressing a gas. The plunger is solid, and in the bottom of the cylinder are two valves, one opening inward and the other outward. When the piston moves upward, the gas is admitted through the left-hand tube, the valve being lifted by the pressure of the gas below it. When the piston descends, this valve closes and the right-hand one opens, affording an exit for the confined gas.

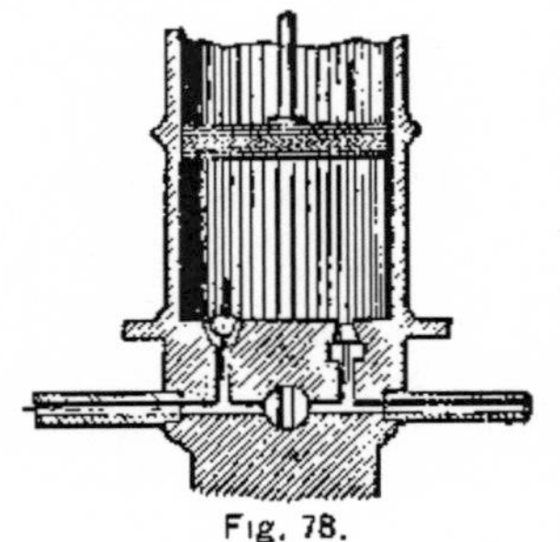
Fig. 78.

This machine is evidently an air-pump when the left-hand tube is connected with a receiver, but it is not capable of producing very high exhaustion, on account of the heavy valves required to give it strength for compression purposes.

156. Applications. — Both the air-pump and the condenser are extensively used in the arts. Sugar refiners employ the air-pump to reduce the boiling point of the syrup (§ 337); manufacturers of soda water use a condenser to charge the water with carbon dioxide; in pneumatic despatch tubes, now extensively employed for rapidly transporting small packages, both pumps are used, the one to exhaust the air from the tubes in front of the closely fitting carriage, and the other to force compressed air into the tube behind it, so as to propel it with great velocity. The condensing pump is also employed to improve the draft of furnaces, to facilitate the ventilation of buildings and mines, to operate pneumatic clocks, Westinghouse brakes on cars, and machinery in places difficult of access.

157. The Siphon, in its simplest form, is a U-shaped tube employed to convey liquids from one vessel to another at a lower level by means of atmospheric pressure (Fig. 79). To set it in action, the usual way is to fill the tube with the liquid, close the ends, place the shorter branch in the liquid, and open the ends. The flow will continue as long as the liquids in the two vessels are at different levels, and the shorter arm dips into the liquid. The siphon may also be started by suction; in the case of corrosive liquids, a suction tube (Fig. 80) is attached in a manner to prevent contact of the liquid with the mouth. The vertical distance of the highest part of the siphon above the surface of the liquid in the vessel being emptied equals the length of the *short arm* of the siphon, as *cd*, Fig. 79; the vertical distance of this highest point above the outlet of the tube equals that of the *long arm*. When the outlet is within the liquid, the measurement must be made to the plane of the surface of the liquid, as *ab*, and not to the end of the tube.

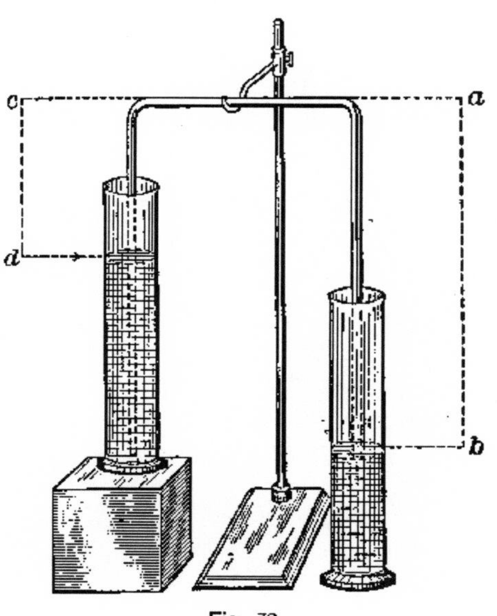

Fig. 79.

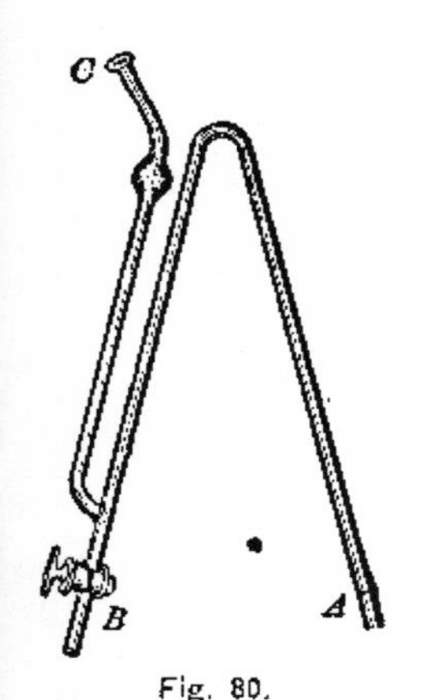

Fig. 80.

158. Its Action Explained. — **Experiment.** — Connect a piece of rubber tube to the long arm of a siphon, so that the length of that arm may be varied by raising or lowering the end of this tube. When the siphon is set in operation, it will be found that the rate of flow will increase as the outer arm is lengthened, and will decrease as this arm is shortened, the flow stopping entirely when the arms are of equal length.

Fig. 81.

Experiment. — Make a siphon of the form shown in Fig. 81, where the short arm is provided with a jet-tube opening within a bottle. If the length of the long arm is increased, the force of the fountain jet within the bottle will increase.

Experiment. — Make a glass siphon, the bore of the tube not to exceed 2 mm. in diameter. Set it in action under a bell-jar on the air-pump table, with mercury as the liquid. When the air is exhausted from the jar, the flow in the siphon will stop, but it will resume on admission of air. If the pump has a pressure gauge, it can be shown that the siphon stops when the pressure of the air in the bell-jar is not sufficient to support a mercury column as high as the top of the siphon.

The following explanation accords with these experiments: —

Let p represent the upward atmospheric pressure at the end of the tube d (Fig. 79). The pressure h of the liquid in that arm is downward. Hence the resultant pressure acting upward in the tube is $p-h$. Similarly if h' is the pressure of the liquid in the long arm, then the upward pressure in that arm is $p-h'$. The difference in resultant pressures at the two ends of the arms is therefore the difference between $p-h$ and $p-h'$, or $h'-h$, a force acting toward b. Hence the force causing the liquid to flow is measured by the pressure of a column of liquid whose height is the difference between the lengths of the arms. It follows also that the elevation over which a liquid can be siphoned cannot exceed the height of a column of that liquid which atmospheric pressure will support.

159. The Suction Pump. — In the *suction pump* a piston *c*, in which there is a valve opening upward, moves practically air-tight in a cylinder, at the lower part of which is an opening fitted with a valve *v*, also opening upward (Fig. 82). From this opening a pipe *s* leads down to a point below the surface of the water. When the piston is drawn upward, the valve in it closes by the pressure of the air above it, and a vacuum is formed in the cylinder below. The pressure of the air in the tube *s* opens the valve *v*, and the space between the piston and the water is filled with air under reduced pressure. Hence, the pressure of the atmosphere on the water in the well forces water up the tube to a height sufficient to produce equilibrium. When the piston descends, the lower valve *v* closes, while the valve *v′* in the piston opens and allows the air in the space below the piston to escape.

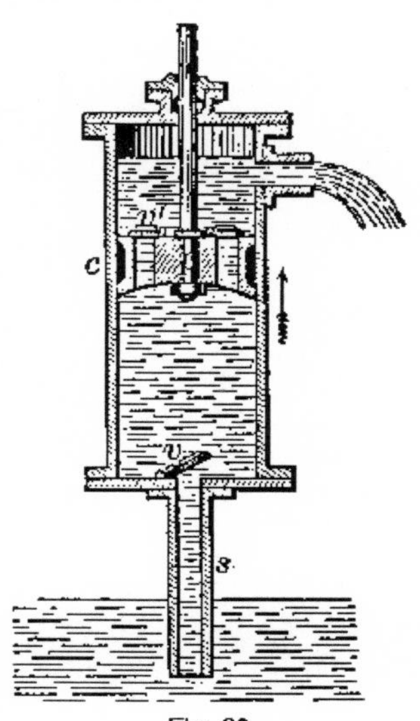

Fig. 82.

It thus appears that each double stroke of the piston removes some air from the cylinder, thereby lessening the pressure in it, while the pressure of the atmosphere on the water in the well causes the water to rise higher and higher in the pipe. If the valve *v′* is distant from the water less than 34 ft., provided the barometric reading is 30 in., the water will be forced by atmospheric pressure through *v′* into the space above the piston, and will be lifted till it flows out of the opening on the side of the pump.

160. The Force Pump. — In this pump (Fig. 83) the piston is solid, and the opening through which the water escapes is between it and the lower valve v, and is closed by a valve v', opening outward from the cylinder. The explanation of the working of this pump is similar to that given for the suction pump. As in the latter, the piston p must be within 34 ft. of the water to be pumped. The height to which the water can be forced in the pipe d depends on the force applied to the piston.

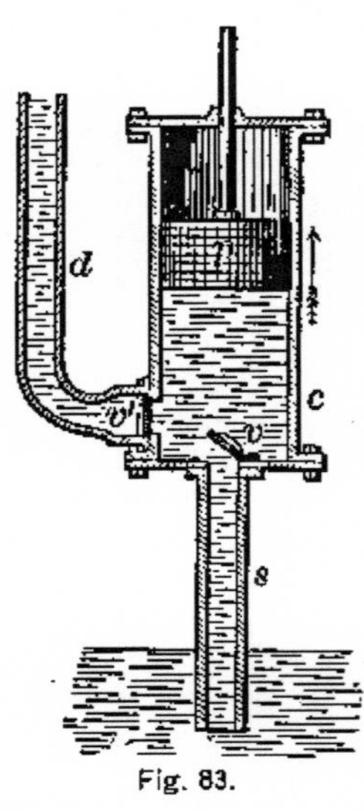

Fig. 83.

In powerful pumps the water usually passes into an air chamber called the *air dome.* Its object is to give steadiness of flow to the water from the delivery pipe. Fire engines and most pumps operated by steam are provided with an air dome.

161. Boyle's Law. — The simple relation existing between the volume of a gas and the pressure applied was first established by Robert Boyle and announced by him in 1662. It is known as *Boyle's Law* among English-speaking peoples, but the French call it Mariotte's Law. It is as follows: —

At a constant temperature the volume of a given mass of gas varies inversely as the pressure to which it is subjected.

If the volume v of a gas under a pressure p becomes v' on changing the pressure to p', then $\frac{v}{v'} = \frac{p'}{p}$, or $pv = p'v'$; that is, the product of the volume of the gas by the corresponding pressure remains constant for the same temperature.

162. The Law Verified. — **Experiment.** — The apparatus (Fig. 84) consists of two glass tubes connected by a stout rubber hose and attached to a wooden support carrying a metric scale. The left-hand tube is closed at the top with an iron cap. Either tube can be fastened at any desired point on the supporting board. Clamp both tubes near the middle of the scale, unscrew the cap, and pour in mercury till the tube with the screw cap is half full. Now screw on the cap, and lower the open tube as far as possible. Note on the scale the position of the mercury in each tube, the position of the top of the capped tube, and also the reading of the barometer. Move the open tube upward a few centimetres, and repeat the readings. Continue in this way till the top of the scale is reached. Since the closed tube is of uniform bore, the volume of air will vary as the length of the column, and hence the length may be used instead of the volume. Find the length of the air column for each set of observations. The difference in the mercury readings, increased by the barometer reading, will give the pressure of the air in the tube in centimetres of mercury. If the temperature of the tube is kept constant, and the air in it is free from moisture, it will be found that the product of the length of each air column by the corresponding pressure is practically constant.

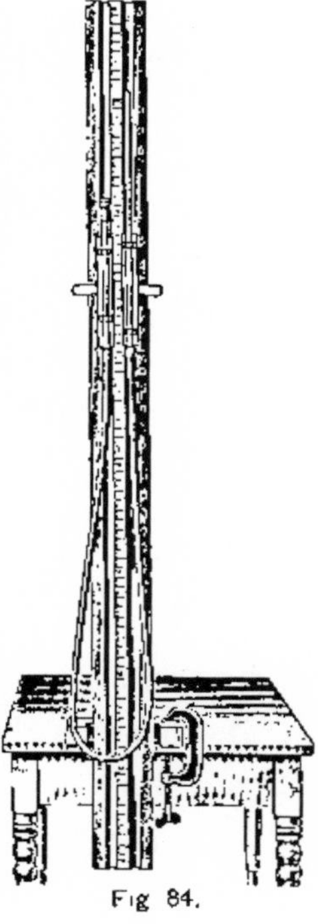
Fig 84.

The following record illustrates the foregoing: —

Top of air column 119.45 cm., Reading of barometer 74.18 cm.

Mercury Readings in Centimetres.		Volume of Air.	Pressure.	Product.
Air Column.	Pressure Column.	v	p	pv
90.50	67.20	28.95	50.88	1473
92.64	73.60	26.81	55.14	1478
94.60	79.90	24.85	59.48	1478
96.40	86.40	23.05	64.18	1479
98.03	92.80	21.42	68.95	1477
99.50	99.45	19.95	74.13	1479

Experiment. — Figure 85 illustrates another form of apparatus for verifying this law. It consists of a cylindrical bottle, through the rubber stopper of which is inserted the capped tube *A*, the open tube *C*, and a short tube terminating in a bicycle valve at *D*. The cork must be tied down. Remove the cap from *A*, attach a bicycle pump at *D*, and pump in air till the mercury rises in both tubes above the stopper. Now screw on the cap, and increase the pressure in the bottle by successive small stages till the mercury in the open tube reaches the top of the scale. Take the readings, as in the first form of the apparatus, with every change of pressure. The readings must not be taken immediately after changing the pressure, because a change of temperature accompanies a change of pressure.

Fig. 85.

163. The Law Inexact. — Extended investigations have shown that Boyle's law is only approximately true even for air at moderate pressures. In general, gases are more compressible than Boyle's law requires. Gases like sulphur dioxide, chlorine, and carbon dioxide, which are easily liquefied by pressure, show the largest variations from the law. Near the point of liquefaction the product pv is much smaller than accords with the law.

Such gases as oxygen and nitrogen show a minimum value of pv; beyond this minimum value an increase of pressure causes the product pv to increase. For hydrogen the value of the product is always higher than the law requires.

Within moderate limits of pressure, however, Boyle's law is extremely useful as a working relation.

Problems.

1. A certain mass of gas under a pressure of one atmosphere has a volume of 5000 cm^3. What will be the volume if the barometer reading is 73 cm.?

2. In a Boyle's law apparatus, when the mercury stands at the same level in both arms, the reading in each being zero, the volume reading of the air in the short arm is 20 cm. What must be the reading of the long arm, if by pouring in mercury, the mercury reading in the short arm is 12 cm., the barometer reading being 74 cm.?

Ans. 123 cm.

3. A coal-gas tank contains 12,500 cu. ft., the pressure being measured by a column of water 2 in. high. If the pressure be increased to 3.2 in., what will be the volume?

4. The barometer reads 740 mm. Into a tube 80 cm. long closed at one end is poured 50 cm. of mercury. The finger is held firmly over the closed end and the tube is inverted in a cistern of mercury as in Torricelli's experiment. If the mouth of the tube is 1 cm. below the surface of the mercury in the basin, at what height will the mercury stand in the tube?

5. If a gas tank of 50 l. capacity contains oxygen gas under a pressure of 10 kgm. per cm^2., what will be the volume under a pressure of one atmosphere?

6. A vessel of 2 cu. ft. capacity has compressed in it 100 cu. ft. of gas. What is the pressure per square inch in atmospheres?

Ans. 50 atmospheres.

7. If an open vessel contains 250 gm. of air when the barometer pressure is 74.3 cm., how much does it contain when the barometer pressure changes to 72.3 cm.?

SUGGESTION. — The mass of air will vary as the pressure.

8. A litre of air at 0° C. and under a barometer pressure of 76 cm. weighs 1.293 gm. What will be the weight of 125 l. of air at the same temperature and under a pressure of 743 mm.?

9. In a Boyle's law apparatus, the reading of the top of the closed arm is 25 cm., the mercury in the closed arm 10 cm., the mercury in the open arm 70 cm., the barometer 74 cm. What will be the reading of the mercury in the closed arm, if by adding more mercury, the mercury reading of the open arm is changed to 125 cm.?

Ans. 14.13 cm.

10. In collecting hydrogen gas over mercury in a graduated cylinder, the volume of gas was 26 cm^3., the mercury standing 15 cm. high in the cylinder, the barometer reading 74 cm. How many cm^3. of gas would there be if the pressure were normal?

CHAPTER IV.

SOUND.

I. WAVE MOTION.

164. Vibration. — **Experiment.** — Suspend a ball by a long thread and set it swinging to and fro like a common pendulum. Notice that the ball returns at regular intervals to the starting-point. Now set the ball moving in a circle, the string describing a conical surface. The ball again returns periodically to the point of departure.

A *vibrating* or *oscillating body* is one which repeats its limited motion at regular short intervals of time. A *complete vibration*, or simply a *vibration*, is the motion comprised between two successive passages of the object in the same direction through any position (§ 70).

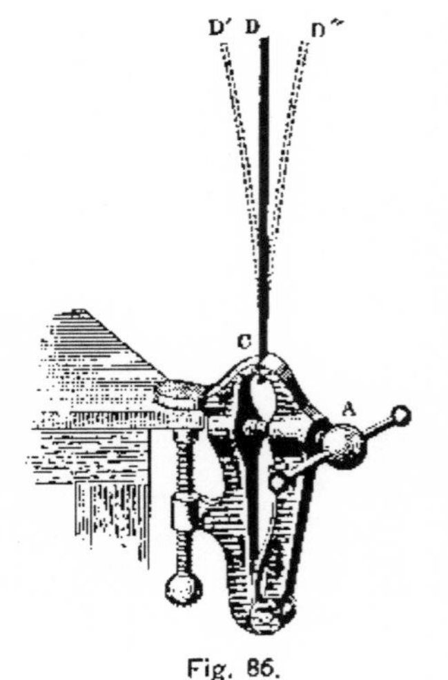

Fig. 86.

165. Vibrations Classified. — **Experiment.** — Clamp one end of a strip of brass or a lath in a vise (Fig. 86). Draw the free end aside and then release it. It moves to and fro like a common pendulum.

Vibrations of this character are called *transverse*, the motion being in a direction at right angles to the length of the vibrating body.

Experiment. — Fasten one end of a long spiral spring[1] to a hook in the wall and

[1] Such a spring may be made by winding No. 18 iron or brass wire on a long rod.

hold the other end in the hand. Crowd together a few turns of the spiral and then release them. A vibratory movement will be started, in which each coil swings to and fro in line with the length of the spiral. Fig. 87 shows the appearance of a portion of the spiral after releasing the compressed coils.

Fig. 87.

Vibrations of this character are called *longitudinal.*

Experiment. — Twist the bob of the torsional pendulum (Fig. 3) part way around. When released it returns periodically to its initial position, as the wire twists and untwists.

Vibrations of this character are called *torsional.*

166. Simple Harmonic Motion. — **Experiment.** — Suspend a ball by a long thread. Set it swinging in a circle (Fig. 88). The string describes the surface of a cone and, consequently, the pendulum is known as a conical pendulum. Place a white screen back of the pendulum and in front a lighted lamp, the light being in the plane of the circle. When the room is darkened the shadow of the pendulum bob will be seen to move to and fro across the screen in a straight line, slowly near the ends of the vibratory motion and rapidly near the middle. This shadow has nearly a *simple harmonic motion.*

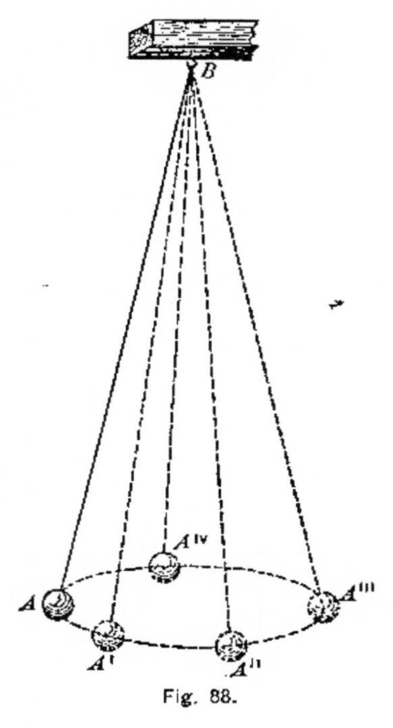

Fig. 88.

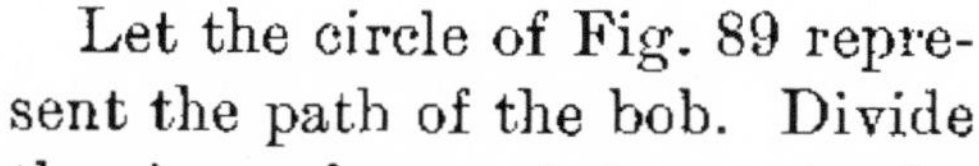

Let the circle of Fig. 89 represent the path of the bob. Divide the circumference into, say twelve, equal parts, as *ab*, *bc*, *cd*, etc. Through the points of division draw perpendiculars to the line AG. Then, the distances AB, BC, CD, etc., are the projections on the straight line of the equal arcs ab, bc, cd, etc., and represent the motion of the ball in successive equal periods of time as viewed from a distance. When a body vibrates to and fro in a straight line, as AG, in such a manner that its position at any moment is the same as the projection on that line of a point moving uniformly in a circle whose diameter is the length of the straight line, it moves with what is known as *simple harmonic motion*. All pendular motions of small amplitude are simple harmonic. The name is due to the fact that musical sounds are caused by bodies vibrating in this manner. The length DA or DG is the *amplitude* of vibration, and the *period* of vibration is the time-interval between two successive passages of the body through any point in the same direction. For example, the time of the body's moving from C to G, back through C to A, and then to C is the period.

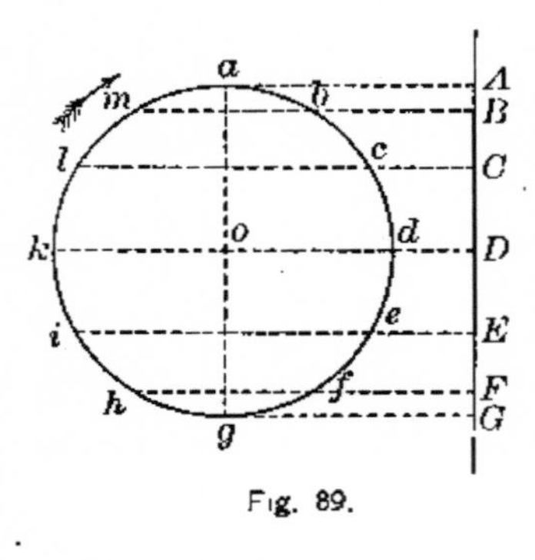

Fig. 89.

167. Waves.—**Experiment.**—Tie one end of a soft cotton clothesline to a rigid support. Grasp the other end and move it up and down quickly. Each point of the cord will be seen to vibrate transversely with a simple harmonic motion, and the disturbance started by the hand will move along the cord from one end to the other.

These curved forms traversing the cord are *waves*, which may, in general, be defined as the configuration of a medium caused by its parts vibrating and passing successively through corresponding positions.

168. The Harmonic Curve or Graphic Wave Form. — **Experiment.** — Make a pendulum, having for a bob a weighted funnel, with a very narrow outlet. Fill the funnel with fine sand, set the pendulum swinging in a plane through a small arc, and slide beneath it a board with constant velocity at right angles to the plane of the arc; the sand will be deposited in a wavy line.

This wave is the result of compounding a simple harmonic motion, with a uniform rectilinear motion at right angles to it. In Fig. 90, the vertical parallel lines *A*, *B*,

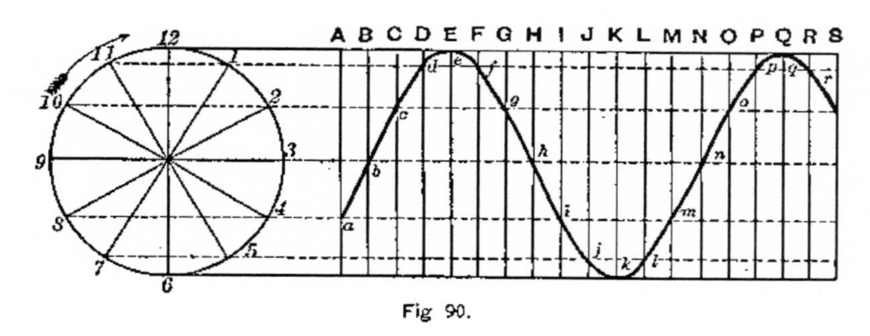

Fig 90.

C, *D*, etc., represent the paths of a series of particles moving with simple harmonic motion. To find the position of these particles in their paths at equal intervals of time, draw a circle whose radius is the amplitude of vibration, divide it into any convenient number of equal parts, say twelve, and through these points draw horizontal lines cutting the vertical ones. These lines will mark off on the lines *A*, *B*, *C*, *D*, etc., spaces which the vibrating particles traverse in intervals of $\frac{1}{12}$ of a period. Now, if the particle in the line *A* has made any number of vibrations and one-third of an additional one, reckoned from the point 12, it will be at *a*; and if each particle in succession is $\frac{1}{12}$ of a period behind the preceding, they will be at *b*, *c*, *d*, etc., respectively. A smooth curve drawn through these points gives the wave form.

If each particle be advanced in its path $\frac{1}{12}$ of a period, and a new curve be drawn (the student should do this), the wave form will be seen to have moved to the right; and if the particle be advanced through a whole period, the wave will have gone through one complete change, the top of the wave or crest having moved from E to Q.

QUERY.—How would you diagram a wave of greater amplitude?

169. Wave Length.—The *length* of a wave is the distance from any particle in the wave to the next one in the same vibration stage, that is, in the *same phase;* as from a to m, c to o, E to Q, etc. (Fig. 90). Art. 168 shows that the wave form travels from a to m, c to o, E to Q, etc., during one period or complete vibration of the particle. Hence, *the wave length is the distance traversed by the wave during one vibration period.*

170. Kinds of Waves.—**Experiment.**—Drop a pebble into a large vessel of water. Circular waves will move outward from the disturbed point to the sides of the vessel. Float a small cork on the surface. It will rise and fall as the waves advance, but will not be carried along with them, showing that the motion of the water is up and down and not forward.

Fig. 91.

Experiment.—Place a lighted candle at the contracted end of a long tin tube (Fig. 91). Over the other end tie a paper membrane. Strike two books together in front of the closed end; the flame is agitated. Tap the membrane with the finger or with a cork mallet; the flame will probably be extinguished. The effect on the candle is not due to a current of air, since the end of the pipe is closed by the membrane. If the tube is filled with smoke, it will not be driven out as by a wind, but it will be agitated by the vibratory movement passing through it.

These experiments illustrate two kinds of waves: the first, *gravitational*, or *waves* of *troughs* and *crests;* and the second, *compressional*, or *waves* of *condensation* and *rarefaction.*

171. Gravitational Waves. — The waves on the surface of water are due to the motion of its particles in closed

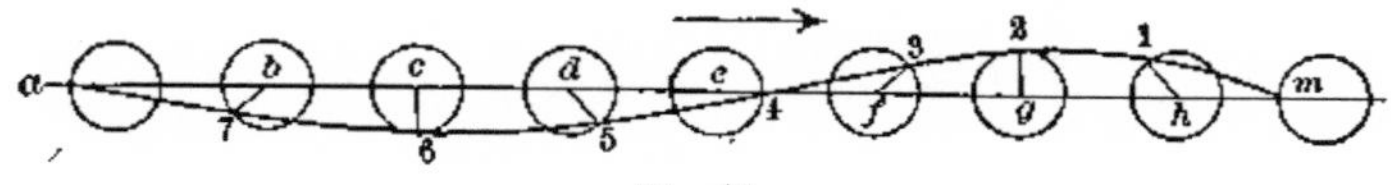

Fig. 92.

curves, which are circular when the amplitude is small. That such a motion will produce a wave form is shown in Fig. 92. Each circle is divided into eight equal parts; a water particle is supposed to move in the circumference of each circle at the same rate, but in any two consecutive circles to be at points separated by $\frac{1}{8}$ of a period. Then when *a* has completed one revolution, *b* will be $\frac{1}{8}$ behind, *c* will be $\frac{2}{8}$ behind, etc. A smooth curve traced through the points found in this manner represents the form of the surface of the water. The figure shows that the crest and trough are not of equal size, the former being narrower. If the circles were larger, that is, if the amplitude were greater, the crests would be still narrower, and would be sharp or looped when the amplitude is very large. The tops of the large waves would then break into foam or white caps.

172. Compressional Waves. — In the second experiment of Art. 170, the air in the tube next to the membrane is compressed. The elasticity of the air forces these air particles apart again, and in turn compresses the air farther along the tube. The continued repetition of this process carries the disturbance through the tube to the flame. Each air particle vibrates longitudinally with a simple harmonic motion, the whole phenomenon being quite similar to the vibrating spiral. Fig. 93 shows the distribution

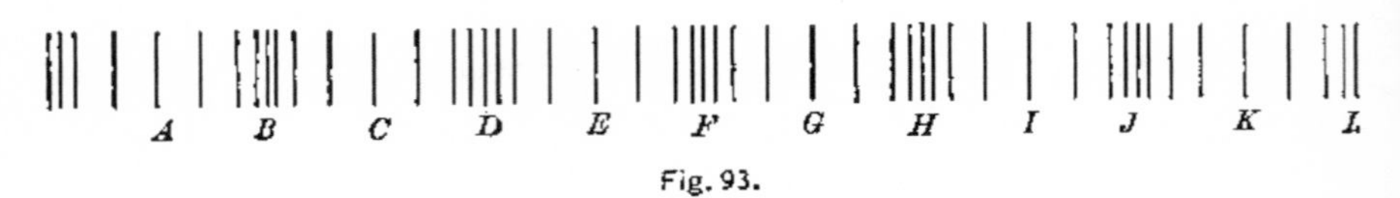

Fig. 93.

of the air particles when disturbed by compressional waves. A, C, E, etc., are regions of *rarefaction*; B, D, F, etc., are regions of *condensation*. Particles A and C, B and D, etc., are in the *same phase*; hence the distance AC is a *wave length, and comprises one rarefaction and one condensation.*

173. Composition of Two Simple Harmonic Motions in the Same Direction. — Let the first two curves (Fig. 94) represent two wave motions in the same medium, having the same amplitude, but differing in wave length. Draw the vertical lines through A'', B'', etc. Lay off $A''a'' = A'a' + Aa$, $B''b'' = B'b' + Bb$, etc.; through the points a'', b'', c'', etc., trace a smooth curve. It will represent the resultant wave form (§ 37). A study of the figure shows that sometimes the two waves act together, resulting in an increased amplitude, while at other times the motions impressed on the particles are opposite in direction, thus reducing the amplitude, and even at times destroying all motion. The figure illustrates the following principle: *If two waves pass simultaneously through the same medium,* *the actual motion of each particle is the resultant of the motions due to each system separately.*

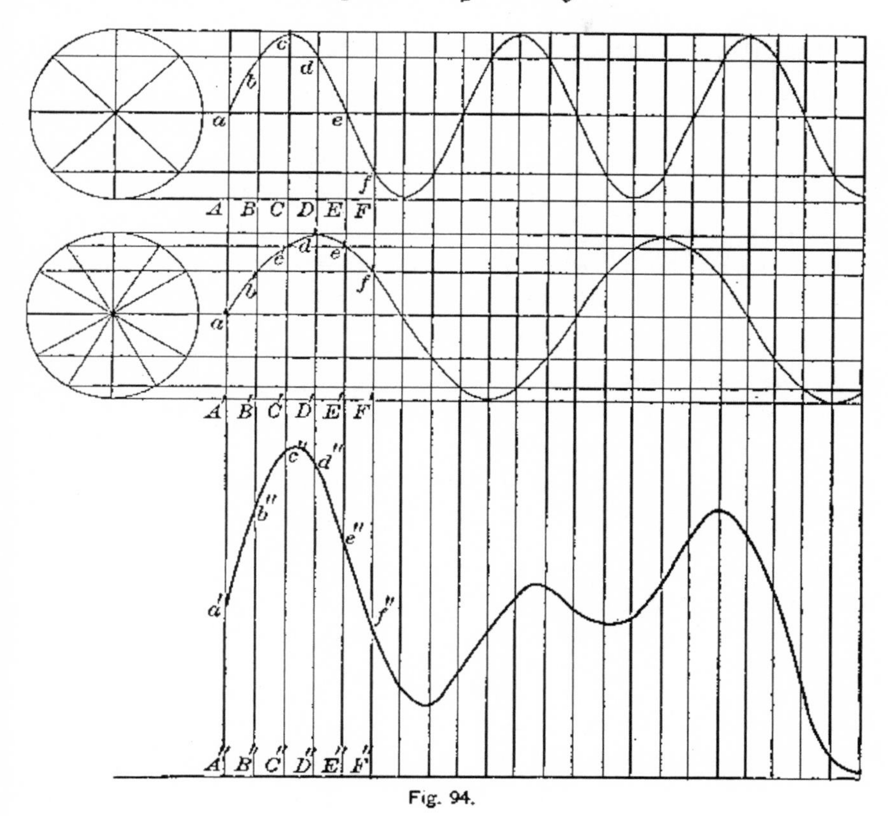

Fig. 94.

Problems.

1. Account for the waves which are often seen moving across fields of grain.

2. Diagram two harmonic curves of equal amplitude, the wave lengths having a ratio of two.

3. Diagram two harmonic curves having equal wave lengths, the amplitudes having a ratio of two.

4. Combine two harmonic curves of equal wave lengths, but of different amplitudes.

5. Combine two harmonic curves of equal wave lengths and of equal amplitudes, but differing in phase by half a period.

II. SOUND AND ITS TRANSMISSION.

174. Sound, as distinguished from the sensation of hearing, is that vibratory disturbance in an elastic medium which is capable of affecting the ear.

175. Source of Sound a Vibrating Body. — **Experiment.** — Suspend a small ball by a thread so that it just touches the edge of an inverted bell-jar. Strike the edge of the jar with a felted or cork mallet. The ball will be repeatedly thrown away from the jar so long as the sound is heard. What must be the condition of the jar?

Fig. 95.

Experiment. — Stretch a piano wire over the table and a little above it. Draw a violin bow across the wire, and then touch it with the suspended ball of the previous experiment. So long as the wire emits sound, the ball will be repeatedly thrown away from it. Inference?

Experiment. — Tap one prong of a tuning-fork (Fig. 95) against a block of soft wood, and, while sounding, touch one prong to the surface of water. In what condition is the fork shown to be?

Experiment.—Insert a whistle in one end of a glass tube (Fig. 96). Distribute evenly within the tube a little cork dust, made by filing cork. Close the other end of the tube, and blow the whistle, holding the tube in a horizontal position. The cork dust will rise in parallel vertical layers, falling back into ridges transverse to the axis of the tube when the sound ceases. Does the behavior of the cork dust indicate a vibratory motion of the air in the tube, or a current?

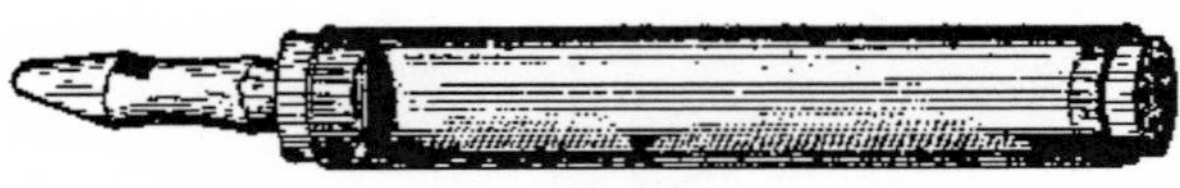

Fig. 96.

These experiments prove that the sources of sound are bodies in a state of vibration, the energy of the motion being sufficient to affect the ear.

176. Air a Medium. — **Experiment.** — Suspend an electric bell in a receiver on the air-pump table (Fig. 97). Set the bell ringing and exhaust the air from the receiver. The bell is heard less and less distinctly as the exhaustion proceeds, and would become inaudible in a perfect vacuum were it not that the suspension wires conduct the sound (§ 178). Readmit the air and the sound is restored. If, after exhausting the air, hydrogen is admitted and then exhausted, the diminution of the sound of the bell will be more marked (§ 197).

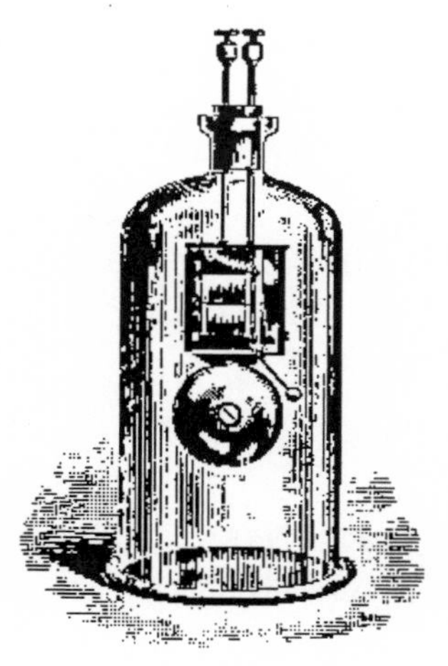

Fig. 97.

The experiment shows that air transmits sound, and also that sound cannot traverse a vacuum. By filling the jar with any kind of gas it may be shown that any gas transmits sound.

177. Liquids as Media. — **Experiment.** — Fill a tumbler with water, or any other liquid, and set it on the table. Insert the stem of a tuning-fork in a thin disk of wood about 3 cm. in diameter. Set the fork in vibration and hold it with the wooden disk resting on the liquid in the tumbler. The fork, which could scarcely be heard when held in the hand, will now be heard distinctly, the sound seeming to come from the table.

The vibrating fork, through the agency of the wooden disk on its stem, throws the liquid into vibration. These vibrations are transmitted by the liquid to the table, and thence to the air of the room.

178. Solids as Media. — **Experiment.** — Hold one end of a long slender wooden bar against the door of the room. Rest the stem of a vibrating tuning-fork against the free end. The sound of the fork will appear to come from the door.

The wood, like the water of the previous experiment, transfers the energy of the fork's vibrations to the door, and the door in turn to the air of the room. It is a familiar fact that, by placing the ear in contact with the metal rail of a railway track, two sounds can be heard, if the rail be struck at some distance away, one sound coming through the rail and the other through the air. The report of a cannon has been heard more than 250 miles by applying the ear to the ground. The great eruption of Cotopaxi in 1744 was heard distinctly 500 miles away, although several gigantic mountains and numerous deep valleys intervened.

The acoustic telephone, familiarly known as the *string telephone*, is a practical application of the sound-transmitting qualities of solids. It was invented in 1667 by Robert Hooke. It consists of a string or wire attached to the thin elastic bottoms of two small conical boxes. By speaking into either of these boxes, one listening at the other can hear distinctly, even for a considerable distance. The membrane vibrating transversely sets up longitudinal vibrations in the wire. These are transmitted to the membrane of the receiving instrument, and reproduce the sound actuating the transmitting instrument.

179. Sound Waves. — When a body, as a tuning-fork, is set in vibration, the disturbances produced in the air around it are known as *sound waves*. These waves consist of a series of condensations and rarefactions, succeeding each other at regular intervals, and forming concentric spherical shells of air of different densities. Each air particle vibrates harmonically and longitudinally in a short path along the radius of the expanding sphere. A *ray* of sound is the line which marks the direction of propagation; it is a radius of the spherical shell, and hence is a perpendicular to the wave front.

III. VELOCITY OF SOUND.

180. Velocity in Air.—In 1738 a commission of the French Academy, and again in 1822 a second scientific commission, experimented to determine the velocity of sound. The method of procedure was to divide into two parties, and by firing a cannon alternately at the two stations to determine the interval between the observed flash and the report. The mean of an even number of results eliminates very nearly the effect of the wind. The final result obtained was 331 m. per second at 0° C. The defect in this method is that the perception of sound and of light are not equally quick, and vary with different persons. Stone determined the velocity of sound in 1871 by stationing two observers three miles apart to give signals by electricity on hearing the report of a cannon. This method employs the sense of hearing only. After correcting as far as possible for all sources of error, the value obtained was 332.4 m. or 1090.5 ft. per second at 0° C. At 20° C. the velocity is about 1130 ft. per second.

181. Velocity in Gases.—It was shown by Newton that the velocity of propagation of a wave through any medium varies directly as the square root of the coefficient of elasticity of volume (§ 14), and inversely as the square root of the density $\left(v = \sqrt{\frac{e}{d}}\right)$. Since the density of oxygen is sixteen times that of hydrogen, it follows that sound will travel in hydrogen four times as fast as in oxygen. Subjecting a gas to pressure increases its coefficient of elasticity and its density at the same rate, and hence does not affect the velocity of sound in it. Heating a gas, however, increases the coefficient of elasticity, and hence increases the velocity of sound. Experiment and calculation agree in showing that the correction is 0.6 m., or nearly 2 ft. for 1° C.

182. Velocity in Liquids.—In 1827 Colladon and Sturm, by a series of experiments in Lake Geneva, found that sound travels in water at the rate of 1435 m. per second at a mean temperature of 8.1° C. Subsequent experiments show that the velocity is affected by changes of temperature and by the intensity of the vibration producing the sound. The velocity of sound in liquids is greater than in gases, owing to the fact that their coefficient of elasticity in proportion to their density is much greater.

183. Velocity in Solids.—The velocity of sound in solids is generally greater than in liquids on account of their high coefficient of elasticity as compared with their density. The velocity in iron is 5127 m. per second; in glass 5026 m. per second; but in lead, on account of its low elasticity, it is only 1228 m. per second, the temperature in each case being 0° C.

Problems.

1. At which place has sound the greater velocity, at the foot of a mountain or at the top? Why?

2. The flash of a cannon was seen, and 10 sec. later the report was heard; how far off was the cannon, the temperature being 16° C.?

3. How long will it take the sound of a signal gun to reach an observer 3.5 mi. away if the temperature of the air is 20° C.?

Ans. 16.35 sec.

4. Two stations, *A* and *B*, are found to be 3.5 km. apart. A person at *B* finds that 10.1 sec. elapse between seeing the flash of a gun at *A* and hearing the report. A person at *A* finds this interval to be 10.3 sec. Calculate the velocity of sound and that of the wind from the data.

5. What is the velocity of sound in coal gas at 0° C., the specific gravity being 0.5 that of air?

6. What must be the temperature of air in order that sound may have a velocity double that at 0° C.?

7. Carbonic acid gas is about 1.5 times more dense than air. Find the velocity of sound in it at 0° C.

8. How long will it take sound to traverse a distance of 5 km. in air, the temperature being 20° C.?

9. Air is about 14.5 times as dense as hydrogen. Calculate the velocity of sound in hydrogen at 0° C.

10. A stone dropped into the mouth of a mine was heard to strike the bottom in 2 sec. How deep was the mine?

11. The Eiffel Tower is 300 m. high. What time will elapse between the dropping of a stone from the top and hearing it strike the bottom, the temperature being 16° C.?

IV. REFLECTION AND REFRACTION OF SOUND.

184. Reflected Sound Waves.—**Experiment.**—Suspend a loud-ticking watch a little in front of the focus (§ 250) of a large concave reflector, as at *W* in Fig. 98. A place will be found at some distance in front where the watch can be heard with great distinctness, as at *E*; but if the reflector be removed, the ticking is nearly, if not quite, inaudible.

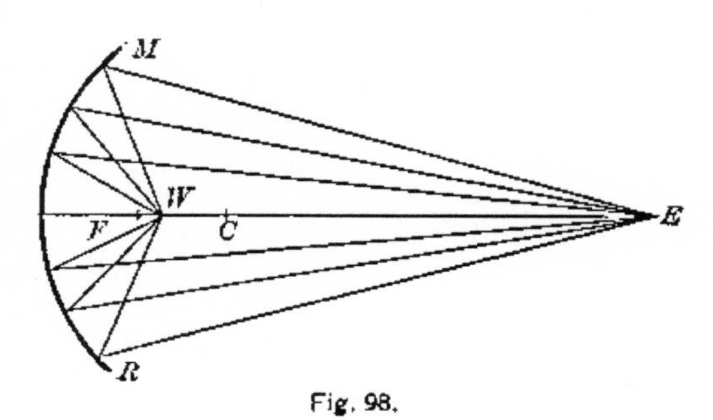

Fig. 98.

When sound waves strike against a smooth surface they are reflected in the same manner as when an elastic body strikes such a surface. If the surface be concave, they are reflected to a point. In Fig. 98 the rays of sound are used instead of waves for simplicity. These, diverging from *W*, are focussed at *E*, just as in the case of light (§ 252).

The ear trumpet is an instrument whose action depends upon the reflection of sound. The sides of the bell-shaped mouth reflect the sound into the tube which conveys it to the ear. Sounding-boards are sometimes placed back of speakers in large halls to reflect those portions of the sound waves that pass back of the speaker.

185. Echoes.—An *echo* is the repetition of a sound caused by the reflection of sound waves from some distant surface, like that of a building, or from cliffs, clouds, trees, etc. The interval between the production of a sound and the perception of its echo is the time that sound takes to travel from the source to the reflecting body and back again. The sensation of sound lasts for about one-tenth of a second, and during that time the sound wave travels $\frac{1}{10} \times 1130$, or 113 ft. If, then, the reflecting surface be about 56 ft. distant, a short sound will be followed immediately by its echo, since the first sound wave will travel to the reflecting surface and back to the ear to renew the sensation just as the first one ceases. If the distance be much less than 56 ft., the reflected sound tends to strengthen the original one, as illustrated by the distinctness of sounds in an ordinary room. The poor acoustic properties of many large halls and churches are due to the confusion of echoes from the large flat walls. Rooms draped with bunting or hangings of some sort have the echo deadened, because of the diffused reflection from the folds of the drapings.

Multiple echoes are caused either by independent reflections from bodies at different distances, or by successive reflections, as in the case of parallel walls at a suitable distance apart. The roll of thunder is partly due to multiple reflection. Multiple echoes following one another very rapidly produce *reverberations.*

186. The Whispering Gallery. — A curious effect of sound reflection is met with in the *whispering gallery*, where a faint sound produced at one point of a very large room is heard distinctly at some distant part, but is inaudible at points between; or where it is heard all round near the wall, but at no other place. In the first case, the walls act as curved reflectors (§ 184), and concentrate the sound waves to a point. In the second case, the sound is reflected from point to point along the curved wall, travelling in a series of equal chords, making the sound audible all round the wall. This is the case with the circular gallery in the dome of St. Paul's Cathedral in London, where a whisper is perfectly audible when the speaker and listener are exactly opposite each other.

187. Refraction. — **Experiment.** — Fill a large toy balloon with carbon dioxide and hang it up. Suspend a loud-ticking watch near it (Fig. 99). There will be a point on the opposite side, but farther away, at which the ticking can be heard very distinctly.

Fig. 99.

The balloon acts on the sound waves in the same manner that a convex lens acts on light waves, converging them to a focus (§ 250). The explanation is to be found in the fact that the middle portions of the waves on striking the balloon are retarded, the outer portions, so to speak, getting ahead and curving around, so that the waves are converted from a convex into a concave form. A balloon filled with hydrogen gas would act in the opposite manner, the middle of the waves getting ahead.

188. Wind Refraction. — It is a matter of common observation that sounds heard with the wind are louder than those heard against it. The reason is to be found in the fact that the velocity of the sound in the first case is increased by that of the wind. But the velocity of the wind is less near the earth's surface than a little above. Hence, the part of the sound waves touching the earth travels more slowly than that above; the waves are thus deflected downward and the sound is condensed along the earth's surface. In the second case, the lower part of the sound wave will be less retarded by the wind than the upper part and hence will get ahead of it. The wave is thus deflected upward.

Problems.

1. Why are there echoes in a very large empty room? Why do they disappear when the room is filled with people?

2. Why does a partially deaf person place his open hand back of his ear when listening to a speaker?

3. A man standing before a high wall shouts and hears the reflected sound in 4.5 seconds. How far away is the wall, the temperature of the air being 25° C.?

4. The report of a gun returns to the gunner in $2\frac{3}{4}$ seconds; how far away is the reflecting surface?

5. If a speaker can articulate distinctly but five syllables a second, what is the least distance from which a reflecting surface will send back a single syllable by echo, the temperature being 20° C.?

V. FORCED AND SYMPATHETIC VIBRATIONS.

189. Forced Vibrations. — **Experiment.** — Suspend a heavy weight by a cord or wire so that it will vibrate nearly as a seconds pendulum. From this weight suspend by a short thread a small weight, as a bullet. Set the system vibrating. The large ball by its superior energy impresses its own period of vibration on the small one and forces it to vibrate with it.

The experiment is an illustration of *forced vibrations*, which in general may be defined as vibrations not agreeing in period with the natural period of the vibrating body, but with that of the periodic force acting on it.

190. Illustrations. — The sounding-board of a piano and the membrane of a banjo are forced into vibration by the strings stretched over them. Two clocks which have nearly the same rate when on separate stands, will keep exact time together when they are placed on the same shelf, because each pendulum supplies a periodic force which acts on the other. They therefore exercise mutual control. The two prongs of a tuning-fork naturally vibrate at slightly different rates on account of unavoidable differences between them; but since they are connected at the stem, the faster one tends to accelerate the slower, and the slower to retard the faster, with the result that they agree in rate. The top of a wooden table may be forced into vibration by pressing against it the stem of a vibrating tuning-fork, and the loudness of the sound is greatly increased. This is a case of forced vibrations, and the table will respond to a fork of any pitch.

191. Sympathetic Vibrations. — **Experiment.** — Place near each other on the table two mounted tuning-forks tuned to exact unison. Keep one of them in vibration for a few seconds, and then stop it. The other one will be heard to sound loudly.

This experiment illustrates *sympathetic vibrations* in bodies having the same natural vibration period as the periodic force. In the case of the forks, the pulses in the air reach the second fork at intervals corresponding to its vibration period, and their effect is cumulative, each impulse arriving just in time to add to the movement. If the forks differ in period, the impulses from the first will not produce cumulative effects on the second, and the second fork will fail to respond.

In *forced vibrations* a vibrating body is compelled to surrender its preference for a particular mode and rate of vibration, and to adopt with more or less accuracy those imposed upon it by some external periodic force. But when there is equality of period between the periodic force and the natural vibration of the body, the co-vibration of the two is known as *resonance.*

192. Illustrations. — Resonance may be mechanical as well as sonorous. A heavy weight suspended by a rope may be set swinging through a wide amplitude by tying to it a thread and pulling gently on it when the weight is moving in the direction of the pull. Each effort then adds to the accumulated motion; the series of small impulses at the right intervals unite to produce a large movement.

If two heavy pendulums, suspended side by side on knife-edges on the same stand, are carefully adjusted to swing in the same period, and one of them is set swinging, it will cause the other one to swing, and will give up to it nearly all its motion.

Many years ago a suspension bridge at Manchester, in England, was destroyed by its vibration reaching an amplitude which exceeded the limits of safety. The cause was the regular tread of troops keeping time with what proved to be the natural rate of vibration of the bridge. Since then the custom has always been observed of breaking step when bodies of troops cross a bridge.

Release the wires of a piano by pressing the loud pedal; a note sung near it will be echoed by the wire which gives a tone of the same pitch. The "sound of the sea" heard when a sea shell is held to the ear is a case of resonance. The mass of air in the shell has a vibration rate of its own, and it amplifies any faint sound of the same period. A vase with a long neck will also exhibit resonance.

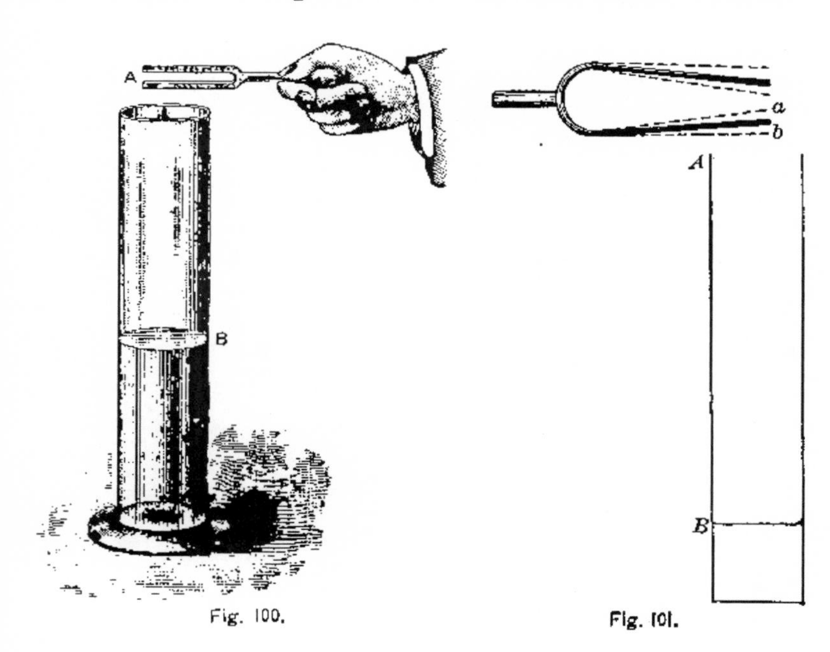

Fig. 100. Fig. 101.

193. Air Resonators. — **Experiment.** — Hold a vibrating fork over the mouth of a cylindrical jar (Fig. 100). Pour in water slowly, and note that, as the air column becomes shorter, the sound grows louder till a certain length is reached, after which it becomes weaker. If forks of different pitch are tried, each will be found to have a different length of air column for reënforcing its sound.

When the prong at *a* (Fig. 101) moves to *b*, it makes half a vibration, and generates half a sound wave. The pulse it sends down the tube AB is reflected from the bottom. Now, if AB is one-fourth of a wave length, then the distance down the tube and back will be half a wave length, and the pulse will return to A in time to strengthen the pulse sent out by the prong in its motion from b to a. If AB be three-fourths, five-fourths, or any odd multiple of one-fourth of a wave length, the effect will be the same; but for other lengths the reflected pulse will not return to A at the proper time to combine with those of the same phase produced by the fork in the air outside the tube, and so to increase their amplitude, but will reduce it instead.

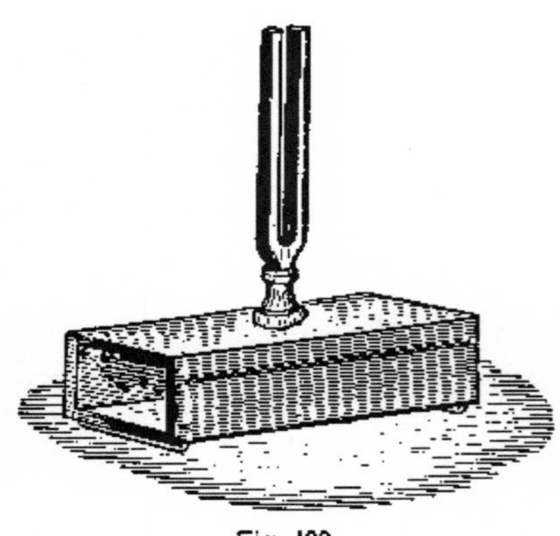

Fig. 102.

Unless the length of the air column is large in comparison with its diameter, it will be somewhat less than one-fourth the wave length of the sound reënforced. The box on which a tuning-fork is mounted (Fig. 102) is a resonator, designed to increase the volume of sound.

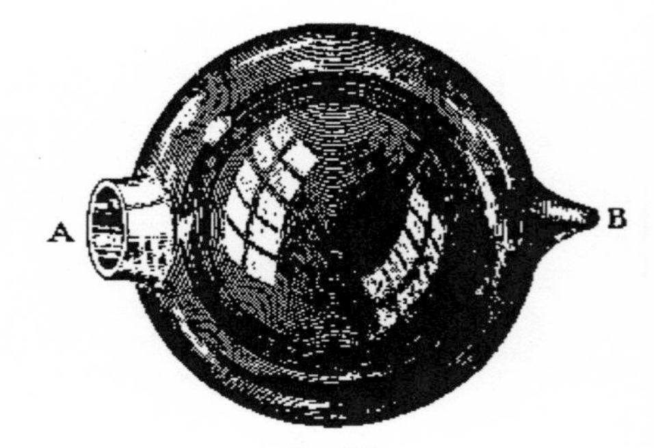

Fig. 103.

194. The Helmholtz Resonator. — The resonator devised by Von Helmholtz, for the purpose of picking out the overtones (§ 215) in a composite sound, is spherical in form, with two short tubes on opposite sides (Fig. 103). The larger opening, A, is the mouth of the resonator; the smaller one, B, fits in the ear. These resonators are made of thin brass or glass, and their pitch is determined by their size. When one of them is held to the ear, it strongly reënforces any sound agreeing with it in pitch, but is silent to others.

VI. INTENSITY AND LOUDNESS.

195. The Physical Intensity of a sound varies as the energy of the vibrating particles of the medium. The *loudness* of a sound depends on the individual sensitiveness of the ear, and on the extent of the physical disturbance reaching the ear drum. The loudness of a sound also involves the pitch. Intensity refers to the mechanical action that gives rise to the sound, while loudness refers to the sensation produced.

196. Effect of Amplitude. — **Experiment.** — Strike one of the prongs of a tuning-fork a slight blow. The sound emitted is feeble, and the prongs when touched to water disturb it but little. Now strike the fork a sharp blow; the sound is much louder, and when the prongs touch the water, it is thrown about more vigorously.

When the amplitude of vibration of the fork is large, a correspondingly large amplitude of vibration is imparted to the air. Since a vibrating body, like a pendulum, has a constant period nearly independent of amplitude, the mean velocity of the vibrating air particles must vary as the amplitude. But the energy of the movement varies as the square of the velocity (§ 85); hence, the intensity of sound varies as the square of the amplitude of vibration.

197. Effect of Density of Medium. — **Experiment.** — Fill a large bell-jar with hydrogen or coal gas. Raise the jar, keeping the mouth downward, and ring within it a small bell. The sound is much feebler than when the jar is filled with air. Now fill a large jar with carbon dioxide, and ring the bell in it. The sound is louder than in air.

The loudness of a sound depends, therefore, on the density of the medium at the place where the vibration is imparted to it.

The energy of the wave motion set up by the bell in the light gas is less than that in the dense one, and there is a corresponding difference in loudness. If a tuning-fork were used in conducting the experiment, the duration of vibration would be found to be longer in the rarer gas, so that the total amount of energy absorbed by the medium from the fork would be the same in each case, as required by the doctrine of Conservation of Energy (§ 87). On high mountains, where the air is quite rare, conversation is carried on with difficulty, and the firing of a gun produces little noise ; while one fired below may be heard as a loud report, even at great elevations.

198. Effect of Distance. — As the sound waves move outward from the vibrating body, each spherical layer of air imparts its energy to the enveloping one. Since these layers are surfaces of spheres, the number of particles composing them increases as the squares of their radii. Hence, the energy of the individual particles must decrease in like ratio, that is, *the intensity of sound varies inversely as the square of the distance from the source.*

During the vibratory movement of the air, some of the mechanical energy is transformed into heat by friction and viscosity and is dissipated. Hence, the actual decrease in intensity of sound is greater than that given by the theoretical law of inverse squares.

199. The Speaking Tube. — The weakening of sound from the enlargement of the sound waves as they recede from the source would evidently not take place if they were confined within a tube. Under such a condition the sound waves would not be propagated as concentric spheres, but the successive layers of air affected would be of equal mass, and the sound would be conveyed with little loss of intensity. Tubes used in this way are called *speaking tubes*. Long galleries, water pipes, and sewers act as speaking tubes.

200. Effect of Area of Vibrating Body. — **Experiment.** — Compare the sound of a small tuning-fork with that of a large one of the same pitch (§ 205). The large one produces the louder sound.

In order that a vibrating body may be a source of sound, the condensations and rarefactions in the air must be well marked. When the object is small, its surface is insufficient to affect a large quantity of air. Hence, *the intensity of sound depends on the area of the sonorous body.*

Illustrations of this fact are found in many stringed musical instruments, where the sound is intensified by placing two or more strings side by side when they are of small diameter; and, secondly, by placing a sounding-board beneath them to be set in motion by the string. The loud sound produced by many wind instruments is explained by the fact that the air within the broad aperture opposite the mouthpiece is a vibrating body of large area.

VII. INTERFERENCE AND BEATS.

201. Interference. — **Experiment.** — Hold a vibrating tuning-fork over a cylindrical jar, serving as a resonator. Turn it slowly around its axis and notice that when the edge of the prong is toward the jar the sound is nearly inaudible. When in one of these positions cover one prong with a pasteboard tube (Fig. 104). The sound is restored to nearly maximum intensity.

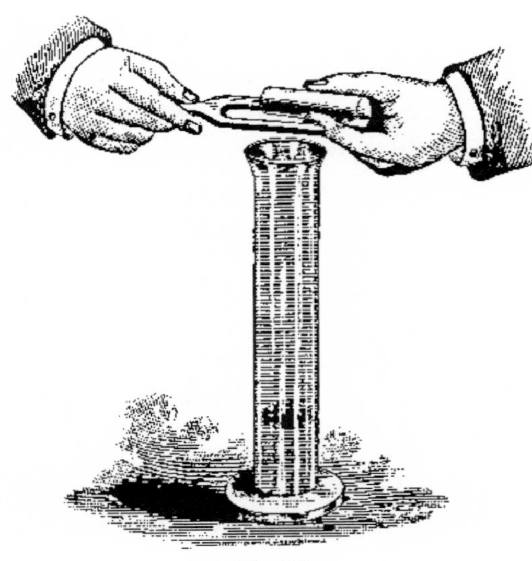

Fig. 104.

The explanation of this experiment is found in the fact that when the two prongs of the fork approach each other a condensation is produced in the air between them, and at the same time two rarefactions are started from the backs of the prongs. These opposite movements communicated to the air meet along surfaces extending outward from the edges of the fork and there neutralize each other. This explanation is supported by the fact of the restoration of the sound on cutting off one set of waves by the paper cylinder.

Interference is the superposition of two similar sets of waves traversing the medium at the same time. If two sound waves of equal length and amplitude meet in opposite phases, the condensation of one corresponding with the rarefaction of the other, the sound at the place of meeting is extinguished by destructive interference; if their phases are not precisely opposite or their amplitudes not equal, the extinction of the sound is not quite complete, or the interference is partial. One of the two series of similar waves may be direct and the other reflected.

202. Beats. — Experiment. — Select two large tuning-forks of the same pitch. When they are set vibrating, the sound is smooth as if only one fork were vibrating. Stick a piece of wax to a prong of one of the forks; the sound will be pulsating or throbbing.

Experiment. — With glass tubes and jet-tubes set up the apparatus of Fig. 105. Provide one tube with a paper slider so that its length may be varied. When the gas flame is turned down to proper size, the tube gives off a continuous sound, and we have what is known as a *singing flame.* By moving the slider, the tubes may be made to yield the same tone, the combined sound being smooth and steady. Now change the position of the slider, and the sound throbs and pulsates in a very disagreeable manner.

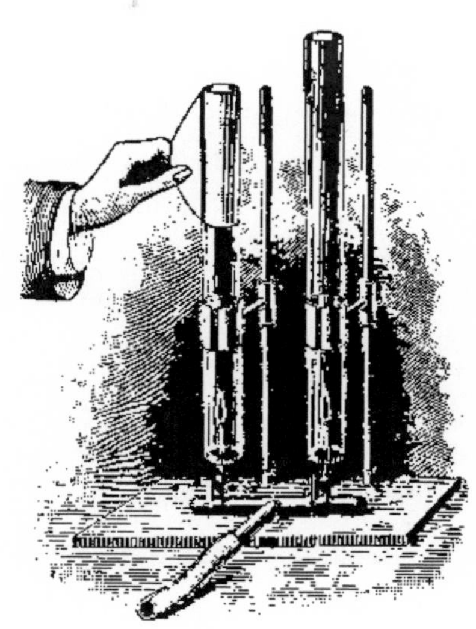

Fig. 105.

Both of these experiments are illustrations of the interference of two sets of sound waves. The outbursts of sound, followed by comparative silence, are called *beats.*

203. Number of Beats. — Let two sounds be produced by forks making, for example, one hundred and one hundred and twenty vibrations per second respectively. Then, in each second the latter fork gains twenty vibrations on the former; and there must be twenty times during each second when they are vibrating in the same phase, and twenty times in opposite phase. Hence, interference and subsidence of sound must occur twenty times during the second, and twenty beats are produced. Therefore, *the number of beats per second is equal to the difference of the vibration rates of the two sounds.*

VIII. PITCH.

204. Musical Sounds are those which are pleasant to the ear, and are caused by regular periodic vibrations. A *noise* is a disagreeable sound, either because the vibrations producing it are not periodic, or because it is a mixture of discordant sounds (§ 223), like the clapping of the hands.

Experiment. — Attach a rose burner to a metal pipe about 15 in. long, and connect it with the gas service by a rubber tube. Light the gas and notice the rustling sound attending its burning. Now hold a large tin tube, several feet long, over the burner. At a certain position of the flame within the tin tube, a sound like that of an organ pipe will be obtained. With tubes of different lengths, the pitch will be different.

The experiment shows that the rustling of the flame is caused by the mixing of many different sounds. If these sounds were not present they could not be reënforced by the different air columns.

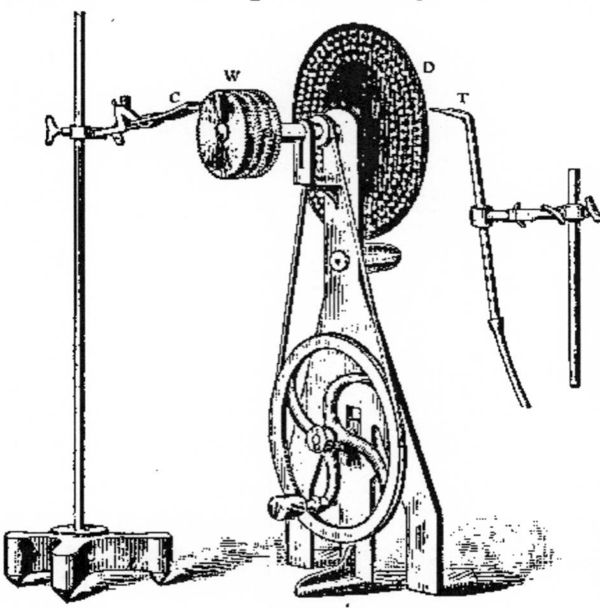

Fig. 106.

205. Pitch. — **Experiment.** — Mount on the axle of a whirling machine (Fig. 106) or on the armature of a small electric motor a cardboard disk (Fig. 107) provided with several concentric rows of equidistant holes differing in number, or several toothed wheels differing in the number of teeth. When rotating rapidly blow a stream of air from the tube T against one of the circles of holes in D, or press a thin card C against one of the toothed wheels W. In either case a distinct note is heard, different for each series of holes, or for each toothed wheel, or for any change in speed of rotation.

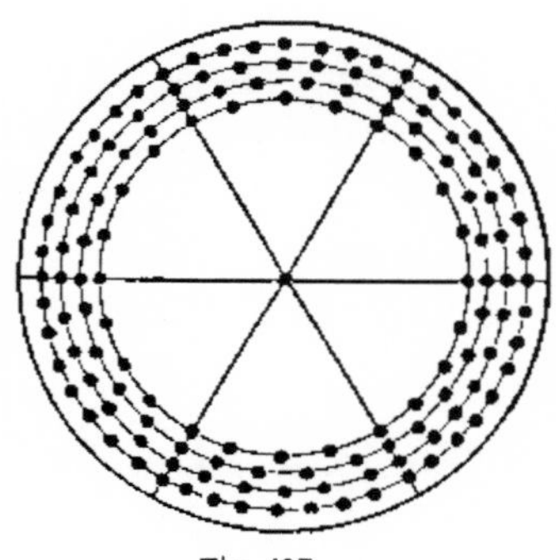
Fig. 107.

In each case, waves are produced in the air which, following each other with definite rapidity, give that characteristic to the sound which is called *pitch*. The perforated disk is called a *siren*, and the toothed wheel is a form of a *Savart's wheel*.

Either of these devices may be used in measuring the pitch of a note. If the number of holes in the circle or of teeth in the wheel be multiplied by the number of revolutions made per second, the product will be the frequency or vibration rate of the sound produced. Audible sounds have a lower limit of about sixteen vibrations a second, and an upper limit of about forty thousand. Most musical sounds are comprised between twenty-seven and four thousand vibrations a second.

206. Relations between Pitch, Wave Length, and Velocity.— If a tuning-fork makes 256 vibrations per second, and in that time a sound wave travels in air 344 m., then the first wave formed will be 344 m. from the fork on the completion of the 256th vibration. Hence, in 344 m. there would be 256 waves, and the length of each is $\frac{344}{256}$ m., or 1.344 m. In general, if l = wave length, v = velocity, and n = vibration rate, then

$$l = \frac{v}{n},\ v = nl, \text{ and } n = \frac{v}{l}. \qquad (25)$$

207. Intervals. — A *musical interval* is the relation between two sounds expressed by the ratio of their frequencies. Many of these intervals have definite names. When the ratio is 1 it is called *unison*, 2 an *octave*, $\frac{3}{2}$ a *fifth*, $\frac{4}{3}$ a *fourth*, $\frac{5}{4}$ a *major third*, $\frac{6}{5}$ a *minor third*, $\frac{25}{24}$ a *chromatic semitone*. Any three notes whose frequencies are as 4 : 5 : 6 form a *major triad*, and together with the octave of the lowest a *major chord*. Any three notes whose frequencies are as 10 : 12 : 15 form a *minor triad*, and together with the octave of the lowest a *minor chord*.

208. The Diatonic Scale or Gamut. — This is a series of eight notes which succeed each other with gradually increasing pitch, the two extremes being an octave apart. The first, or lowest note, is called the *keynote*, and the last is regarded as the keynote of another set of eight notes. In this way the series is repeated till the limit of pitch is reached, or a sufficiently extended scale is obtained. The tones comprised in each octave are named C, D, E, F, G, A, B, C′. The keynote may be given any pitch at pleasure. Physicists have agreed to assign to C, known as "middle C," 256 vibrations per second. In music the standard of pitch is variable; in the United States piano manufacturers agreed in 1892 to adopt as their standard A = 435.

The relative values of the notes composing the gamut is shown in the following table: —

Name . . .	C	D	E	F	G	A	B	C′
Vibration No.	256	288	320	$341\frac{1}{3}$	384	$426\frac{2}{3}$	480	512
Vibration ratio	C	$\frac{9}{8}$C	$\frac{5}{4}$C	$\frac{4}{3}$C	$\frac{3}{2}$C	$\frac{5}{3}$C	$\frac{15}{8}$C	2C
Intervals . .		$\frac{9}{8}$	$\frac{10}{9}$	$\frac{16}{15}$	$\frac{9}{8}$	$\frac{10}{9}$	$\frac{9}{8}$	$\frac{16}{15}$

An inspection of this table shows that the intervals between the successive tones are not equal, but are of three kinds, — $\frac{9}{8}$, called a *major tone;* $\frac{10}{9}$, a *minor tone,* and $\frac{16}{15}$, a *major semitone,* — and that they succeed each other in a definite order. If a note be raised by a chromatic semitone, $\frac{25}{24}$, it is said to be *sharpened*, and if lowered by $\frac{24}{25}$, to be *flattened.*

209. The Tempered Scale. — If C were always the keynote, the diatonic scale would be sufficient for all purposes except for minor chords; but if some other note be chosen for the keynote, in order to maintain the same order of intervals new and intermediate notes will have to be introduced. For example, let D be chosen for the keynote, then the next note will be $288 \times \frac{9}{8} = 324$ vibrations, a number differing slightly from E. Again, $324 \times \frac{10}{9} = 360$, a note differing widely from any note in the series. In like manner, if other notes are taken as keynotes, and a scale is built up with the order of intervals of the diatonic scale, many more new notes will be needed. This interpolation of notes for both the major and minor scales would increase the number in the octave to seventy-two.

In instruments with fixed keys such a number is unmanageable, and it becomes necessary to reduce the number by changing the value of the intervals. Such a modification of the notes is called *tempering.* Of the several methods proposed by musicians, that of *equal temperament* is the one generally adopted. It makes all the intervals from note to note equal, interpolates one note in each whole tone of the diatonic scale, and thus reduces the number of notes in the octave to twelve. Each interval is a semitone and equals $\sqrt[12]{2}$ or 1.05946. The only accurately tuned interval in this scale is the octave; the thirds are sharp, and the fifths flat. The following table shows the differences between the diatonic and the equally tempered scales: —

	C	D	E	F	G	A	B	C′
Diatonic . .	256	288	320	341.3	384	426.7	480	512
Tempered . .	256	287.3	322.5	341.7	383.6	430.5	483.3	512

Problems.

1. Explain why after a bell is struck the sound gradually "dies away."

2. Why is the pitch of the sound emitted by a phonograph raised by increasing the speed of the cylinder?

3. When soldiers are marching over a bridge of doubtful strength they are ordered to "break step." Why is this?

4. If a wave travels at the rate of 200 m. per second, and the wave length is 0.5 m., what is the vibration rate?

5. Calculate the wave length of a tone produced by 288 vibrations per second when the temperature of the air is 18° C.

Ans. 1.19 m.

6. What length of tube, closed at one end, will reëuforce a sound produced by a tuning-fork giving 320 vibrations per second, the temperature being 16° C.?

7. A pendulum whose length is 80 cm., and a tuning-fork are arranged so as to record their movements side by side on a smoked paper wound on a revolving cylinder. It is found that the line made by the fork indicates 203 double vibrations between each two marks made by the pendulum. Find the rate of the fork.

Ans. 226.2 per second.

8. The wheel of a Savart's apparatus has 40 teeth, and makes 1023 revolutions per minute. Find the pitch of the resulting tone.

Ans. 682.

9. A siren has 48 holes in the disk; how many revolutions must it make per minute in order that the sound emitted by blowing a stream of air against it may be in unison with "middle C" of the piano (258.7)?

10. A siren has 20 holes in the plate, and the indicator records 1200 revolutions in $1\frac{1}{2}$ min.; find the pitch of the note emitted.

Ans. 266.67.

11. A tuning-fork is held over a tall glass jar into which water is gradually poured until the maximum reënforcement of the sound is produced. This is found to be the case when the length of the column of air is 63.8 cm. What is the vibration number of the fork, the temperature being 16° C.?

12. How many beats are produced in a second by two notes whose rates of vibration are respectively 371 and 388?

13. How many beats per second between "middle C," physicists' standard pitch, and "middle C," equally tempered scale, international pitch ($A = 435$)?

Ans. 2.7.

14. What is the wave length of the note due to 538 vibrations per second at 16° C.?

15. Calculate the vibration numbers of the major triad, whose keynote is E (diatonic scale, physicists' pitch).

16. Calculate the wave length of the note G, which is a fifth above the middle C, in air at 15° C., physicists' pitch.

Ans. 88.9 cm.

IX. VIBRATIONS OF STRINGS.

210. Mode of Vibration. — Strings when used for the production of sound, are fastened at their ends, stretched to the proper tension, and made to vibrate transversely either by drawing a violin bow across them, striking them with a light hammer, or plucking them with the fingers. An examination of any stringed musical instrument, as a violin, will make it evident that by varying the tension, the length, or the mass per unit length of the wires or strings, tones of any desired pitch may be secured.

211. Laws of Strings. — In order to study the laws governing the vibration of strings, an instrument called a *sonometer* is used. It consists of a thin wooden box (Fig.

108), near the ends of which are fixed bridges, *A* and *D*. Wires or strings may be stretched lengthwise of the box by attaching them to the pins set in the frame at one end and to the weights at the other, the wires passing over

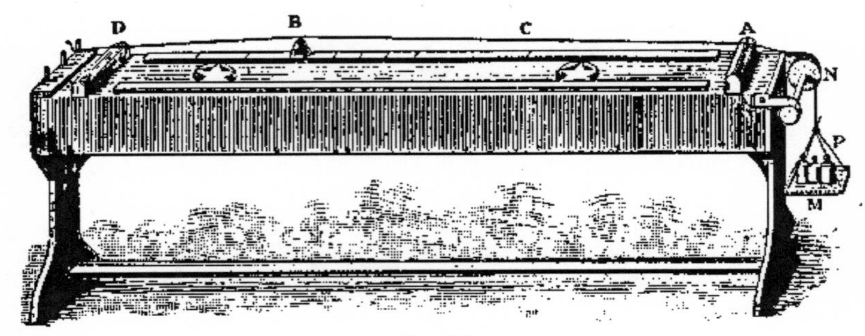

Fig. 108.

pulleys, as at *N*. By means of a movable bridge, *B*, the length of a wire may be shortened at pleasure. Below the wires there is a scale of equal parts.

Experiment. — Stretch two similar wires on the sonometer and tune them to unison by varying the weights. With the movable bridge shorten one of them successively to $\frac{8}{9}$, $\frac{4}{5}$, $\frac{3}{4}$, $\frac{2}{3}$, etc. The successive intervals between the notes given by the long wire and the shortened one will be $\frac{9}{8}$, $\frac{5}{4}$, $\frac{4}{3}$, $\frac{3}{2}$, etc., and the notes emitted by the wire of variable length will be those of the diatonic scale. Hence,

THE LAW OF LENGTHS. — *The tension and the diameter being constant, the vibration number varies inversely as the length.*

Experiment. — Stretch equally two wires of different known diameters. Shorten the larger one until it is in unison with the smaller. The ratio of the lengths will be inversely as that of the diameters. Hence,

THE LAW OF DIAMETERS. — *The tension and the length being constant, the vibration number varies inversely as the diameter.*

Experiment. — Stretch two similar wires with unequal known tensions. Shorten the one of lower pitch till it is in unison with the other. The ratio of the lengths will be that of the square root of the tensions. Hence,

THE LAW OF TENSIONS. — *The length and the diameter being constant, the vibration number varies as the square root of the tension.*

Experiment. — Stretch equally two wires of about the same diameter but of different material. Bring them to unison with the movable bridge. The ratio of the lengths will be the inverse of that of the square roots of the masses per unit length. Hence,

THE LAW OF DENSITIES. — *The length, tension, and diameter being constant, the vibration number varies inversely as the square root of the mass per unit length.*

212. Applications. — In the piano, violin, harp, and other stringed instruments, the pitch of each string is determined partly by its length, partly by its tension, and partly by its size. The tuning is done by varying the tension.

X. OVERTONES AND HARMONIC PARTIALS.

213. Fundamental Tone. — **Experiment.** — Fasten a silk thread to one prong of a large tuning-fork (Fig. 109). Set the fork vibrating and apply tension to the thread till it vibrates as a single spindle.

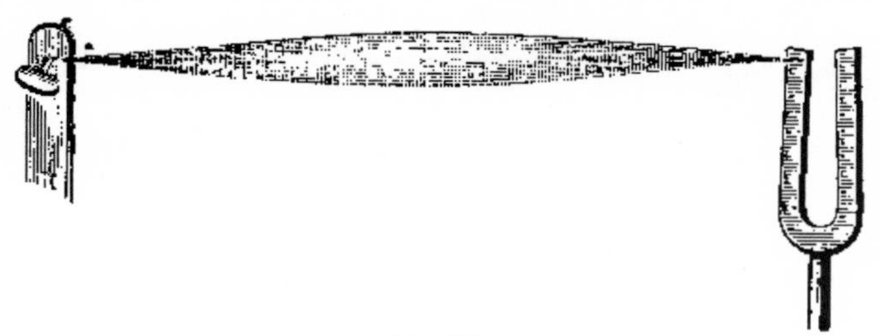

Fig. 109.

The experiment illustrates the manner in which a string or wire vibrates when emitting its lowest tone. The *fundamental tone* of a vibrating body is the lowest tone that it can yield. It is produced when the body vibrates as a whole, or in the smallest number of segments possible.

214. Nodes and Ventral Segments. — **Experiment.** — Proceeding as in the last experiment, it will be found that under a suitable

Fig. 110.

tension the thread can be made to vibrate in a number of parts (Fig. 110), giving it the appearance of a succession of spindles.

Experiment. — Stretch a wire on the sonometer with a thin slip of cork strung on it; then touching the cork lightly at one-third or one-fourth, or any aliquot part from one end (Fig. 111), bow or pluck the

Fig. 111.

shorter portion. The wire will vibrate in equal segments. This may be made more evident by placing narrow V-shaped pieces of paper, or *riders*, on the wire before bowing it. Some of them will be thrown off, and others will remain on, marking the places of maximum and of minimum vibration respectively.

The intermediate points of minimum vibration and the ends are called *nodes;* the vibrating portions between the nodes are called *loops* or *ventral segments;* and the middle points of the loops are called the *antinodes.* The experiment also illustrates *stationary waves.* The nodal points are caused by the bowed segment sending out waves along the wire, which interfere with similar waves reflected from the opposite end. The distance between two nodes is half a wave length.

215. Overtones and Harmonics. — **Experiment.** — Stretch a wire on the sonometer and set it in vibration by plucking or bowing it near one end. The tone heard most distinctly is its fundamental. Touch the wire lightly at its middle point. Instead of stopping the sound, a note an octave higher will be heard, showing that the wire is vibrating in two parts. If the wire be again plucked, both sounds can be heard together. Touching the wire one-third from the end brings out a tone an octave and a fifth higher, showing that the wire vibrates in thirds at the same time that it is vibrating as a whole. With a long string, it is possible to prove that a still further subdivision of a vibrating string takes place. In conducting such experiments, care must be exercised in selecting the point at which the string is plucked, for it is evident that there can be no node at that point.

The tones produced by sounding bodies vibrating in parts are called *overtones* or *partial tones.* If the vibration rate of an overtone is an exact multiple of the fundamental it is called a *harmonic partial.* In strings the overtones are usually harmonics, but in vibrating plates and membranes they are generally not. The overtones are named *first*, *second*, *third*, etc., in the order of their vibration rates as compared with that of the fundamental. The frequency of an overtone is found by multiplying the fundamental by a number one greater than the number of the overtone. For example, the frequency of the first overtone of $C = 256$ is $256 \times 2 = 512$, that of the second is $256 \times 3 = 768$, and so on.

XI. VIBRATION OF AIR IN PIPES.

216. Gases as Sources of Sound. — It was seen in the use of the resonator that gases can be thrown into vibration when they are confined in tubes or globes, and that they thus become sources of sound. Such a column of gas can be set in vibration in two ways: by a vibrating tongue, as in reed instruments; or by a stream of air striking against the edge of a lateral opening in the tube, as in the whistle, flute, etc.

217. Laws for Air Columns. — **Experiment.** — Fit a cork piston in a glass tube whose length is about 30 cm. and diameter 2.5 cm. With a piece of brass tubing flattened at one end direct a stream of air across the mouth of the tube. If the position of the piston, as well as the force of the blast, be right, the tube will yield a pure tone. If we shorten or lengthen the air column by moving the piston, the pitch of the tone will rise or fall accordingly. If we determine by trial the different lengths necessary to give the gamut, a comparison of them will give the continued ratio, $1 : \frac{8}{9} : \frac{4}{5} : \frac{3}{4} : \frac{2}{3} : \frac{3}{5} : \frac{8}{15} : \frac{1}{2}$ (§ 208), showing that,

The pitch varies inversely as the length of the air column.

Experiment. — Prepare two glass or paper tubes, 20 and 10 cm. long, respectively, and about 2 cm. in diameter. Hold the hand over one end of the shorter tube, and blow across the open end so as to produce its lowest pure tone. A comparison of this tone with that obtained by blowing across one end of the longer open tube will show that the pitch is the same. Hence,

For the same pitch, the open pipe is twice the length of the closed one.

218. State of Air in Sounding Tubes. — **Experiment.** — Employing an organ pipe, made either of glass or with one glass side (Fig. 112), lower into it, as it emits its fundamental, a light membrane covered with fine sand. The sand will be agitated the least at the middle of the tube, and most at the ends. The vibration of the air in the pipe is longitudinal. A *node* is a place of least motion and greatest change of density; an *antinode* is a place of greatest motion and least change of density. The closed end of a pipe is necessarily a node, and the open end an antinode. Hence,

In open pipes, for the fundamental tone, there is a node at the middle and an antinode at each end; in stopped pipes there is a node at the stopped end and an antinode at the other end.

The pipe or tube acts as a resonator, and for the fundamental tone the length of the closed pipe is one-fourth of a wave length; since the open pipe has a node in the middle, its length must be half a wave length.

Fig. 112.

219. Overtones. — **Experiment.** — Blow a strong blast across the end of the long tube used in Art. 217, and notice that notes of higher pitch than the fundamental are produced. These are *overtones*, caused by the air column vibrating in parts or segments. In proof of this, insert a piston in the tube, and by means of it shorten the air column till it gives as its fundamental the overtone previously obtained.

Since there can be no motion of the air at the surface of the piston, it must mark a node; and since the tone is unchanged by the presence of the piston, there must have been a node at that point before the introduction of the piston.

When an open pipe yields its fundamental there is a

node at the middle; hence to yield higher tones there must be added one, two, three, or more nodes so placed that there will be an antinode at each end. *A*, *B*, and *C*, of Fig. 113, illustrate the division into segments of an open pipe. If *A* gives the fundamental, then *B* must give a tone an octave higher or the first overtone, since the first node is one-fourth from the end. In *C*, the node is one-

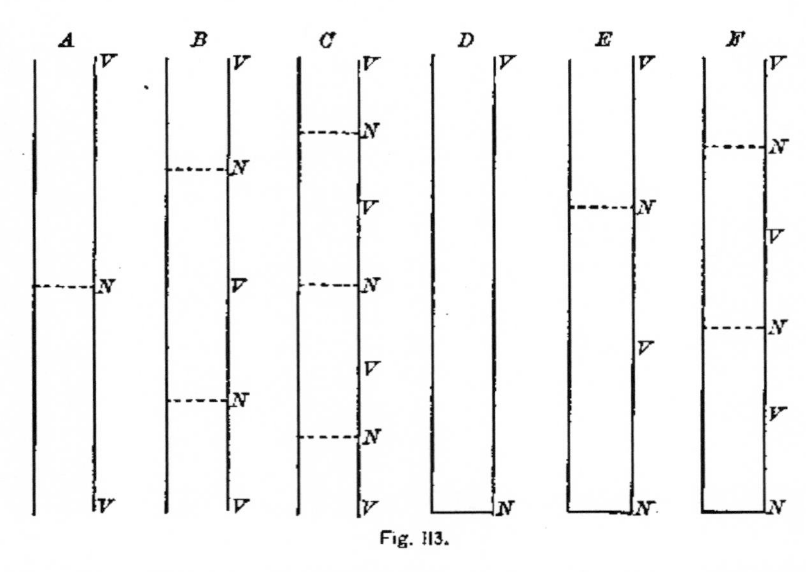

Fig. 113.

sixth from the end, and the frequency is therefore three times that of *A*, or the overtone is the second. Hence,

In open pipes the complete series of overtones is possible.

For stopped pipes, the open end is always an antinode, and the closed end a node; then by adding successively a node we have the conditions shown in tubes *D*, *E*, *F*, of Fig. 113. If *D* is the fundamental, then the vibration rate of *E* is three times that of *D*, giving the second overtone, since the node is one-third from the end. In like manner *F* is the fourth overtone, the node being one-fifth from the end. Hence,

In closed pipes only those overtones are possible whose vibration rates are 3, 5, 7, etc., times the fundamental.

Problems.

1. How can any particular overtone be omitted from the complex tone of a vibrating string?

2. What notes are the overtones of a "middle C" stopped pipe? *Ans.* G′, E″.

3. A string sounding C′ is 18 in. long; how much must it be lengthened to sound the note D?

4. A string stretched by a weight of 4 lbs. sounds a certain note C. What weight will make it sound F of the scale?

5. A stretched wire 10 ft. long is in unison with a tuning-fork marked 256; the wire is shortened 4 ft. How often will it now vibrate in a second?

6. A string is fastened at one end to a peg in a horizontal board, and the other end passes over a pulley and carries 16 lb. The string gives the note C. What weight must be used instead of 16 lb. so that the string will give the octave below?

7. A wire stretched by a weight of 13 kgm. sounds a certain note. What must be the stretching-weight to produce the major third?

8. A string whose length is 120 cm. vibrates 288 times per second under a tension of 50 kgm. What must be the tension, if a movable bridge be placed at the middle of the string, to maintain the same pitch?

9. A wire whose length is 100 cm. when under a tension of 60 kgm. has a frequency of 320. What must be the relative diameter of a second wire 70 cm. long and under a tension of 50 kgm. to give a fundamental an octave higher? *Ans.* 0.652.

10. What effect will it have on the pitch of a wire to double the length, diameter, and tension? *Ans.* Multiplies frequency by 0.354.

11. A steel and a brass wire of the same diameter and length are stretched with equal tensions. The density of the steel wire is 7.8 and that of the brass 8.5. Calculate the frequency of the brass wire, that of the steel being 256. *Ans.* 245.21.

12. Find the relative frequency of two notes yielded, one by steel ($d = 7.8$) wire 1 m. long, 1 mm. diameter, and stretched by a force of 30 kgm., and the other a brass ($d = 8.8$) wire 75 cm. long, ½ mm. diameter, and stretched by a force of 20 kgm. *Ans.* 2.05.

13. A closed organ pipe is 3½ ft. long. Find the wave length of the sixth overtone.

14. Find the frequency of the fundamental note of a closed organ pipe 6 ft. long, the temperature of the air being 18° C.

15. An open organ pipe is 4 ft. long, the temperature of the air is 20° C. Calculate the frequency of the first three overtones.

Ans. 282.62, 423.94, 565.25.

16. Compute the lengths of open pipes necessary to give the tones of a major chord at a temperature of 16° C., calling C 256.

Ans. 2.192 ft., 1.754 ft., 1.462 ft., 1.09 ft.

17. What overtone of E (320) forms an interval of a major third with C′ (512)?

18. What is the interval between the fourth overtone of a note whose fundamental is 256, and the third overtone of one whose fundamental is 320?

19. Calculate the length of a closed pipe that will give the note C (256) when the temperature is 10° C. Calculate the frequency when the temperature is 20° C. *Ans.* 13 in.; 260.5.

20. The lowest note on a piano is A (27). Compute the length of an open pipe that will give this note when blown with air at 15° C.

XII. QUALITY OF SOUND.

220. Definition of Quality.—Two of the essential characteristics of musical sounds have already been considered, namely, pitch and loudness. There is a third important difference between musical sounds. We easily perceive that one sound differs from another not only in being more acute or grave, louder or softer, but also in respect to the character of the sound. We have no difficulty in distinguishing the notes of a piano from those of a violin, even though they may be of the same pitch and apparent loudness. Similar differences enable us to distinguish one voice from another in speech and in song. Even the untrained musical ear can readily appreciate differences in the character of the music produced by different instruments of the same class. All such differences, not assignable to pitch or loudness, are included under the term *quality*. Those characteristics of a sound that enable one to refer it to its source are called its *quality*.

221. Quality due to Overtones.— Experiment. — Pluck the wire of a sonometer at the middle and compare the sound with that given when the wire is plucked near one end. In the first case the first and third overtones are necessarily lacking (§ 215), while in the second the whole series is probably present.

Although these two notes are of the same pitch, yet there is a marked difference in their quality, caused by the difference in their overtones. The sound waves in each case are the result of compounding the fundamental with the overtones present. Pitch depends on the wave length; loudness on the amplitude; and quality depends upon the only other physical difference between aerial sound waves, namely, their vibrational form, due to the relative phase and intensity of the overtones. When the fundamental is relatively strong, the overtones being few and weak, the tone is full and mellow; but when the fundamental is weak, and the overtones numerous and strong, the tone is metallic.

Differences in quality are to be referred to the series of overtones present in each case. Voices differ for this reason. Violins differ in sweetness of tone, because the sounding-boards of some bring out the overtones differently from others. Voice culture consists in training and developing the vocal organs and resonant cavities, to the end that purer overtones may be secured, and greater richness may thereby be imparted to the tones. Often in the reflection of sound by distant objects its character is greatly changed, on account of the partial or complete suppression of the fundamental or some of the overtones in the process of reflection.

222. Analysis and Synthesis of Sound. — Helmholtz constructed a series of resonators of different sizes, like those of Fig. 103, one to reënforce a certain fundamental tone and one for each of its overtones. By placing these resonators successively to the ear, when the fundamental tone is sounded, the presence of any of its overtones is indicated by the resonator responding. Proceeding in this way, Helmholtz demonstrated that quality of sound is determined by the overtones present; that to give a tone richness and sweetness the first four or five overtones are essential.

Helmholtz also proved that a sound of any quality may be built up by combining a fundamental tone with overtones. To do this he employed a set of tuning-forks, kept in vibration by electromagnets (§ 455). The set consisted of ten forks, nine of which sounded the overtones of the tenth. Each fork was provided with a resonator to strengthen its sound. By means of a set of keys any particular resonator could be brought into action, and the sound of the fork made audible at a distance.

XIII. HARMONY AND DISCORD.

223. Consonance and Dissonance. — **Experiment.** — Tune two wires on the sonometer to unison. By the movable bridge raise the pitch of one wire a fifth. When the wires are plucked, the combination of sound is smooth and agreeable. Now move the bridge till the interval is a second; the sound of the two wires will be disagreeable.

When two notes differ in pitch and their combination is agreeable to the ear, they are said to be *consonant;* when disagreeable, *dissonant.*

Helmholtz concluded that dissonance is due to beats. He even expressed their relation quantitatively to the effect that dissonance occurs when the number of beats a second is between 10 and 70, and that maximum discord is caused by 30 or 32 a second. Later investigators have shown that the number of beats that produce the greatest discord is not constant, but depends upon the pitch. Beats may be regarded as the physical aspect of discord. They are still recognized as one cause of it, but psychologists are now agreed that discord cannot be defined by beats.

Problems.

1. Point out three important differences between two musical sounds.

2. Why is the quality of the sound of a "middle C" stopped pipe different from that of a "middle C" open pipe?

3. Why is a minor triad more dissonant than a major one?

4. Explain why a major third is less pleasant on the equally tempered scale than on the diatonic scale.

5. Write out the vibration rates of the fundamentals and the first four overtones of the notes C, D, and E.

XIV. VIBRATING RODS, PLATES, AND BELLS.

224. Vibration of Rods. — Rods of metal, of wood, and of glass may be made to vibrate either transversely or longitudinally. A rod may be vibrated transversely by fixing it and drawing a violin bow across the free end, or by striking it with a suitable hammer; it may be set in longitudinal vibration by clamping at the middle and stroking lengthwise with a cloth dusted with powdered resin. A moist cloth is better for glass. The jews'-harp, the music-box, and the coiled-wire gong of a clock are illustrations of the transverse vibration of rods or plates clamped at one end. The tuning-fork may be regarded as an elastic bar free at the ends, and supported in the middle by a stem which is subject to all the motion of the middle of the ventral segment, giving it an up and down movement which is transmitted to the supporting sounding-board (Fig. 114). The overtones are of high pitch and feeble intensity, and soon vanish, leaving the tone pure.

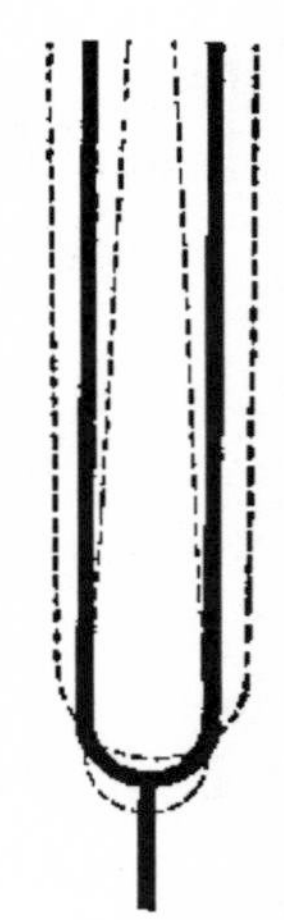
Fig. 114.

225. Vibration of Plates. — **Experiment.** — Support a brass or glass plate, as shown in Fig. 115. Scatter a little fine sand evenly over it, touch the plate at some point with the finger, while a violin bow is drawn across the edge. The plate is thrown into vibration, the sand arranging itself in symmetrical figures whose complexity increases as the pitch becomes higher.

Fig. 115.

These sand figures make it clear that the plate vibrates transversely in segments, the sand being thrown to places of least vibration, which lie between parts having opposite motions. The arrangement of nodal lines is determined by the relative position of the point bowed to that pressed by the finger. This method of studying vibrating bodies was first employed by Chladni, and the figures are called Chladni's figures.

Vibration of Bells. — **Experiment.** — Draw a violin bow across the edge of a large bell or goblet half full of water. It will yield a musical sound, and at the same time the surface of the water will be greatly agitated in sections corresponding to the several segments into which the vibrating body is divided.

For the fundamental tone the bell divides into four segments, the pitch rising with the number of segments. Powdered sulphur sifted evenly on the water will make the position of the nodes more conspicuous.

XV. GRAPHIC AND OPTICAL METHODS.

226. Graphic Methods of studying sound are of service in determining the vibration rate of sounding bodies. In one of the simplest a small style is attached to the vibrating body and traces its movements upon a piece of smoked paper or glass, which moves uniformly beneath it. Generally, the paper is wrapped

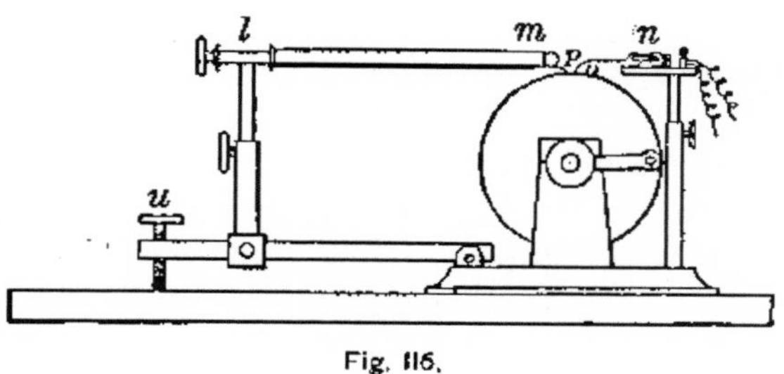

Fig. 116.

around a cylinder, mounted on an axis, one end of which has a screw thread on it, so that when the cylinder turns it also moves axially (Fig. 116). If the vibrating body is a fork, the beats of a seconds pendulum may be marked on the paper by electric sparks from the style, and the number of sinuosities in the line traced on the cylinder between the marks may be counted. The rate of the fork is thus easily measured.

227. Manometric Flames.—**Experiment.**—A box with mirror faces is mounted so as to turn on a vertical axis (Fig. 117). In front of these revolving mirrors is supported a short cylinder which is divided into two parts by a partition of gold-beater's skin. Illuminating gas is admitted to one compartment; a small gas jet is connected with the same one and a speaking-tube with the other.

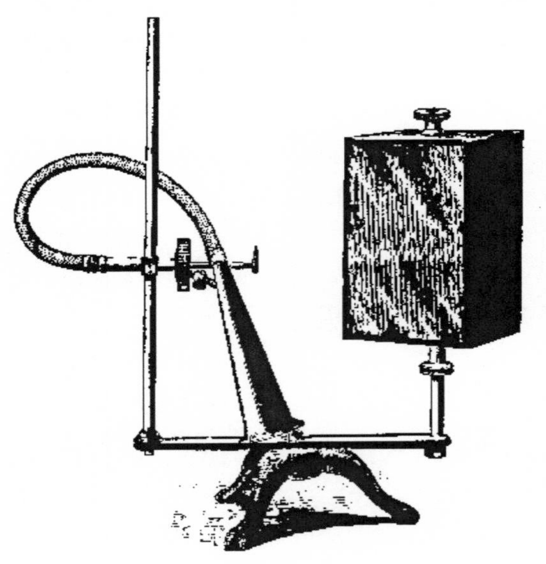

Fig. 117.

In a darkened room, the image seen in the revolving mirror will be a smooth band of light. Now sound a heavy C-fork in front of the mouthpiece, or produce there any pure tone; the appearance in the rotating mirror will be that of the upper band of Fig. 118. A condensation entering the box acts on the membrane, compressing the gas, thereby extending the flame; a rarefaction entering it produces an opposite effect. Hence a serrated band is seen in the mirror. Now sound a C′-fork, a strong tone an octave higher than that first used; the appearance is the second band of Fig. 118, differing from the last in having teeth half as wide. If we connect two mouthpieces with the box, using a T-pipe, and sound the C-fork in front of one, and the C′-fork in front of the other, we obtain the last band of Fig. 118, the short tongues of which are due to

C', the octave of C. The same figure is obtained by singing the vowel sound *o* on the note $B♭$, showing that this sound is composed of the fundamental combined with the first overtone.

Fig. 118.

The experiment shows the possibility of analyzing sounds by the flame pictures they produce. This method was invented by the late Rudolph Koenig of Paris, and it has the great advantage of being independent of hearing. If this box, or manometric capsule, as it is called, be attached to a Helmholtz resonator, the flame will respond whenever any sound is produced that affects the resonator. With a complete set of resonators, each with its manometric capsule, a most efficient apparatus is provided for the analysis of sounds.

CHAPTER V.

LIGHT.

I. NATURE AND PROPAGATION OF LIGHT.

228. Light, as distinguished from the sensation of seeing, is a periodic or undulatory disturbance in a medium which is assumed to exist everywhere in space, even penetrating between the molecules of ordinary matter. This medium is known as the *ether*. Light waves do not consist of alternate condensations and rarefactions, as in sound, but of periodic *transverse* disturbances. These disturbances are probably not transverse movements of the ether itself, but transverse alterations in the electrical and magnetic condition of the ether. But whatever may be the nature of the medium, light is a wave motion in it, and the vibrations are transverse.

229. Definition of Terms. — In general, when light falls on a body, a part of it is reflected, a part is transmitted, and the rest is absorbed. *Transparent* bodies allow light to pass through them with so little loss that objects can be easily distinguished through them, as glass, air, pure water. *Translucent* bodies transmit light so imperfectly that objects cannot be seen distinctly through them, as horn, oiled paper, very thin sheets of metal or wood. *Opaque* bodies transmit no light, as blocks of wood or iron. This classification is one of degree; no sharp line of separation between these classes can be drawn.

No body is perfectly transparent. If several layers of glass are put together, the distinctness of vision through them diminishes with the increase in the number of layers; stars which are invisible at the foot of a mountain are often visible at the top. (Why?)

230. Speed of Light. — Until 1676 it was believed that light travelled instantaneously. In that year Roemer, a Danish astronomer, was engaged at the Paris observatory in observing the eclipses of Jupiter's moons. Confining attention to the one nearest the planet, because its eclipses occurred most frequently, he found that the time of the observed eclipses did not correspond with the computed values derived from observations extending over a long period. He found the interval between two successive eclipses at E and E'' (Fig. 119) the same; but for intermediate positions this interval was greater while the earth was receding from Jupiter in going from E through E' to E'', and less while approaching Jupiter again. It was greatest at E' when the earth was moving directly away from the planet. The sum of all these excesses from E to E'' amounts to 16 min. 38 sec., or 998 sec. In accounting for these facts Roemer advanced the theory that the increase is caused by the time taken by light to traverse the added distance when the earth is moving away from Jupiter, and that 998 sec. represent

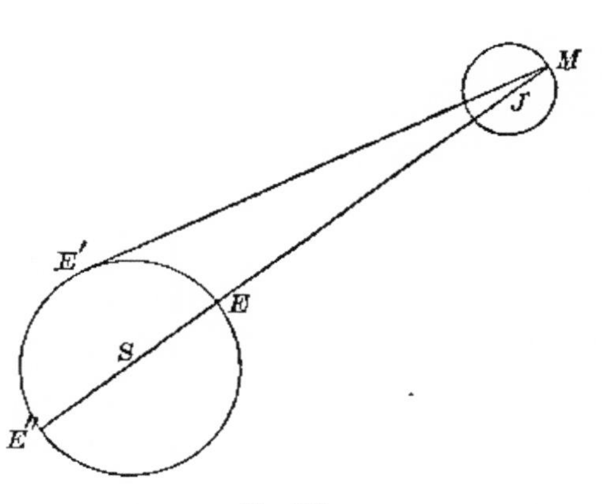

Fig. 119.

the time required by light to traverse the diameter of the earth's orbit. This gives for the speed of light about 186,000 miles a second.

In 1849 Fizeau determined the time required for light to travel a known distance on the earth's surface. In 1850 Foucault showed not only that light takes a measurable time to travel short distances, but that the speed of light varies with the medium. These experiments have been repeated in a modified form by Michelson and Newcomb in our own country, and the results, as summarized by Professor William Harkness, show that the velocity of light is about 186,337 miles (299,877 km.) a second.

231. Light travels in Straight Lines. — If an opaque object, as a book, is between the eye and a lamp, it hides the lamp from view. From such facts as this we learn that *light is propagated in straight lines*. Other facts to be considered in a subsequent article (§ 265) make it necessary to add the restriction, in a medium having the same physical properties in all directions. Rays of light are the directions in which light is propagated. These directions are radii of the spherical waves and normals to the wave fronts. When the source is at a great distance, the rays are sensibly parallel, and a number of them taken together constitute a *beam of light;* for example, in the case of light from the sun or stars, the distance is so great that the rays are considered to be parallel. Rays of light proceeding outward from a point form a *diverging pencil;* and rays proceeding toward a point, a *converging pencil.*

232. Shadows. — **Experiment.** — Hold a ball or disk between the flame of a lamp and a white screen. From a part of the screen the light will be wholly cut off, and surrounding this area is one from which the light is excluded in part. If three small holes be made in the screen, one where the screen is darkest, one in the part where the screen is less dark, and one in the lightest part, it will be found, on looking through them, that the flame of the lamp is wholly invisible through the first, part is visible through the second, and the whole flame through the third.

The space behind the object from which the light is excluded is called the *shadow*. The figure on the screen is a section of the shadow. The darker part of the shadow, called the *umbra*, is caused by the total exclusion of the light by the opaque object; and the lighter part, the *penumbra*, by its partial exclusion.

When the luminous object is a point, as *L* (Fig. 120), then the shadow will be bounded by the cone of rays,

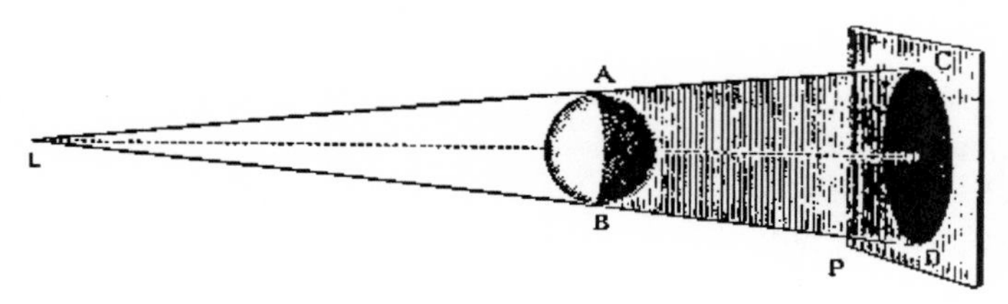

Fig. 120.

ALB, tangent to the object, and will have but one part, the umbra. When the luminous body has magnitude, as *LL* (Fig. 121), then the space *ABDC* behind the opaque body receives no light, and the parts between *AC* and

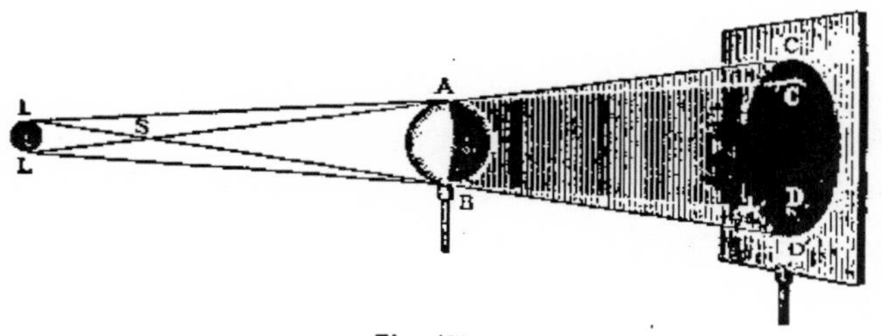

Fig. 121.

AC', and between BD and BD', receive some light, the amount increasing as AC' and BD' are approached. [The student should draw figures for the cases when the luminous body is larger than the opaque body, and when of the same size.]

233. Images by Small Apertures. — **Experiment.** — Support in vertical planes two sheets of cardboard, *A* and *B* (Fig. 122). In the centre of *A* cut a round hole about 2 mm. in diameter; and place in front of it, at a distance of a few centimetres, a lighted candle, *CD*. An inverted image of the flame will be found on the other sheet. If a triangular or a square hole is used, an equally distinct image of the candle will be formed, showing that the image is independent of the shape of the opening. Again, if the aperture is gradually enlarged, the image loses in distinctness of outline, gains in brightness, and gradually assumes the shape of the aperture.

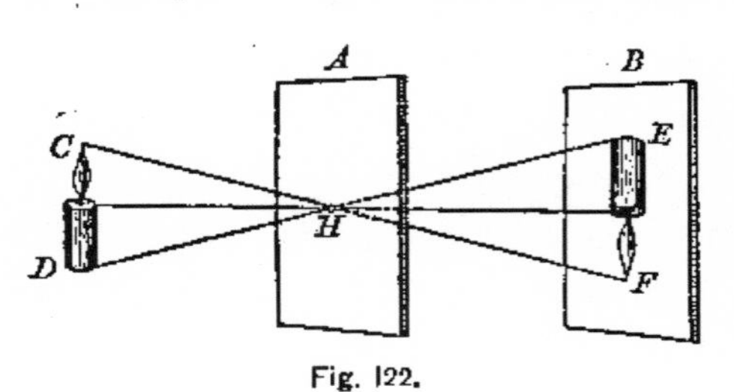

Fig. 122.

The experiment shows that the definition and the brightness of the image are independent of the shape of the aperture, but are affected by its size. To understand the origin of this image it must be borne in mind that each point of the object is the vertex of a cone of rays passing through the aperture and forming an image of the aperture on the screen. These images will be symmetrically placed with reference to the points emitting the light, and consequently will build up a figure of the same form as the luminous object. Now it is evident that these numerous pictures of the aperture overlap in forming a picture of the object, the number at any one place determining the brightness. The edge of the picture will, therefore, be less bright than the other portions, and these differences will be more noticeable, the larger the opening. In the case of a very large opening, the overlapping of the images of the aperture destroys all resemblance of the image to the object, the resulting image having the shape of the aperture.

234. Illustrations. — When the sun shines through the small chinks in the foliage of a tree, there may be seen on the ground a number of spots of light, either round or oval. These are images of the sun. During partial eclipses of the sun such figures assume a crescent shape.

In the photographer's camera, and in the eye, we have further examples of the formation of images by small apertures. In these cases, as will be seen in subsequent articles, the definition is improved by means of a lens (§ 269).

II. PHOTOMETRY.

235. The Intensity of illumination is the quantity of light received on a unit of surface. Everyday experience shows that it varies, not only with the nature of the source, but also with the distance at which the source is placed.

236. Law of Intensity. — **Experiment.** — Cut from cardboard three squares, 4 cm., 8 cm., and 12 cm. on a side, respectively, mounting them on wire supports. The centres of these screens should be at the same distance above the table as the source of light. Use a lamp giving a small, flat flame, and set it with the edge of the flame toward the surface of the largest screen, and distant about one metre. Interpose the medium-sized screen so that it exactly cuts off the light from the lateral edges of the largest. If the source were a point, the light would now be cut off from the whole screen. In like manner place the smallest screen with reference to the intermediate one. If these screens are placed with care, it will be found that their distances from the light are as 1 : 2 : 3. Now as each screen exactly cuts off the light from the one next to it in the series, it follows that each receives the same amount of light from the source when it is not intercepted. The surfaces of these screens are as 1 : 4 : 9, and hence the amount of light per unit of surface must be inversely as 1 : 4 : 9, the squares of 1, 2, and 3, respectively.

Any disturbance propagated in spherical waves must have its energy distributed over a continually increasing area as the radius increases; and since this area increases as the square of the radius, it follows that the energy per unit of area must decrease at the same rate. Therefore, *the intensity of the illumination varies inversely as the square of the distance from the source of light.*

237. A Photometer is an instrument for comparing the intensity of one light with that of another assumed as a standard. The principle applied is that the ratio of the intensities of two lights equals the square of the ratio of the distances at which they give equal illuminations. The standard in general use is the light emitted by a sperm candle of the size known as "sixes," when burning 120 grains per hour. The illuminating power of a light is expressed by stating the number of times it is greater than the standard candle.

238. The Bunsen Photometer. — In this photometer, a screen of paper, *A* (Fig. 123), having a translucent spot, made by applying a little hot paraffin, is supported on a graduated bar between the standard candle, *B*, and the light to be measured, *C*. When the two surfaces are equally illuminated, the spot is scarcely distinguishable from the surrounding paper. By moving *B*, this condition is easily secured. Then the candle power of *C* equals $\overline{AC}^2 \div \overline{AB}^2$.

Since the spot never appears exactly like the surrounding paper, it is better to place a pair of mirrors, forming a V, opening toward the edge of *A*. By looking

Fig. 123.

in them, both sides of *A* can be seen at once. *B* is then adjusted to give these images the same appearance.

Problems.

1. In what respects does light differ from sound?

2. What principle is recognized in aiming a rifle?

3. Why is the image formed by a small aperture inverted?

4. Oculists often introduce atropine into the eye to enlarge the pupil. Why is one then unable to see objects distinctly?

5. When has the umbra a finite length?

6. If the distance of an object from the source of light is doubled, how is the size of the shadow affected? Show by figure.

7. If the shadow of a flag-pole, as cast on the ground by the sun, is 20 ft. long, how high is the pole if a vertical rod, 5 ft. long, casts a shadow 1.2 ft. long?

8. By means of a small hole in a window shutter of a darkened room an image of a factory chimney, 50 ft. away, is thrown on a screen 6 in. from the hole. The image is 8 in. high; how high is the chimney?

9. In measuring the candle-power of a lamp with a Bunsen

photometer the standard candle was 10 cm. from the screen, and the lamp was 54.5 cm. Calculate their relative illuminating power.

Ans. 29.7.

10. A lamp and a candle are 200 cm. apart. The intensity of the lamp is 20 times that of the candle. Where must a screen be placed between them in order that the two sides may be equally illuminated?

Ans. 36.549 cm. from the candle.

11. A circular uniform source of light, 10 cm. in diameter, is placed parallel to an opaque disk, 5 cm. in diameter, and distant from it 1 m. Find the length of the umbra.

12. Two lights, A and B, are in intensities as 10 to 7, and are 100 m. apart. Find a point C that will be equally illuminated by A and B.

13. A light of 500 candle-power is placed 100 m. from a wall. At what distance must a light of 100 candle-power be placed to produce the same illumination?

III. REFLECTION OF LIGHT.

239. Regular Reflection. — When a beam of light falls on a polished surface AC (Fig. 124), the greater part of it is reflected in a definite direction. The angle that the incident ray makes with the normal PB to the reflecting surface at the point of incidence is called the *angle of incidence*, as IBP; and the angle between the reflected ray and this normal is the *angle of reflection*, as RBP.

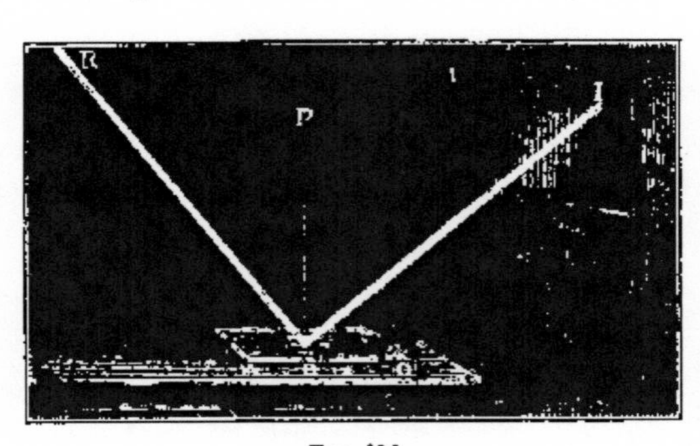

Fig. 124.

240. Law of Reflection. — **Experiment.** — A semicircular board is provided with two arms pivoted at the centre, one carrying a lighted candle and a convex lens (§ 269), the other an oiled-paper screen and lens (Fig. 125). A plane mirror is mounted at the centre of the semicircle, with its reflecting surface parallel to the diameter. A scale on the edge of the semicircle has its zero in a normal to the mirror. The first lens, with a wire bent across it, is so placed that the shadow of

Fig. 125.

the wire, after reflection from the mirror, is focussed sharp and clear on the screen by the second lens. Now give the candle-arm any desired position and move the screen-arm till the shadow of the wire falls across the middle of the screen. The arms will be found on examination to make equal angles with the normal to the mirror.

Hence, *the angle of reflection is equal to the angle of incidence, and the two angles lie in the same plane.*

241. Diffused Reflection. — **Experiment.** — Fill a large glass jar with smoke. Cover the mouth with a piece of cardboard, in which is a hole about 1 cm. in diameter. With a small hand-glass reflect a beam of sunlight into the jar through the hole in the cover. The whole interior of the jar will be illuminated.

The small particles of smoke floating in the jar furnish a great many surfaces. The light falling on them is reflected in as many different directions, the result being seen in the diffusion of the beam.

All reflecting surfaces, to a greater or less extent, scatter light in the same way as these smoke particles, on account of the irregularities of their surfaces. Figure 126 illustrates the difference between a perfectly smooth reflector

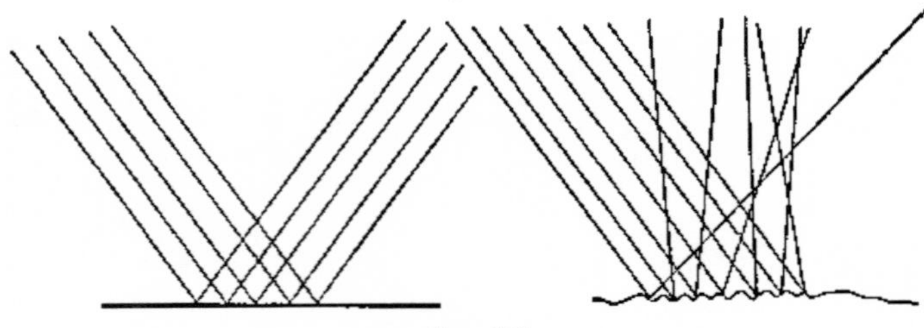
Fig 126.

and reflectors as they actually are, more or less irregular according to the degree of polish.

It is by diffused reflection that objects become visible to us. Perfect reflectors would be invisible. The trees, the ground, the grass, and particles floating in the air, reflect the light from the sun in every direction, and thus fill the space about us with light. Aeronauts tell us that when they reach very high altitudes, the sky grows black, owing to the absence of floating particles to diffuse the light.

242. A Mirror is any smooth surface. *A plane mirror* is one whose reflecting surface is plane. *A spherical mirror* is one whose reflecting surface is a portion of the surface of a sphere.

243. Image of a Point in a Plane Mirror. — Let A be a luminous point in front of a plane mirror MN (Fig. 127). Any ray AB incident on the mirror is reflected in the direction BD, making the angle of reflection FBD equal to that of incidence FBA. (See Appendix.) In like manner, a second ray, as AC, is reflected along CE. If BD and CE are produced, they meet at A'. Join A and A'. Then AA' is perpendicular to MN,[1] and A' is as far back of MN as A is in front. Since AB and AC are any two rays diverging from A, and incident on the mirror, it follows that all rays from A, incident on MN, must be reflected from MN as if they came from a point as far back of MN as A is in front. Hence, the eye placed at DE will receive the rays as if they came from A'. The point A' is accordingly called the image of A in the mirror MN, and is known as a *virtual image*, because the light only *apparently* comes from it. Therefore, *the image of a point in a plane mirror is virtual, and is as far back of the mirror as the point is in front;* it may be found by drawing from the point a perpendicular to the mirror, and producing it till its length is doubled.

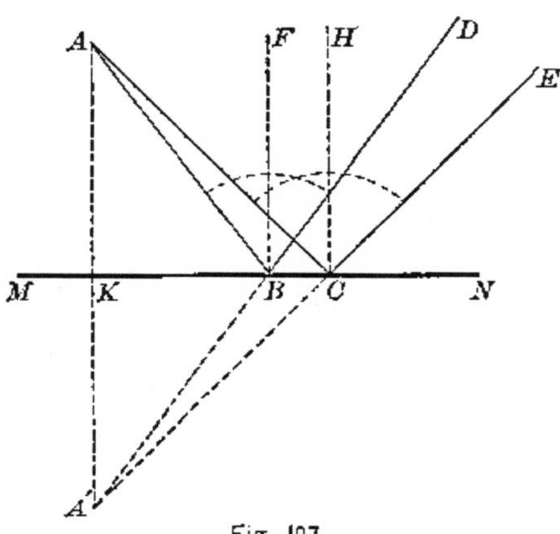

Fig. 127.

244. Image of an Object. — Since the image of an object is an assemblage of the images of its points, it follows that the image of an object can be located by finding those of its points. Let AB (Fig. 128) represent an object in front of the mirror MN. Drop perpendiculars from the

[1] Angle $ABK = DBN = A'BK$. $\therefore ABN = A'BN$. Angle $ACK = ECN = A'CK$. Hence, the triangles ABC and $A'BC$ are equal, and $AB = A'B$. In the triangles AKB and $A'KB$, it follows that the angles at K are right angles, and $AK = A K$.

points of the object to the mirror, and produce them till their length is doubled. Then $A'B'$ is the image of AB. *It is evidently virtual, erect, and of the same size as the object.*

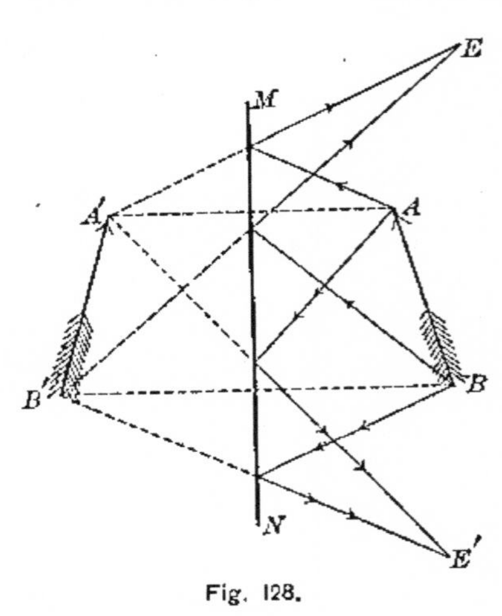

Fig. 128.

The rays which form the image for one observer are not those which form it for another. Let E and E' represent two different observers. To find the path of the rays which enter the eye at E, draw lines from A' and B' to E. The intersections of these lines with MN are the points of incidence of the rays from A and B which are reflected to E. In like manner we may trace the path of the rays for the position E'.

245. Experimental Proof of the Position of the Image. — **Experiment.** — Support a pane of window glass in a vertical position. On one side of it place a lighted candle, and on the other a tumbler of water, each at the same distance from the glass, and in a line perpendicular to it. An image of the candle will be seen in the tumbler of water, showing that the image is as far back of the mirror as the object is in front.

The experiment also explains how many optical illusions, such as Pepper's ghost, are produced. A large sheet of unsilvered plate glass, with its edges hidden from view by curtains, is so placed that the audience have to look obliquely through it to see the actors on the stage. Other actors, strongly illuminated and out of view by the audience, are seen by reflection in the glass and appear as ghosts on the stage. The magic cabinet and the head without a body are also illusions produced by the aid of mirrors.

246. Multiple Reflection. — **Experiment.** — Support two mirrors so that their reflecting surfaces form an angle. If a lighted candle be placed between them, several images may be seen in the mirrors; three when at right angles, the number increasing with the diminution of the angle. When the mirrors are parallel, all the images are on a straight line perpendicular to the mirrors.

These several images are caused by the successive reflection of the light from the mirrors; the image in one mirror serves as an object for the second mirror, and the image in the second becomes in turn an object for the first mirror. In Fig. 129 the two mirrors are at right angles. O' is the image of O in AB, and is found as directed in § 244. O''' is the image of O' in AC, and is found by the perpendicular $O'O'''$. O'' is the image of O in AC, and since the mirrors are at right angles, O''' is also the image of O'' in AB. O''' is situated behind the plane of both mirrors, and no images of it are formed. All the images are situated in the circumference of the circle whose centre is A and radius AO. If E is the position of the eye, then O' and O'' are each seen by one reflection, and O''' by two reflections, and for this reason it is less bright. To trace the path of a ray from O''', draw $O'''E$, cutting AB at b, and from the intersection b draw bO'', cutting AC at a. Join aO; the path is $OabE$.

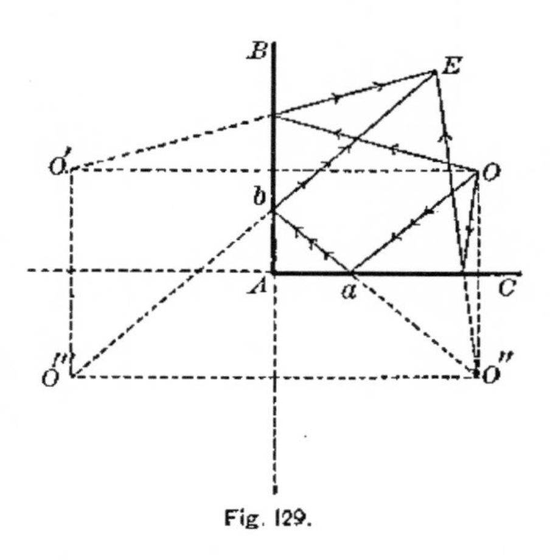

Fig. 129.

247. Illustrations. — The double image of a bright star and the several images of a gas-jet in a thick mirror (Fig. 130) are examples of multiple reflection, the front surface of the mirror and the metallic surface at the back serving as parallel reflectors. Theoretically the number of images is infinite; but on account of their faintness only a limited number is visible. The *kaleidoscope*, a toy invented by Sir David Brewster, is an interesting application of the same principle. It consists of a tube containing three mirrors extending its entire length, the angle between any two of them being 60°. One end of the tube is closed by ground glass, and the other by a cap with a round hole in it. Pieces of colored glass are placed loosely between the ground glass and a plate of clear glass parallel to it. On looking through the hole at any source of light, multiple images of these pieces of glass are seen, symmetrically arranged around the centre, and forming beautiful figures, which vary in pattern with every change in the position of the objects.

Fig. 130.

248. Deviation by Revolving Mirror. — If a ray of light is incident on a plane mirror, and the mirror is turned through an angle, the angular deviation of the ray will be double that of the mirror. Let the ray AM be normal to the mirror (Fig. 131); it will then be reflected back on itself. Turn the mirror through the angle θ (theta); the normal AM is turned through the same angle, so that the angles of incidence and reflection are now both equal to θ. The deviation of the reflected ray is the angle AMB or 2θ.

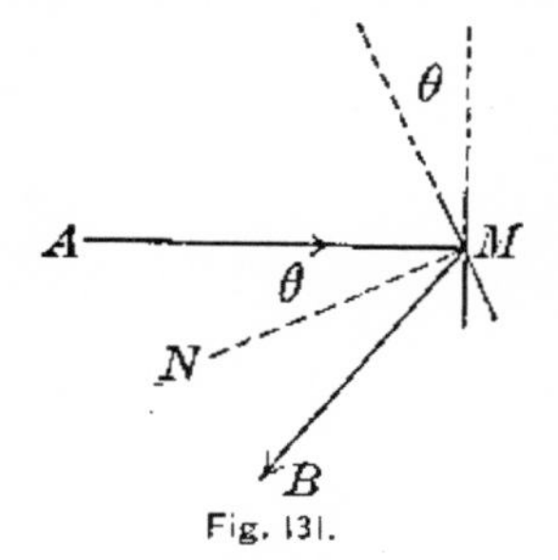

Fig. 131.

This principle is applied in measuring small movements of rotation, as in reflecting galvanometers; for not only is the deviation doubled, but the ray of light acts as a long weightless pointer, thus adding greatly to the sensitiveness of the instrument.

249. Spherical Mirrors. — A mirror is *spherical* when its reflecting surface is a portion of the surface of a sphere. If the inner surface is polished for reflection, the mirror is *concave;* if the outer surface, it is *convex.* Only a small portion of a spherical surface cut off by a plane is used as a mirror, and its boundary is therefore circular. The *centre* of the mirror is the centre of curvature of the sphere of which the reflecting surface is a part. The middle point of the reflecting surface is the *pole* or *vertex* of the mirror, and the straight line passing through the centre of curvature and the pole of the mirror is its *principal axis.* Any other straight line through the centre and intersecting the mirror is a *secondary axis.*

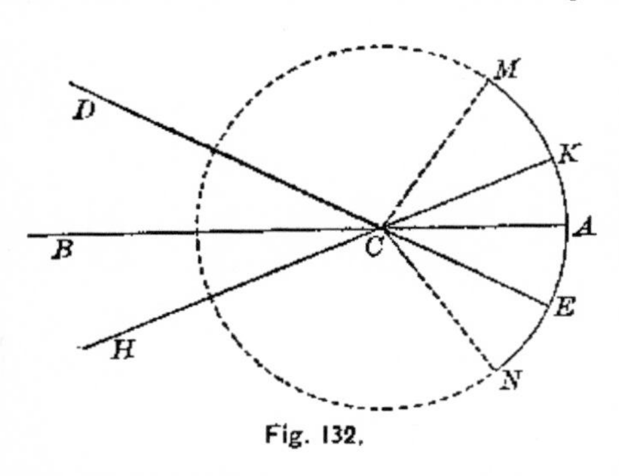

Fig. 132.

In Fig. 132, MN is the spherical mirror,[1] C is the centre, A is the pole, BA the principal axis, and DE and HK secondary axes. The angle MCN measures the *aperture* of the mirror.

The difference between a plane mirror and a spherical one is that the normals to a plane mirror are all parallel lines, while those of a spherical mirror are the radii of the surface, and all pass through the centre of curvature.

250. A Focus is the point common to the paths of all the rays of a pencil of light after incidence. It is a *real* focus if the rays and waves of light actually pass through the point, and *virtual* if they only appear to do so.

251. Principal Focus of Spherical Mirrors.—Experiment.—Let the rays of the sun fall on a concave spherical mirror. Hold a graduated ruler in the position of its principal axis, and slide along it a small strip of cardboard. Find the point where the image of the sun is smallest. This will mark the principal focus, and it is a real one. If a convex spherical mirror be used, the light will be reflected as a broad pencil diverging from a point back of the mirror. The focus is then a virtual one.

When a pencil of parallel rays is incident on a concave spherical mirror, parallel to its principal axis, the point to which they converge after reflection is called its *principal focus*. In the case of a convex spherical mirror, the principal focus is the point on the axis behind the mirror from which the reflected rays diverge. The distance of the principal focus from the mirror is its *principal focal length.*

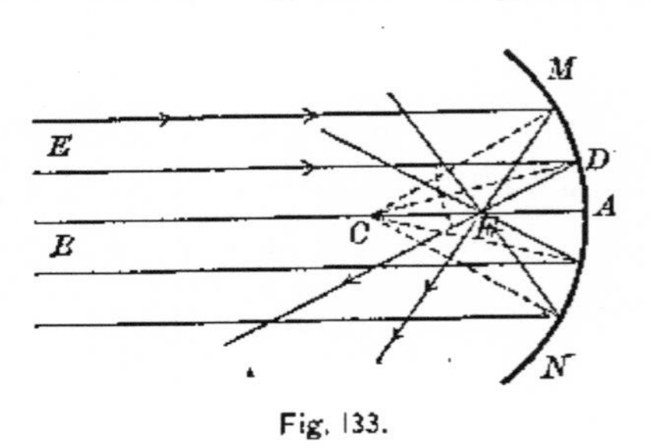

Fig. 133.

---

[1] The figures of mirrors in this chapter are sections made by a plane passing through the principal axis.

Let *MN* (Fig. 133) be a concave mirror whose centre is at *C* and principal axis is *AB*. Let *ED* be a ray parallel to *BA*. Then *CD* is the normal at *D*; and *CDF*, the angle of reflection, must equal *EDC*, the angle of incidence. Since the ray *BA* is normal to the mirror, it will be reflected back along *AB*. The reflected rays *DF* and *AB* have a common point *F*, which is the principal focus. The triangle *CFD* is isosceles with the sides *CF* and *FD* equal. (Why?) But when the point *D* is near *A*, *FD* is sensibly equal to *FA*; *F* is therefore the middle point of the radius *CA*. Other rays parallel to *BA* will pass after reflection nearly through *F* (§ 255). Hence, *the principal focus of a concave spherical mirror is real and is half way between the centre of curvature and the vertex.*

Let *MN* (Fig. 134) be a convex spherical mirror. *ED* and *BA* are rays parallel to the principal axis. Their common point *F*, after reflection, is back of the mirror and halfway between *A* and *C*. (Why?) Hence, *the principal focus of a convex spherical mirror is virtual and halfway between the centre of curvature and the mirror.*

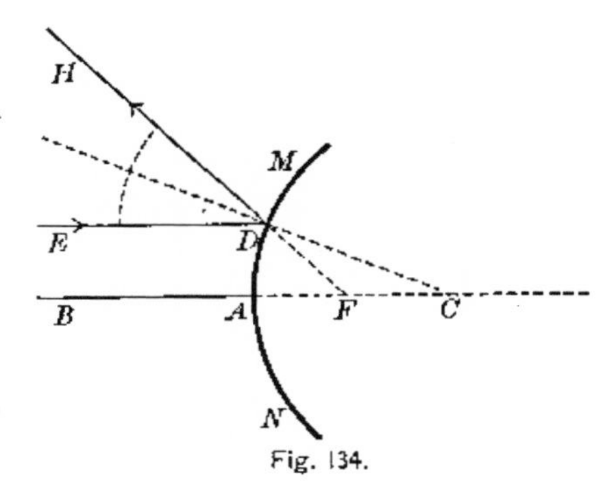

Fig. 134.

252. Conjugate Foci. — If a diverging pencil of light is incident on a spherical mirror, it is focussed after reflection at a point on the axis which passes through the radiant point or source of light; after reflection the rays diverge from this focus as a new radiant point. Thus, in Fig. 135, the rays *BA* and *BD* diverge from *B*; *BA* is reflected back along its own path (why?), and *BD* is reflected along DH, making the angles of incidence and reflection equal to each other. After reflection they both pass through B' and diverge from it. B and B' are *conjugate foci*. Rays diverging from either point will converge to the other.

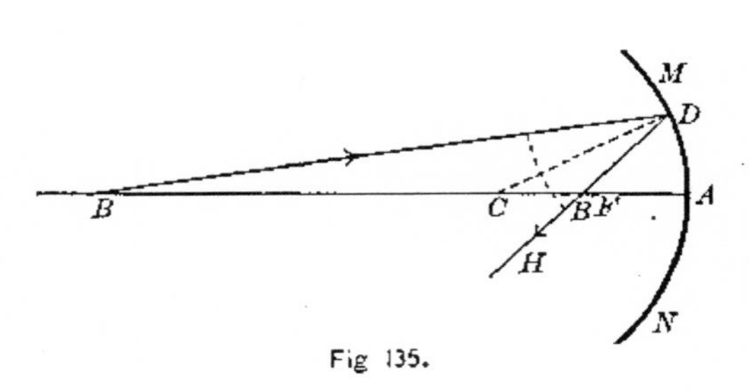

Fig 135.

In Fig. 136, the rays BA and BD diverge from B as the radiant point; after reflection they diverge from B' behind the reflecting surface, and B' is a virtual focus. B and B' are again conjugate foci.

In the first case the source of light is farther from the mirror than the centre of curvature, and the focus is real; in the second case it is nearer the mirror than the principal focus, and the focus is virtual.[1]

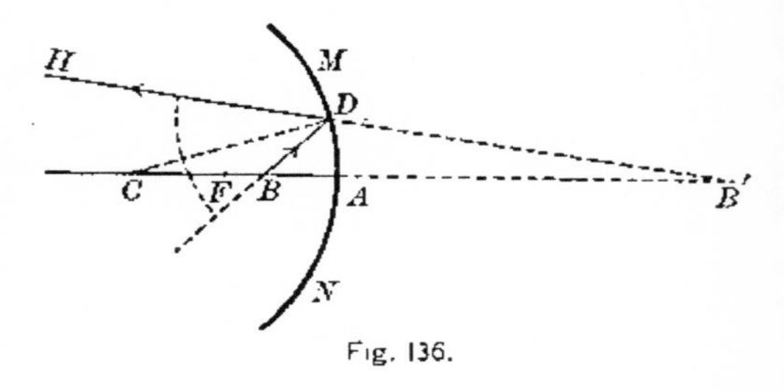

Fig. 136.

253. Formation of Images. — **Experiment.** — Darken the room and support on the table a spherical mirror, a lamp, and a small screen. The lamp and screen must be placed so that the screen will not cut off too much light from the mirror. Place the lamp anywhere beyond the

---

[1] In Fig. 135, CD bisects the angle BDH. Hence, $\frac{BD}{B'D}=\frac{BC}{B'C}$. If D is close to A, we may, without sensible error, place $BD = BA$ and $B'D = B'A$. Put $BA = p$, $B'A = p'$, $CA = r = 2f$. Then $BC = p - r$, $B'C = r - p'$, and $\frac{p}{p'}=\frac{p-r}{r-p'}$, from which $\frac{1}{p}+\frac{1}{p'}=\frac{2}{r}=\frac{1}{f}$. By measuring p and p', we may compute r and f. For the convex mirror, p' and r are negative.

focus, and move the screen till a clear image of the flame is formed on it. Notice the size and position of the image, and whether it is erect or inverted. When the lamp is between the focus and the mirror, an image of it cannot be obtained on the screen, but it can be seen by looking into the mirror. The same is true for the convex mirror, whatever be the position of the lamp; in these last cases the image is a virtual one.

The experiment illustrates the several relative positions of the object and its image for a concave mirror, all depending on the position of the object with respect to the mirror: —

First. — When the object is at a finite distance beyond the centre of curvature, the image is real, inverted, smaller

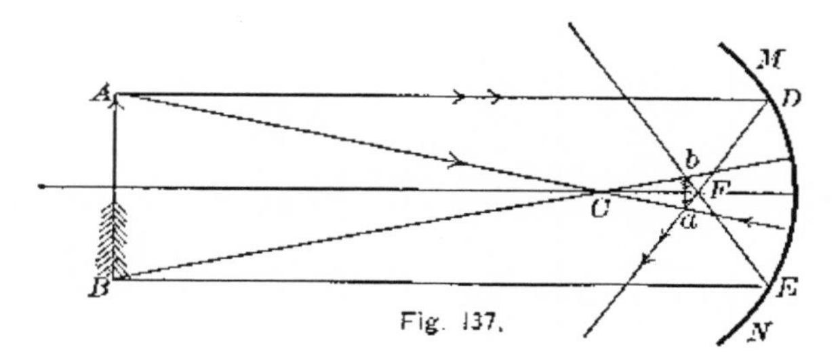

Fig. 137.

than the object, and between the centre of curvature and the focus (Fig. 137).

Second. — When a small object is at the centre of curvature, the image is real, inverted, of the same size as the object, and at the centre of curvature.

Third. — When the object is between the centre and the focus, the image is real, inverted, larger than the object, and is beyond the centre.

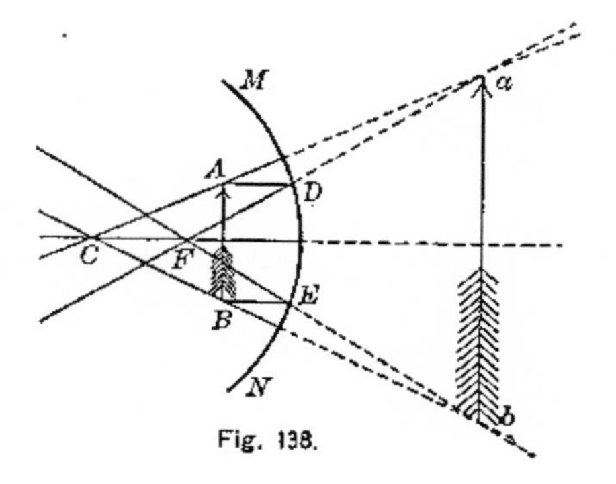

Fig. 138.

Fourth. — When the object is at the principal focus, the rays are reflected parallel to the principal axis, and no image is formed.

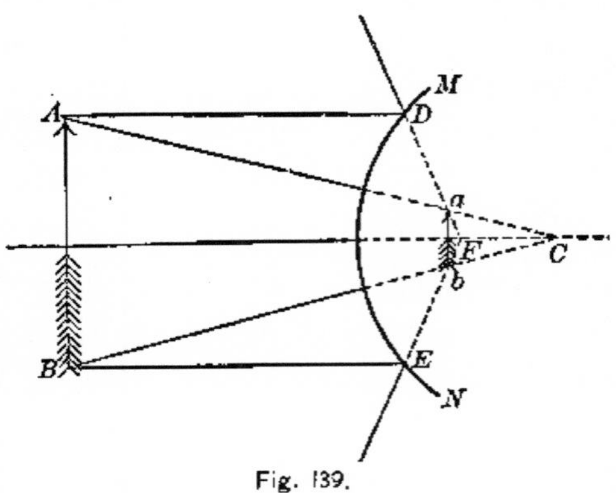

Fig. 139.

Fifth. — When the object is between the principal focus and the mirror, the image is virtual, erect, and larger than the object (Fig. 138).

When the mirror is convex, the image is always virtual, erect, and smaller than the object (Fig. 139).

254. Construction for Images. — The geometrical construction for images in spherical mirrors consists in finding conjugate focal points. For this purpose it is necessary to trace only two rays for each point of the object, one along the secondary axis through it, and the other parallel to the principal axis. The first ray is reflected back on itself, and the second through the principal focus. The intersection of the two reflected rays from the same point of the object locates the image of that point.

To illustrate: In Fig. 137, *AC* is the path of both the incident and the reflected ray, while the ray *AD* is reflected through the principal focus *F*. Their intersection is at *a*. The rays *BC* and *BE* are reflected similarly through *b*. Hence, *ab* is the image of *AB*. In Fig. 138, the ray *AC* along the secondary axis, and *AD* reflected back through *F* as *DF*, must be produced to meet back of the mirror at the virtual focus *a*. *A* and *a* are conjugate foci; also *B* and *b*, and *ab* is a virtual image.

The construction for the convex mirror (Fig. 139) is the same. From the point A draw AC along the normal or secondary axis, and AD parallel to the principal axis. The latter is reflected so that its direction passes through F. The intersection of these two lines is at a. The image ab is virtual and erect.

255. Spherical Aberration. — **Experiment.** — Bend a strip of bright tin or polished brass into as true a semicircle as possible. Place it on a sheet of paper with its concave surface toward a candle or lamp (Fig. 140). The light focuses on curved lines.

Fig. 140.

Experiment. — Project the image of a candle or a lighted lamp on a screen with a concave spherical mirror. The edge of the image will be indistinct; that is, not sharply defined. Now cover up the reflecting surface, exposing only the central portion. The image will be less bright (why?), but the definition will be sharper.

Rays incident near the margin of the mirror cross the principal axis at points nearer the mirror than F, the principal focus.

This distribution of the focus is known as *spherical aberration*, and the curve obtained in the first experiment is called the *caustic by reflection*. This curve may also be exhibited by allowing sunlight to fall on a cup partly full of milk, or on a plain gold ring supported on a white surface. If parallel rays incident on a concave mirror of wide aperture are traced (Fig. 141), it will be seen that the caustic is formed by their intersection after reflection. If the aperture of the mirror does not exceed 10°, this confusion of focus is small, and may be neglected. By decreasing the curvature of the mirror from the vertex outward, the aberration may be corrected. This is accomplished in the parabolic mirror, a form used as a reflector in lighthouses, in the headlights of locomotives, and for astronomical purposes.

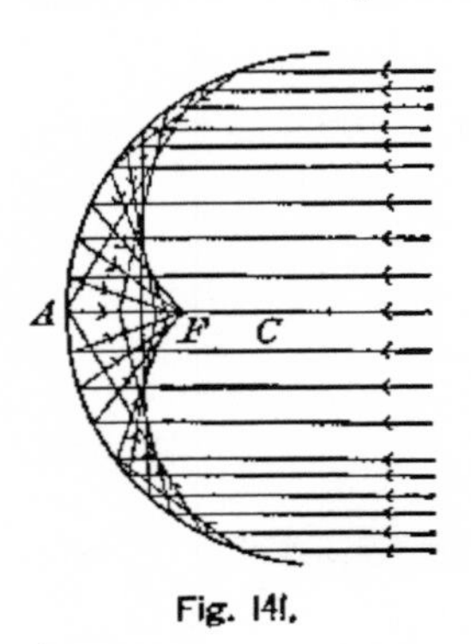

Fig. 141.

Problems.

1. A plane mirror is inclined to a horizontal plane at an angle of 45°. Find the position of the image of a vertical object in such a mirror.

2. A ray of light is successively reflected from two mirrors inclined at right angles to each other. What is the relative position of the first incident ray and the last reflected one?

3. Curved mirrors are sometimes made by silvering one face of a lens. How can you distinguish a concave mirror from a convex one?

4. Use a straight line for an object. Mark on it five equidistant points. Find, by construction, the conjugate foci to these points in a concave mirror. Join them, and compare the image obtained with the object. Account for the distortion in the image.

5. Find, by construction, the number of images of a point placed between two plane mirrors, whose angle of inclination is 60°.

6. Find, by construction, the number of images formed of a point by two plane mirrors inclined to each other at an angle of 40°. First, place the point on the bisector of the angle; second, to one side of the bisector. *Ans.* 8; 9.

7. The radius of curvature of a concave spherical mirror is

20 cm. If rays of light diverge from a point 60 cm. in front of it, at what point will they focus?

8. The principal focal length of a convex mirror is 30 cm. Locate the image of an object placed 10 cm. in front of the mirror.

Ans. 7.5 cm. behind the mirror.

9. The focal length of a concave spherical mirror is 12 in. An object is placed 16 in. in front of the mirror; find the position of the screen on which the image is received.

10. Find the focal length of a concave spherical mirror, whose radius of curvature is 24 in., and find the position of the image (*a*) of a point 18 in. in front of the mirror; (*b*) of a point 36 in. in front of the mirror; (*c*) of a point 8 in. in front of the mirror. *Ans.* 12 in., 36 in., 18 in., in front of the mirror, and 24 in. back of the mirror.

11. An object is placed 30 cm. in front of a concave mirror. The image is found to be 90 cm. in front of the mirror. Calculate the focal length.

12. The flame of a candle is placed on the axis of a concave spherical mirror, at the distance of 160 cm., and its image is formed at the distance of 40 cm. What is the radius of curvature of the mirror?

IV. REFRACTION OF LIGHT.

256. Refraction. — **Experiment.** — Place a rectangular tank, provided with a glass face, so that the light from a candle passing over the upper edge of one end just illuminates the whole of the opposite end (Fig. 142). The bottom of the tank lies wholly in the shadow cast by the end. Now fill the tank with water. The shadow no longer covers the whole bottom, since the rays are bent at an angle at the edge of the tank, as in *B*.

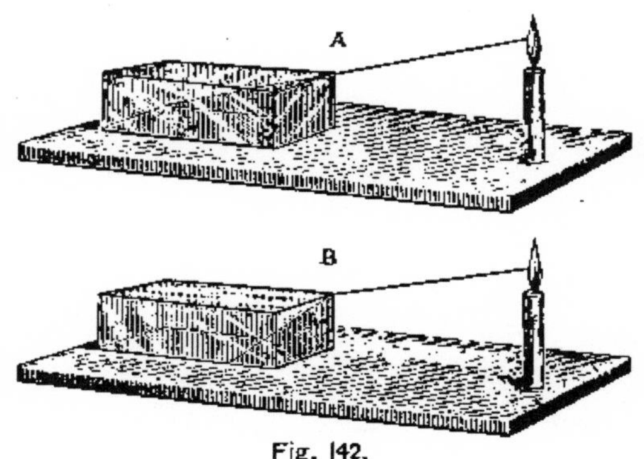

Fig. 142.

This change in the course of light in passing from one medium to another is called *refraction*.

257. Its Cause. — The investigations of Foucault, Michelson, and others, show that light has a less velocity in glass, water, etc., than it has in air. Now if a beam of light is incident obliquely on the surface *MN* of water (Fig. 143), all parts of a wave do not enter the water at the same time. Let the parallel lines perpendicular to *AB* represent the waves. Then one part of *f* will reach the water first and will travel less rapidly. The other portions, on entering, will be retarded in succession, the result being that the wave is swung around; that is, the direction of propagation *BC*, perpendicular to the wave front, is changed; or, in other words, the ray is refracted.

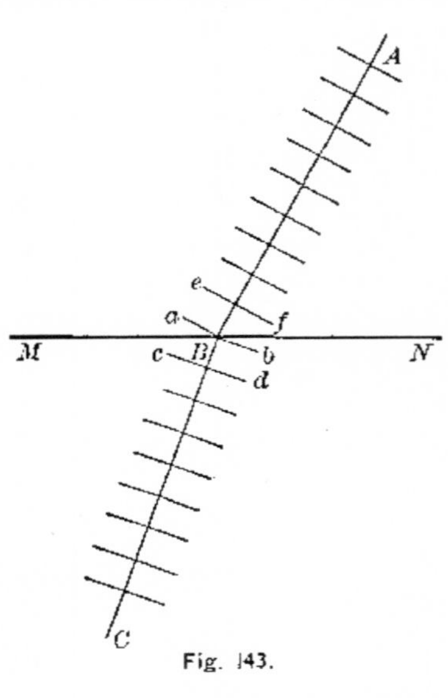

Fig. 143.

258. Terms Defined. — Let *BA* (Fig. 144) represent a ray of light in air incident obliquely at *A* upon the surface *MN* of another medium, as water. Draw the normal *DE* to the refracting surface. The angle *BAD* between the incident ray and the normal to the surface at the point of incidence is the angle of incidence; and the angle *CAE* between

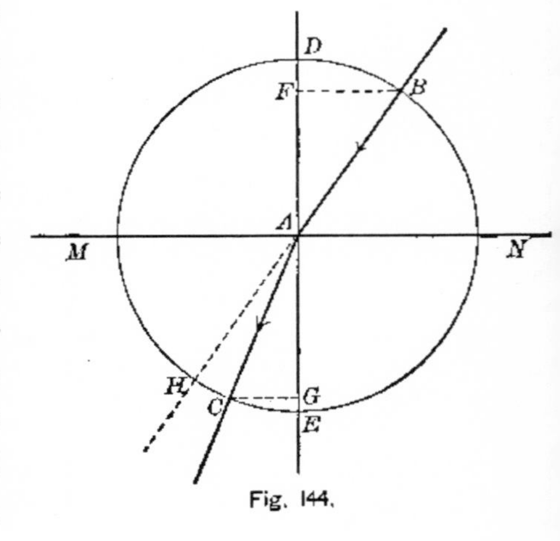

Fig. 144.

the refracted ray and the normal is the angle of refraction. Produce BA to H. The angle CAH measures the deviation of the light from its original course, and is called the *angle of deviation.* With A as a centre, describe a circle. Draw BF and CG, perpendicular to DE. Then the ratio $\frac{BF}{AB}$ is the sine of the angle BAD, and the ratio $\frac{CG}{AC}$ is the sine of CAE. The ratio of these sines is called the *index of refraction;* and, since $AB = AC$, this index equals $\frac{BF}{CG}$.

259. Laws of Refraction. — **Experiment.** — A rectangular glass jar is provided with a circular protractor scale on one face, and a slotted cardboard cover to the top (Fig. 145). Fill the jar with water exactly to the horizontal diameter of the circle. With a plane mirror, reflect a strong beam of light through the slit at such an angle as to be incident on the water exactly back of the centre of the scale. Read the angles of incidence and refraction on the circular scale, and find the ratio of their sines (see Appendix). Move the slit and obtain other angles for comparison. These ratios will be found to be constant.

Fig. 145.

The experiment illustrates the following laws: —

I. *When a beam of light passes obliquely from a less highly to a more highly refractive medium, it is bent toward the normal; when it passes in a reverse direction, it is bent from the normal.*

II. *Whatever the angle of incidence, the ratio of the sines of the angles of incidence and refraction is constant for the same two media.*

III. *The planes of the angles of incidence and refraction coincide.*

These laws were first discovered by Snell, a Dutch physicist, in 1621.

260. Indices of Refraction. — The *absolute index of refraction* is the ratio of the sines of the angles of incidence and refraction when the ray passes from a vacuum into the substance. The *relative index of refraction* is the index for light passing from one substance into another; it is found by dividing the absolute index of the latter by the former. The larger the index of refraction, the greater is said to be the *optical density* of the substance.

The following table gives the indices of a few substances relative to air: —

Water	1.333	Crown glass	1.51
Alcohol	1.36	Flint glass	1.54 to 1.71
Carbon bisulphide	1.64	Diamond	2.47

For the purposes of this book, the refractive index for water may be taken to be $\frac{4}{3}$; for crown glass, $\frac{3}{2}$; for flint glass, $\frac{8}{5}$; and for diamond, $\frac{5}{2}$.

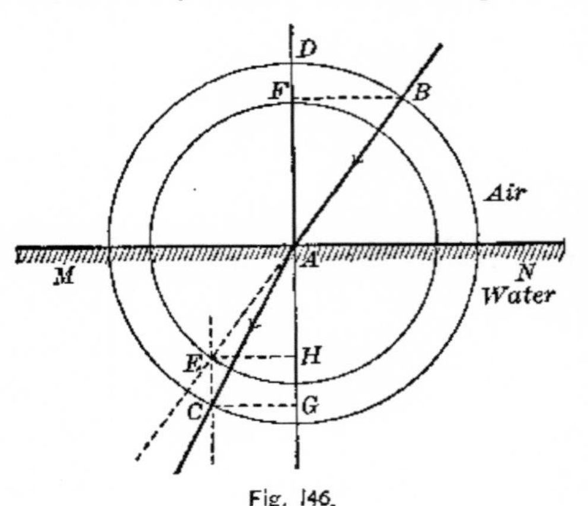

Fig. 146.

261. Construction for the Refracted Ray. — Let MN (Fig. 146) be the surface separating the two media, air and water; and let BA be a ray of light incident at A; it is required to draw the refracted ray. With A as a centre, and with radii whose ratio is $\frac{4}{3}$, the index of refraction, draw two concentric circles. Through E, one of the intersections of BA with the smaller circle, draw EC parallel to the normal AD, cutting the larger circle at C. Draw AC. This will be the refracted ray, because $\frac{BF}{CG} = \frac{4}{3}$, a fact easily shown.[1]

When the ray passes into a medium of smaller optical density, then the intersection of the incident ray with the *larger* circle must be used; through this point draw a parallel to the normal, cutting the *smaller* circle. The line through the intersection with the smaller circle and the point of incidence is the refracted ray.

It should be observed that it does not matter whether the intersections of the incident ray with the circles be taken before the ray enters the medium or afterwards.

Problems.

1. Trace a ray of light from air into crown glass.
2. Trace a ray of light from water into air.
3. Trace a ray of light from air into diamond.
4. Trace a ray of light from glass[2] into air.
5. Trace a ray of light from water into glass.
6. Trace a ray of light from glass into water.

262. Refraction through a Parallel Plate. — **Experiment.** — Draw a straight line on a sheet of paper. Place a piece of thick plate

---

[1] The triangles EAH and ABF are similar. Hence $\frac{BF}{EH} = \frac{AB}{AE}$. But $EH = CG$, and $\frac{AB}{AE} = \frac{4}{3}$, by construction. Therefore $\frac{BF}{CG} = \frac{4}{3}$, the index of refraction.

[2] By "glass" crown glass is to be understood, and the index is taken as $\frac{3}{2}$.

glass over the line, covering a portion of it. Look obliquely through the glass; the line will appear broken at the edge of the plate, the part under the glass appearing laterally displaced.

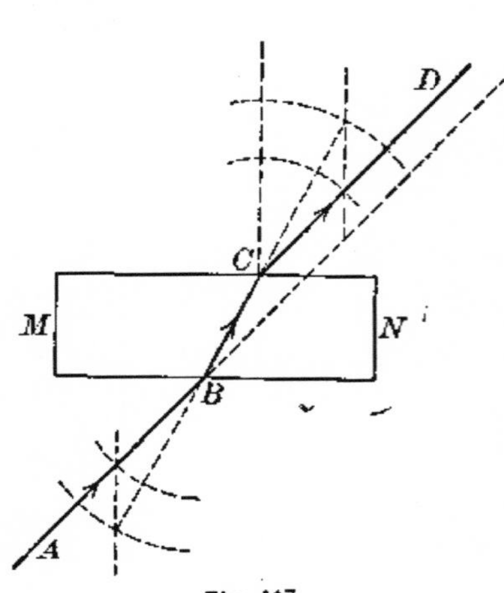

Fig. 147.

To explain this, let *MN* (Fig. 147) represent a thick plate of glass, and *AB* a ray of light incident obliquely upon it. If the path of the ray be determined, the emergent ray will be parallel to the incident ray. Hence, the apparent position of an object viewed through a plate is at one side of its true position.

263. A Prism. — Optically, the portion of a transparent substance lying between two intersecting planes is a *prism*, and the angle between these planes is the *refracting angle* of the prism. A beam of light incident on a prism is bent away from the refracting angle, and consequently the apparent position of an object seen through it is displaced toward the refracting edge.

Let *ABC* (Fig. 148) represent a section of a glass prism

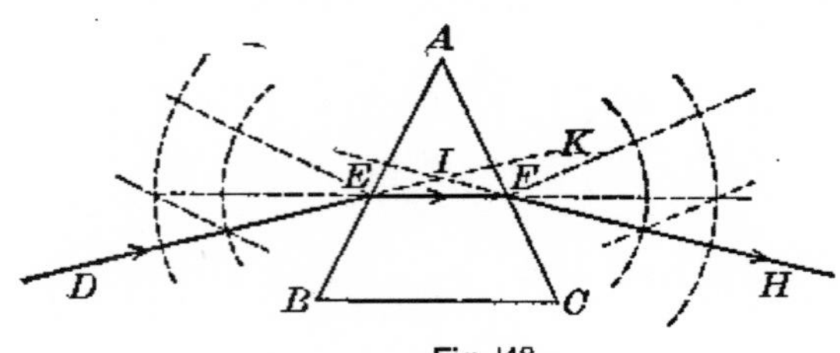

Fig. 148.

made by a plane perpendicular to the refracting edge *A*. Also, let *DE* be a ray incident on the face *BA*. This ray will be refracted along EF, and entering the air at the point F will be refracted again, taking the direction FH.

264. The Angle of Deviation. — **Experiment.** — Make a small hole in the centre of a sheet of cardboard. Arrange a screen, the sheet of cardboard, and a Bunsen flame as shown in Fig. 149. The burner should have a short cylinder of asbestos paper soaked in sodium nitrate at its top; the light will then be of one color. Mount a prism back of the aperture in the cardboard with the refracting edge uppermost. A on the screen will be the illuminated spot before the prism is in place, and B will be the spot when the light passes through the prism. The angle APB is roughly the *deviation.*

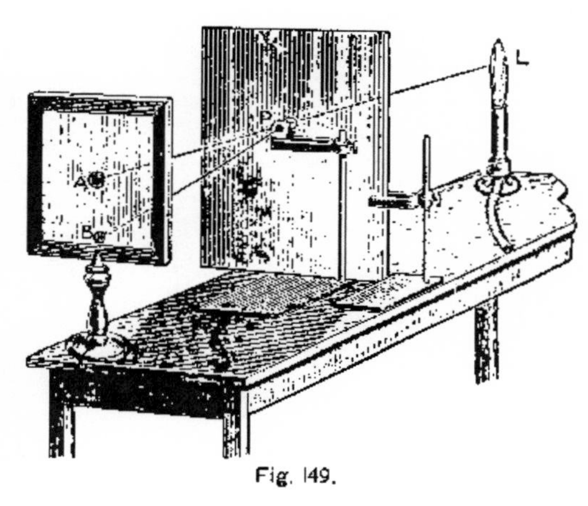

Fig. 149.

The angle of deviation is the angle included between the incident and the emergent ray, as KIH (Fig. 148). By tracing different rays through the prism, and by using prisms with different angles, the deviation will be found to vary with the angle of the prism, with the index of refraction, and with the angle of incidence. The least deviation for any prism occurs when the angles of incidence and emergence are equal.

265. Phenomena of Refraction. — Light in passing from water into air is refracted from the normal, and consequently objects under water appear elevated above their true position. A familiar instance is a coin in an empty cup just hidden from view, becoming visible when the cup is filled with water. The light from the coin is bent on leaving the water so that it reaches the eye. The apparent shoaling of a pond of water is explained in the same way. Light coming to the eye from a star will be gradually bent, describing a curve (Fig. 150), because the refractive index of the air is greater near the earth than higher up. Since the direction in which the star is seen is that of a tangent to the curve at the eye, the effect will be to increase the apparent altitude of the star. For the same reason the sun is visible when it is actually below the horizon.

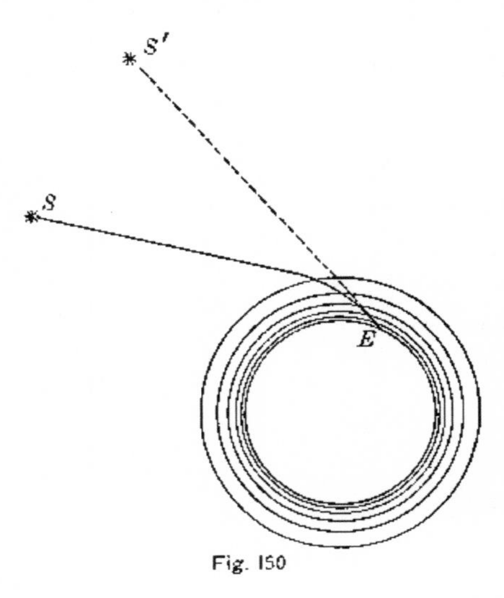

Fig. 150

266. Total Reflection. — **Experiment.** — Using the apparatus of § 259, place the cardboard against the end so that the slit is close to the bottom of the jar (Fig. 151). By means of mirrors, send a beam of light upward through the water and incident on the surface just back of the centre of the circular scale. The light will be reflected from the surface back into the water as by a mirror.

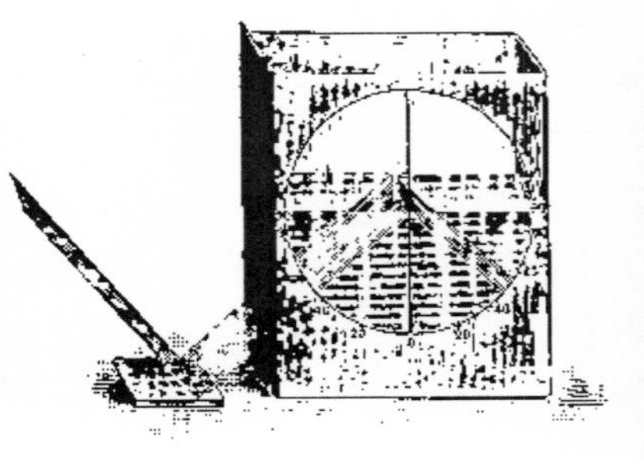
Fig. 151.

When a beam of light passes from one medium into another of smaller optical density, it is refracted from the normal, the angle of refraction being always greater than the angle of incidence. Hence, by gradually increasing the angle of incidence, a value can be found in which the angle of refraction is 90°; that is, the rays pass off in the surface. When the angle of refraction in the rarer medium is a right angle, the angle of incidence is called the *critical angle*. This angle varies with the substance, being $48\frac{1}{2}°$ for water, 41° for crown glass, and 24° for diamond. When the angle of incidence exceeds the critical angle, as in the experiment, the light suffers *total internal reflection*.

267. Construction for the Critical Angle. — Let *MN* (Fig. 152) separate two media, as air and glass. With *A* as a centre, draw two concentric circles, the ratio of their radii being the index of refraction, $\frac{3}{2}$. The issuing ray must lie in *AN*. Hence, draw the normal *CE*, cutting the larger circle at *E*, and the line *BA* through *A* and *E* will be the required incident ray (§ 261), the angle *BAD* being the critical angle.

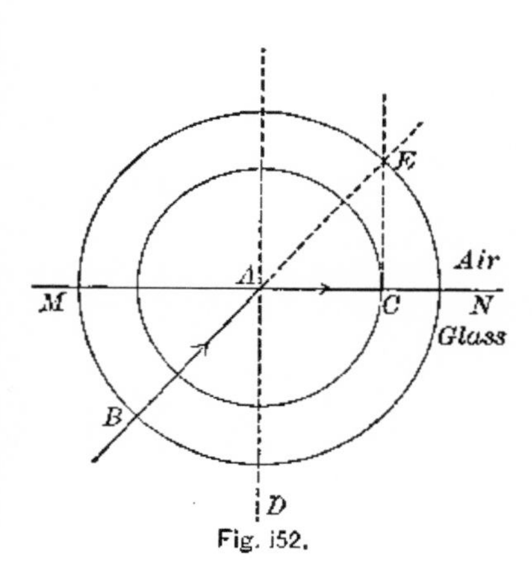

Fig. 152.

268. Illustrations of Total Reflection. — Of all the rays diverging from a point at the bottom of a pond and incident on the surface, only those within the cone whose semi-angle is $48\frac{1}{2}°$ can pass into the air, on account of the separation of the refracted from the totally reflected rays by total internal reflection (Fig. 153). Hence, an observer under water sees all objects out-

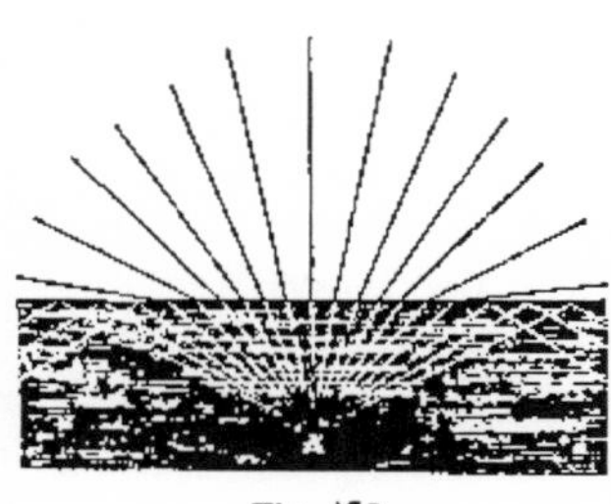
Fig. 153.

side as if they were crowded within this cone, and beyond this cone he sees by reflection objects lying on the bottom of the pond.

Total reflection in glass is best shown by means of a prism whose cross-section is a right-angled isosceles triangle (Fig. 154). A ray incident normally on either face about the right angle enters the prism without refraction, and is incident on the hypotenuse at an angle of 45°, which is greater than the critical angle (§ 266). Hence, the ray suffers total reflection and leaves the prism at right angles to the incident ray. Such a prism makes the best possible reflector in an optical instrument where it is desirable to change the direction of the light by 90°.

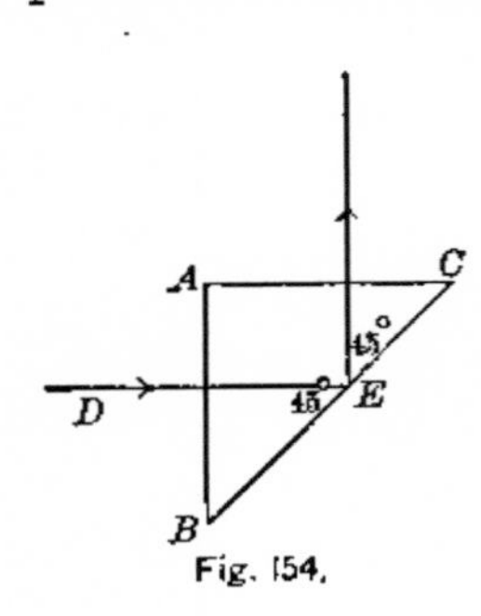

Fig. 154.

Problems.

1. Construct the critical angle for water.

2. Construct the critical angle for diamond.

3. Construct the critical angle for flint glass.

4. The refracting angle of a prism is 60°, and the index of refraction 1.5. Trace a ray of light through the prism, the angle of incidence being 15°.

5. The refracting angle of a prism of diamond is 60°. Trace a ray of light through it, the angle of incidence being 30°.

V. LENSES.

269. A Lens is a portion of a transparent substance bounded by two curved surfaces, or one plane and one curved surface. Those most commonly met with have spherical surfaces (Fig. 155), and are classified as follows: —

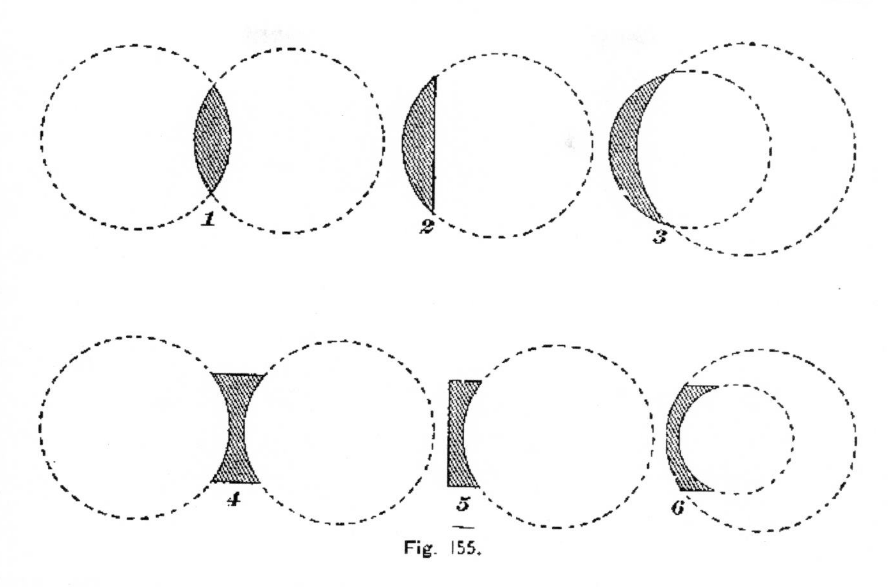

Fig. 155.

1. Double-convex, — both surfaces convex . . 2. Plano-convex, — one surface convex, one plane . . . . 3. Concavo-convex, — one surface convex, one concave	Converging lenses, thicker at the middle than at the edges.
4. Double-concave, — both surfaces concave, . 5. Plano-concave, — one surface concave, one plane . . . . 6. Convexo-concave, — one surface concave, one convex	Diverging lenses, thinner at the middle than at the edges.

The concavo-convex and the convexo-concave lenses are frequently called *meniscus* lenses. The double-convex lens may be regarded as the type of the converging class of lenses, and the double-concave lens of the diverging class.

270. Terms Defined. — The centres of the spherical surfaces bounding a lens are the *centres of curvature*. The *optical centre* is a point such that any ray passing through it and the lens suffers no change of direction. In lenses whose surfaces are of equal curvature, it is their centre of volume, as O, in Fig. 156. In plano-lenses, the optical centre is the middle point of the curved face. Any straight line, as EH, through the optical centre is a *secondary axis*, and the axis through the centres of curvature is the *principal axis*, as CC'. The normal at any point of the surface is the radius of the sphere drawn to that point; thus CD is the normal to the surface AnB at D.

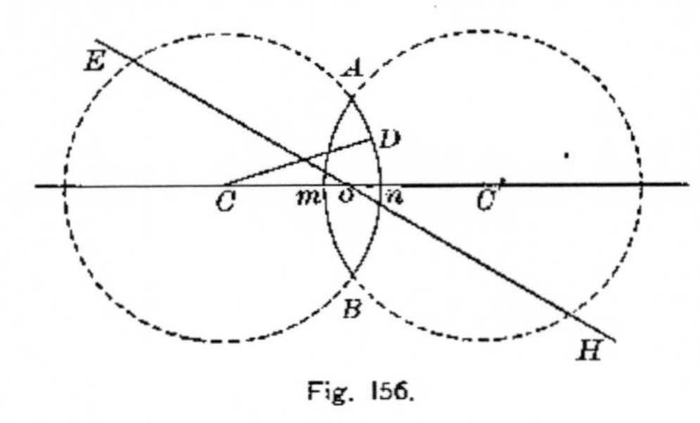

Fig. 156.

271. Tracing Rays through Lenses. — Let MN represent a lens whose centres of curvature are C and C', and AB

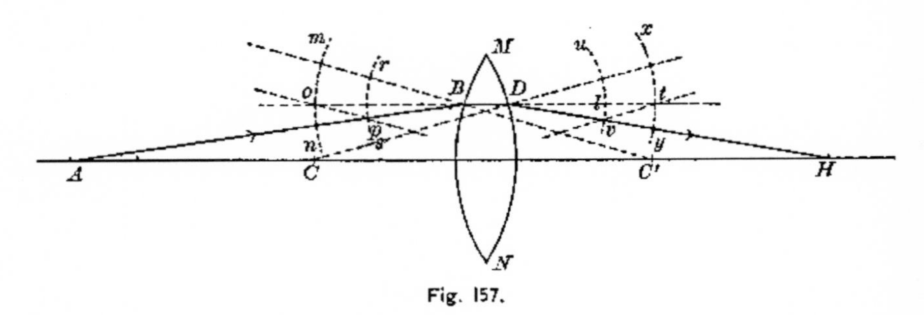

Fig. 157.

the ray to be traced through it (Figs. 157, 158). Draw the normal, $C'B$, to the point of incidence. With B as a centre, draw the arcs mn and rs, making the ratio of their radii equal the index of refraction, $\frac{3}{2}$.[1] Through p, the intersection of AB with rs, draw op parallel to the normal, $C'B$, and cutting mn at o. Through o and B draw oBD;

[1] The figures of lenses in this chapter are sections made by planes through the centre of curvature, and the index is taken at $\frac{3}{2}$.

this will be the path of the ray through the lens (§ 261). At D it will again be refracted; to determine the amount, draw the normal CD and the auxiliary circles, xy and uv, as before. Through the intersection of BD produced with xy, draw lt parallel to the normal CD, cutting uv at l. Through D and l draw DH; this will be the path of the ray after emergence. (Compare this procedure with that of § 263.) It should be noticed that the convex lens bends the ray toward the principal axis, while the concave lens bends it away from it.

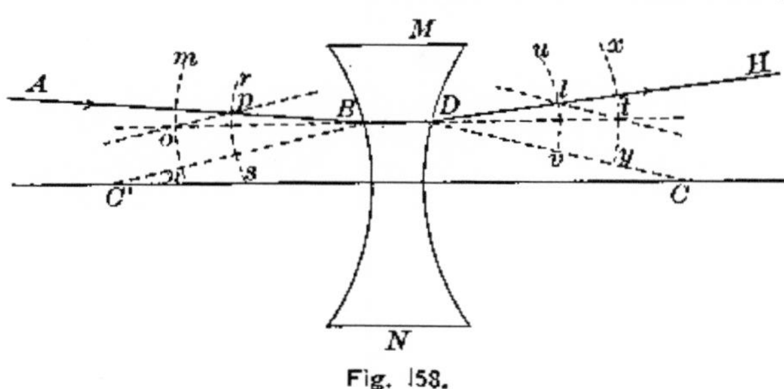

Fig. 158.

272. Principal Focus of a Lens. — **Experiment.** — Let the rays of the sun fall on a convex lens parallel to its principal axis. Hold beyond the lens a sheet of paper, moving it till the round spot of light is smallest and brightest. If held steadily, a hole may be burned in the paper. This spot marks the *principal focus* of the lens, and its distance from the optical centre is the *principal focal length*.

Figures 159 and 160 illustrate the method of finding the focus geometrically. In the double-convex lens, it is real, and at the centre of curvature, *if the index of refraction is* $\frac{3}{2}$. For the plano-convex lens, the focal length is twice the radius of curvature for the same

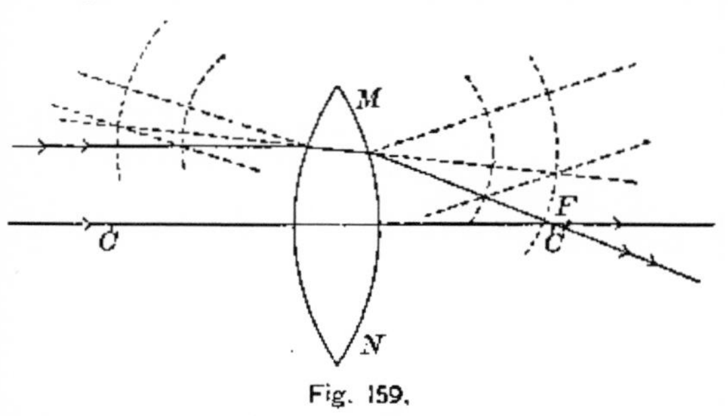

Fig. 159.

refractive index. Convex lenses are often called *burning glasses*, because of their power to focus the *heat rays*, as

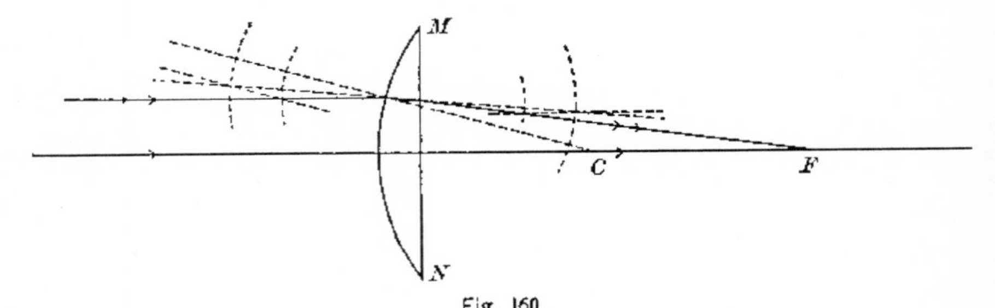

Fig. 160.

shown in the experiment. Figure 161 shows that parallel rays are rendered diverging by a double-concave lens, *the*

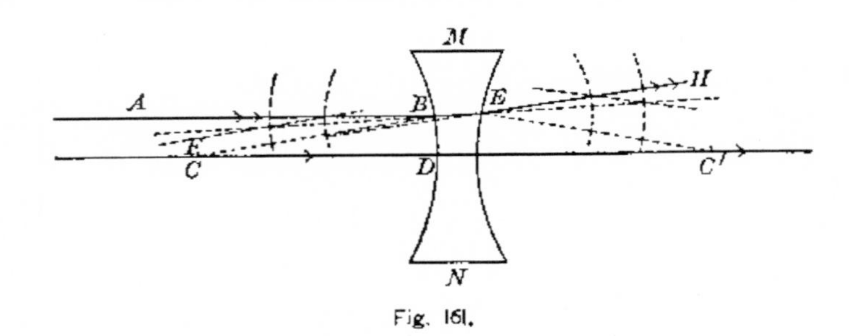

Fig. 161.

principal focus being virtual, and at the centre of curvature, when the refractive index is $\frac{3}{2}$. Concave lenses increase the divergence of light, whereas convex lenses decrease it.

273. Conjugate Foci.—Lenses resemble spherical mirrors in this respect, that if a pencil of light diverges from a point and is incident on the lens, it is focussed at a point on the axis through the radiant point. These points are called *conjugate foci*, for the same reason as in mirrors. If p and p' are the distances of these points from the lens, and f is the principal focal length, then the relation between the three quantities is expressed by the equation

$\frac{1}{p}+\frac{1}{p'}=\frac{1}{f}$.[1] For converging lenses p' must be considered negative when the image is virtual; and for diverging lenses both p' and f are to be treated as negative. By measuring p and p', the focal length of a lens may be computed.

Diverging lenses always increase the divergence of the rays incident upon them, and hence the focus of such lenses is always virtual. With converging lenses the results vary with the angle of divergence. Hence, the following cases arise: —

First. — When the incident rays diverge from a point more than twice the focal distance from the lens.

In Fig. 157 the two rays diverging from A focus at H, less than twice the focal distance.

Second. — When the incident rays diverge from a point at twice the focal distance from the lens.

(The student should draw a figure, and show that the focus is real, and at twice the focal distance.)

Third. — When the incident rays diverge from a point at less than twice the focal distance.

(The student should draw a figure and show that the focus is real and at more than twice the focal distance.)

Fourth. — When the rays diverge from the principal focus.

---

[1] In Fig. 163, from the similar triangles AOB and aOb, we have $\frac{AB}{ab}=\frac{KO}{OL}$. If E and I be connected by a straight line, this line may be taken as approximately equal to AB, and to pass through O. From the similar triangles EFI and aFb we have $\frac{EI}{ab}=\frac{OF}{FL}$. Hence, $\frac{KO}{OL}=\frac{OF}{FL}$. Put $KO=p$, $OL=p'$, $OF=f$. Then $FL=p'-f$, $\frac{p}{p}=\frac{f}{p'-f}$, and $\frac{1}{p}+\frac{1}{p'}=\frac{1}{f}$. By measuring p and p', we may compute f. For the diverging lens f and p' are negative.

(The student should draw a figure and show that the emergent rays are parallel.)

Fifth. — When the rays diverge from a point between the principal focus and the lens.

Figure 162 shows that the divergence of the rays is not

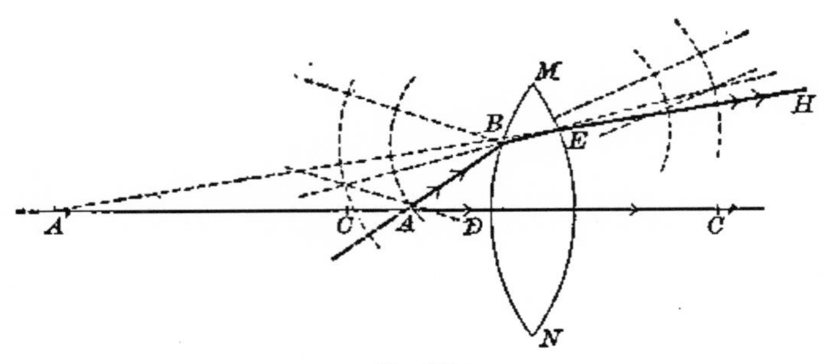

Fig. 162.

wholly overcome by the lens, but that they leave the lens as if they emanated from a point farther from the lens than the actual radiant point. Hence, the focus is virtual.

274. Formation of Images. — The image of an object formed by a lens may be found by means of the images of its points. The work is considerably lessened by observing the following method of construction: —

Draw secondary axes through the extremities of the object. These will be rays of light which suffer no change of direction. (Why?) Also through these extremities draw rays parallel to the principal axis, and find by construction their path in the lens (§ 261). On leaving the lens they will pass through the principal focus. (Why?) The image of each extremity will be the intersection of the two rays drawn from it. The image of any point is always on the secondary axis passing through it.

To illustrate, let AB be the object and MN the lens (Fig. 163). Rays along the secondary axes pass through the lens without deviation. The rays AD and BH, parallel to the principal axis, are refracted on entering the lens along DE and HI respectively, and pass through F, the principal focus, after leaving the lens. The intersec-

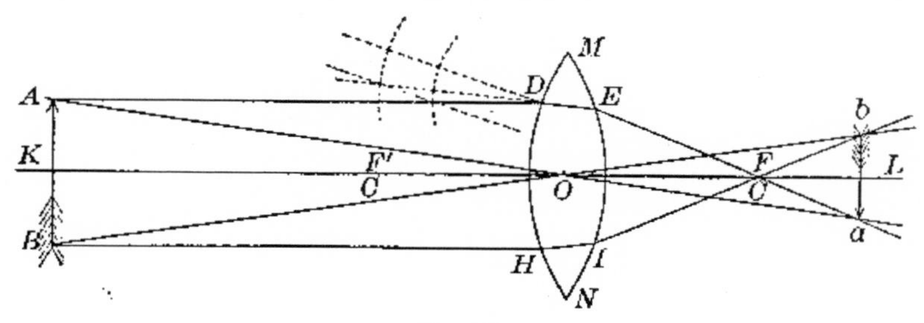

Fig. 163.

tion of Aa with $ADEa$ is the image of A, and that of $BHIb$ with Bb is the image of B. If other rays from A and B be drawn, they will focus at a and b respectively. Hence, ab is the image of AB.

The student should draw figures and establish the following propositions for converging lenses: —

I. When the object is at a finite distance greater than twice the focal distance, the image is real, inverted, situated beyond the principal focus, and is smaller than the object (Fig. 163).

II. When the object is at twice the focal distance, the image is real, inverted, situated at twice the focal distance, and is of the same size as the object.

III. When the object is at less than twice the focal distance, the image is real, inverted, situated beyond the principal focus, and is larger than the object.

IV. When the object is at the principal focus, the light leaves the lens in parallel rays, no image being formed.

V. When the object is between the principal focus and the lens, the image is virtual, erect, and enlarged (Fig. 164).

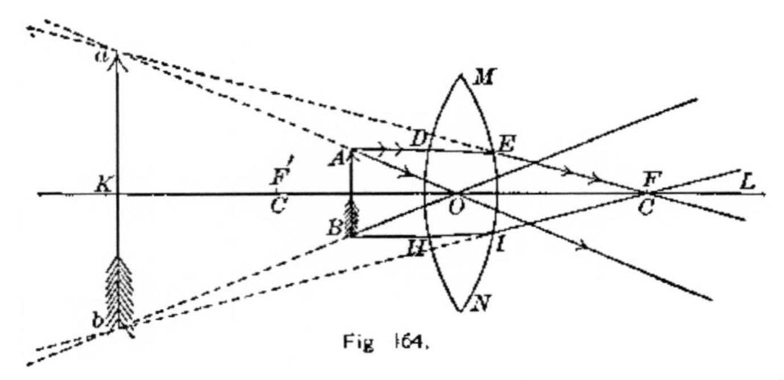

Fig 164.

The images formed by diverging lenses are always virtual, erect, and smaller than the object (Fig. 165).

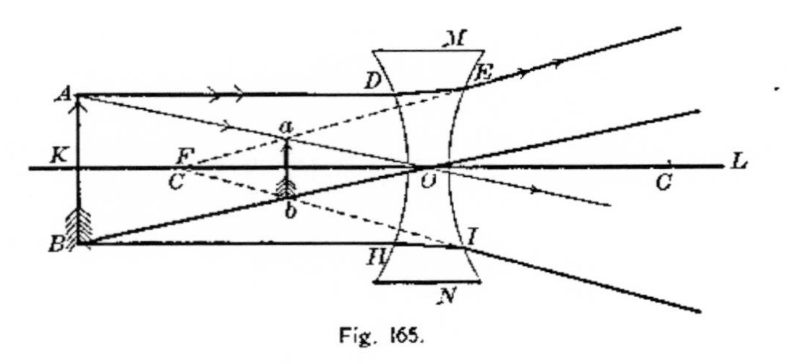

Fig. 165.

The distance of the object from the lens affects only the size of the image.

275. Experimental Illustrations. — **Experiment.** — Arrange in line on the table a lamp, a converging lens, and a screen. Give the lamp successively the positions described in the preceding article, adjusting the screen each time till a sharply defined image is obtained. The results will be in harmony with the preceding propositions. If the lamp is placed at the focus of the lens, only a blurred image is obtained; if between the focus and the lens, no image is formed on the screen, but a magnified image of the lamp may be seen on looking through the lens.

If, after focussing the image on the screen, the eye be placed in line with the object and the lens, and the screen be removed, an inverted image of the object can be seen suspended in mid-air.

The projection of images on a screen by means of a lens has its application in the projection lantern (§ 299), the compound microscope (§ 296), the telescope (§ 297), and the photographer's camera (§ 300). The formation of virtual images by a converging lens is applied in the simple magnifier and in the eyepieces of telescopes and compound microscopes.

276. Spherical Aberration. — If rays from any point be drawn to different parts of a lens, and their directions be determined after refraction, it will be found that those incident near the edge of the lens cross the principal axis, after emerging, nearer the lens than those incident near the middle. (The student should convince himself of this by drawing a figure.) The principal focal length for the marginal rays is therefore less than for central rays. This indefiniteness of focus is called *spherical aberration by refraction*, the effect of which is to lessen the distinctness of images formed by the lens. In practice an annular screen, called a *diaphragm*, is used to cut off the marginal rays; this renders the image sharper in outline, but less bright. In the large lenses used in telescopes the curvature of the lens is made less toward the edge, so that all parallel rays are brought to the same focus.

Problems.

1. Compare by accurate construction the focal length of a lens of crown glass with that of a lens of diamond, the lenses having the same shape and size.

2. The radius of curvature of a double-convex lens of crown glass is 12 in. Find the position of the image of an object distant 2 ft. from the optical centre of the lens.

3. An object 120 cm. distant from a converging lens gave an image 10 cm. from the lens. Calculate the focal length.

4. Rays of light diverge from a point 20 cm. in front of a converging lens. The focal length of the lens is 4 cm. How do the rays behave after refraction through it?

5. An object is placed at a distance of 14 cm. in front of a double convex lens of 40 cm. focal length. Where will the image be formed?

Ans. Virtual, 21.54 cm. from lens.

6. A converging lens of 20 cm. focal length forms an image of an object on a screen. The object is distant 24 cm. from the lens. How far from the optical centre must the screen be placed to receive the image?

7. It is found that when the image formed by a converging lens and the object are of the same size, they are 100 cm. apart. Find the focal length of the lens.

VI. DISPERSION.

277. Analysis of White Light. The Solar Spectrum.— **Experiment.**— Close the opening of a porte-lumière[1] with a piece of tin, in which is cut a very narrow vertical slit. Let the ribbon of sunlight issuing from the slit be incident obliquely on a glass prism (Fig. 166). A many-colored band, gradually changing from red at one end through orange, yellow, green, blue, to violet at the other, appears on the screen. If a converg-

Fig. 166.

[1] The porte-lumière is a device by means of which a beam of sunlight can be reflected horizontally into a darkened room through an opening in the shutter. In the simplest form it is a hinged mirror set outside of the window, and so arranged that its position can be changed from within.

ing lens of about 30 cm. focal length be used to focus an image of the slit on the screen, and the prism be placed near the principal focus, the colored images of the slit will be more distinct.

This experiment, though not original with Sir Isaac Newton, was first explained by him in 1666. It shows that white or colorless light is a mixture of an infinite number of differently colored rays, differing in refrangibility, the red being least and the violet most refrangible. The brilliant band of light consists of an indefinite number of colored images of the slit; it is called the *solar spectrum*, and the opening out or separating of the beam of white light is known as *dispersion.*

278. Synthesis of Light. — Experiment. — Project a spectrum of sunlight on the screen. Now place a second prism like the first behind it, but reversed in position (Fig. 167). There will be formed a colorless image, slightly displaced on the screen.

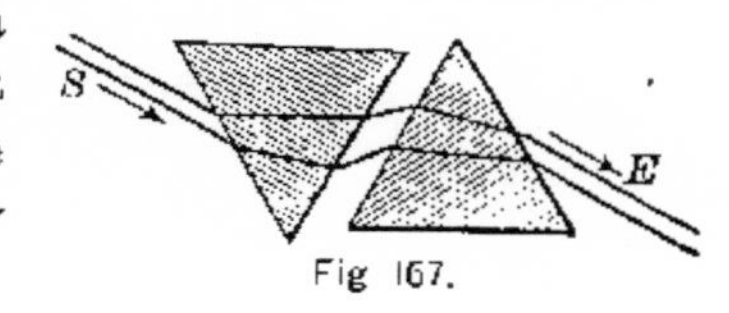

Fig 167.

The second prism reunites the colored rays, making the effect that of a thick plate of glass (§ 262). The recomposition of the colored rays into white light may also be effected by receiving them on a concave mirror or a large convex lens.

279. Chromatic Aberration. — Experiment. — Close the opening of the porte-lumière with a piece of cardboard in which is a small round hole. Project an image of this aperture on the screen, using a double-convex lens for the purpose. The round image will be bordered with the spectral colors.

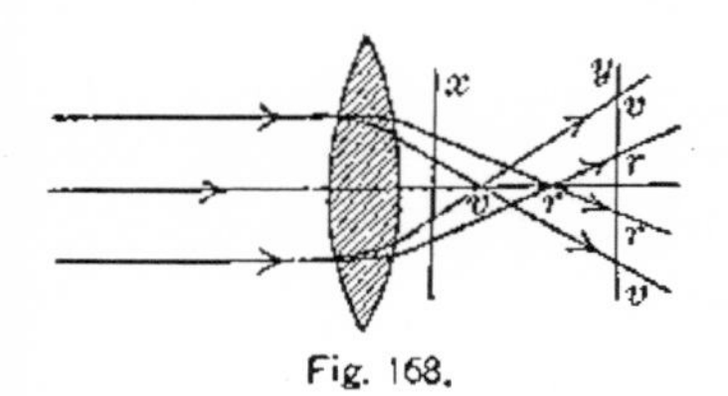

Fig. 168.

This defect in lenses is known as *chromatic aberration.* It is caused by the lens refracting the rays of different colors to different foci. The violet rays, being mor. refrangible than the red, will have their focus nearer to the lens than the red, as shown in Fig. 168, where v is the principal focus for violet light and r for the red. If a screen were placed at x, the image would be bordered with red, and if at y with violet.

280. The Achromatic Lens. — **Experiment.** — Project a spectrum of sunlight on the screen, using a prism of crown glass, and note the length of the spectrum when the prism is turned to give the least deviation. Repeat the experiment with a prism of flint glass having the same refracting angle. The spectrum formed by the flint glass will be about twice as long as that given by crown glass, while the position of the middle of the spectrum on the screen is about the same in the two cases. Now use a flint glass prism whose refracting angle is half that of the crown glass one. The spectrum is nearly equal in length to that given by the crown glass prism, but the deviation of the middle of it is considerably less. Finally, place this flint glass prism in a reversed position against the crown glass one. The image of the aperture is no longer colored, and the deviation is about half that produced by the crown glass alone.

The above facts suggested to Dolland, an English optician, in 1757, that by combining a double-convex lens of crown glass with a plano-concave lens of flint glass the dispersion by the one would neutralize that due to the other, while the refraction would be reduced about half (Fig. 169). Such a lens or system of lenses is called *achromatic*, since images formed by it are not fringed with the spectral colors.

Fig. 169.

281. Dispersion by a Globe of Water. — **Experiment.** — Fill an air-thermometer bulb, about 4 cm. in diameter, with clear water. Cover the opening of the porte-lumière with a large sheet of white cardboard, in which is a circular hole about 3.75 cm. in diameter. Support the bulb a short distance from the opening so that the cylinder of sunlight is incident upon it. There will appear on the cardboard screen one or more circular spectra, resembling rainbows.

To understand this experiment it must be kept in mind that when light passes from one medium to another, part of the light is always reflected. So in the case of the globe of water, part passes through it and part enters and is internally reflected from the back surface, forming the colored image on the cardboard screen.

Let the circle whose centre is O (Fig. 170) represent the globe of water, and SS' rays of sunlight incident upon

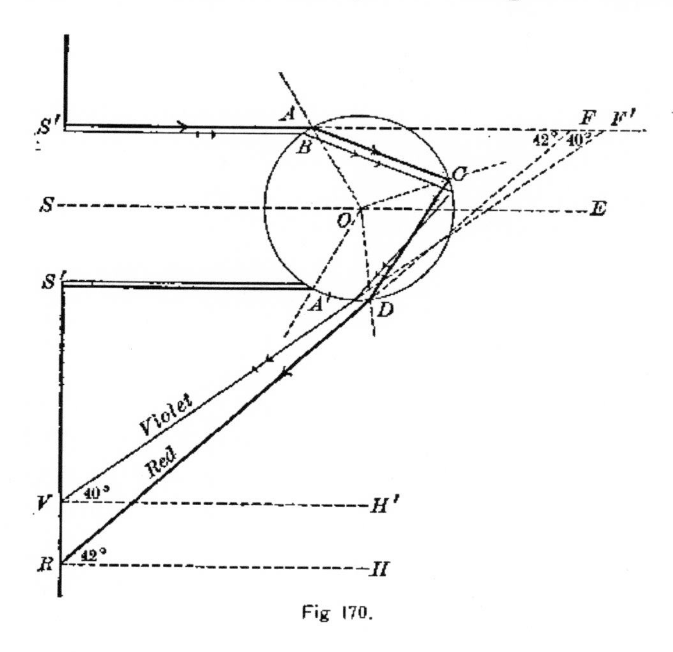

Fig 170.

it. Of all the rays entering the globe, it is shown in mathematical optics that the *red* rays incident in the immediate vicinity of 59° 23′ 30″ from the axis SE, as at A, keep together after reflection and subsequent refraction; that is, they are parallel on leaving the sphere, and hence have sufficient intensity to produce a colored image on the screen. The violet rays at 58° 40′ from the axis, as at *B*, produce a violet image for a like reason. Between these positions the other colors arrange themselves in order. Since the sphere is symmetrical with respect to *SO*, it follows that the colored images will appear as circles. In this way the inner colored ring is produced in the above experiment. The angle *S′FR* is the deviation of the red rays, and equals 42° 1′ 40″; the deviation of the violet rays *S′F′V* is 40° 17′. Hence the red circle on the screen has a radius of about 42°, and the violet one about 40°. The second colored band is caused by rays that are twice internally reflected. The violet part has a radius of about 54° and the red 51°.

282. The Rainbow is a solar spectrum formed by spherical raindrops dispersing the sunlight falling upon them. Usually two bows are visible, the *primary* and the *secondary*. The *primary bow* is the inner and brighter one, and is distinguished by being red on the outside and violet on the inside. The *secondary bow* is much fainter, and has the order of colors reversed. Figure 171 shows the relative position of the sun, the observer, and the raindrops which form the bows. An observer at *E*, with his back to the sun, receives red light from every raindrop situated 42° from the line *SC*, drawn

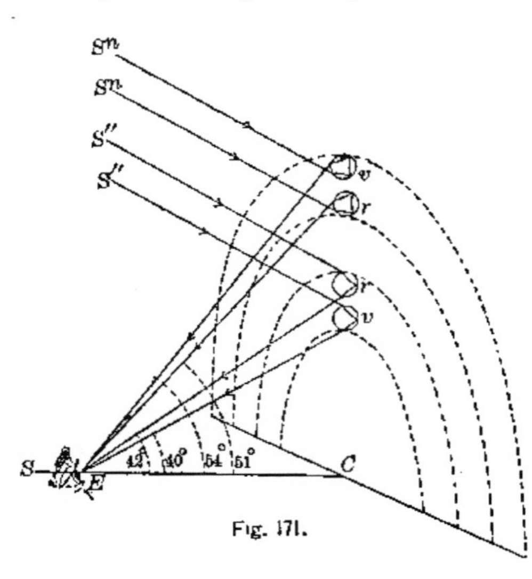

Fig. 171.

through the sun and the eye, and violet from those 40° distant. The drops, being approximately spheres, send out light on all sides, and hence all of those situated in a circle about *C*, and distant 42°, will send red light to the eye and give the appearance of a red circle, while those at 40° will send violet light, the other colors arranging themselves between these limits. Again, those drops which are 51° from *C* will also send red light to the eye, the rays being twice reflected internally, and those 54° will send violet light, thus forming the secondary bow.

283. The Spectroscrope (Fig. 172) is an instrument for viewing spectra. In one of its simplest forms it consists of a prism, a telescope, and a tube called the *collimator*, carrying an adjustable slit at the outer end and a converging lens at the other to render parallel the diverging rays coming from the slit. The slit must therefore be placed at the principal focus of the collimating lens. To mark the deviation of the spectral lines, there is provided on the supporting table a divided circle, which is read by the aid of the vernier attached to the telescope arm.

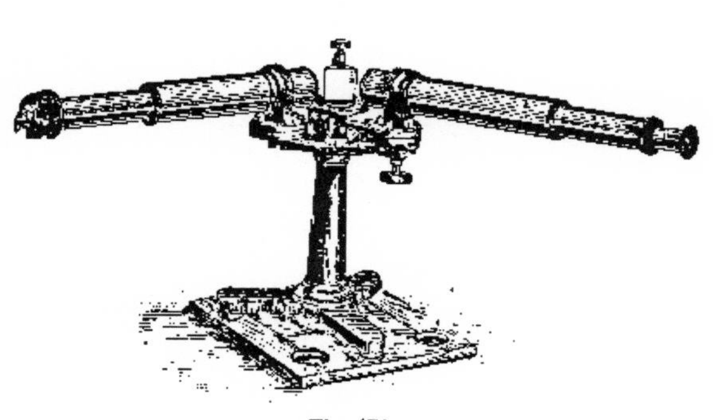

Fig. 172.

284. Kinds of Spectra. — **Experiment.** — Place a lighted candle or lamp in front of the slit of the spectroscope. The spectrum will be seen to pass from red at one end, through the various colors, to violet at the other, without any interruptions.

Experiment. — Over the tube of a Bunsen burner place a short cylinder of asbestos paper into which has been fused a salt of either sodium, lithium, or strontium. The heat of the flame vaporizes these salts. View this flame through the spectroscope, and instead of a bright band there will be seen one or more bright colored lines, depending on the nature of the salt, sodium giving a bright yellow line, lithium a bright carmine line, strontium a cluster of lines in the orange-red, etc. We infer that the lights produced by an incandescent vapor are of certain particular wave lengths only.

Experiment. — Project a spectrum of the electric arc on the screen, using a hollow prism with plane glass sides and filled with carbon bisulphide. The spectrum will be composed of colors from red at one end through all the intermediate spectral tints to violet at the other, without any interruptions or gaps; that is, it will be *continuous*. With a clean glass rod introduce a few crystals of sodium nitrate into the arc, and lengthen it so as to reduce the intensity of the continuous spectrum. The spectrum will now consist of a few bright colored lines, one red, one yellow, three green, and one violet, the yellow being especially prominent. Now substitute for the lower carbon in the lamp a larger one with a cup-shaped depression in its upper end. Reverse the direction of the current through the lamp, making the lower carbon the positive. Place a few crystals of sodium nitrate in the carbon cup immediately after it has been heated by the current, and close the circuit. The heat will vaporize the sodium, and the intensely hot arc will be surrounded with an atmosphere of sodium vapor at a lower temperature. This outer vapor will absorb the light emitted by the hotter vapor within, and dark bands will appear in place of the bright ones in the faint continuous spectrum, particularly in the red and yellow.

These experiments illustrate the three kinds of spectra, namely, the *continuous*, the *discontinuous* or *bright-lined*, and the *absorption* or *reversed* spectra. Solids, liquids, and dense vapors and gases, when heated to incandescence, give continuous spectra; rarefied gases and vapors heated to incandescence give discontinuous spectra; and gases and vapors absorb light of the same refrangibility as they emit at a higher temperature.

285. The Fraunhofer Lines. — If sunlight be analyzed with a spectroscope, a number of dark lines will be seen to cross the spectrum (Fig. 173). This discovery was made by Wollaston in 1802. Fraunhofer

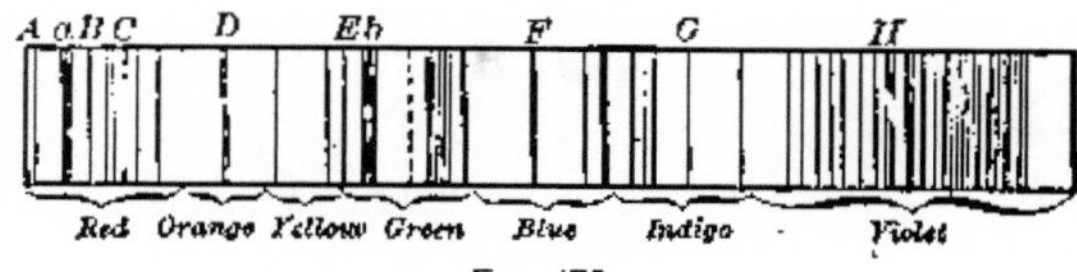

Fig. 173.

was the first to notice that some of these lines coincide in position with the bright lines of certain artificial lights. He mapped no less than 576 of them, and designated the more important ones by the letters *A*, *B*, *C*, *D*, *E*, *F*, *G*, *H*, the first in the extreme red and the last in the extreme violet. For this reason it is customary to refer to them as the Fraunhofer lines. In recent years Lockyer, Rowland, Langley, and many others, employing greatly improved apparatus, have found the number of these lines to be practically unlimited.

The explanation of these dark lines was first suggested by Stokes in 1852, and the theory was fully established by Kirchhoff in 1859. In the last experiment it was shown that sodium vapor absorbs that part of the light of the electric arc which is of the same refrangibility as the light emitted by the vapor itself. Similar experiments with other substances show that every substance has its own absorption spectrum. These facts suggested the following explanation of the Fraunhofer lines: The heated nucleus of the sun gives off light of all degrees of refrangibility. Its spectrum would therefore be continuous, were it not surrounded by an atmosphere of metallic vapors and of gases, which absorb or weaken those rays of which their spectra consist. Hence, the parts of the spectrum which would have been illuminated by those particular rays have their brightness diminished; since the rays from the nucleus are absorbed, and the illumination is due to the less intense light coming from the vapors. These absorption lines are not lines of no light, but are lines of diminished brightness, appearing dark by contrast with the other parts of the spectrum.

286. Spectrum Analysis consists in detecting the presence of substances by the spectra of their heated vapors. The great delicacy of the method is exhibited in the statement made by Professor Swan, that he was able to detect by its spectrum the presence of $\frac{1}{2500000}$th part of a grain of sodium.

The applications of the spectroscope are many and various. By an examination of their absorption spectra, normal and diseased blood are easily distinguished, the adulteration of substances is detected, and the chemistry of the stars is approximately determined.

VII. COLOR.

287. The Color of light depends on its wave length, extreme red being due to the longest waves, and extreme violet to the shortest. The unit employed in measuring wave lengths of light is the *tenth-metre*, of which 10^{10} are required to make a metre. The following are the values for the principal Fraunhofer lines in air at 20° C. and 760 mm. pressure: —

A	Dark Red . . .	7621.31	E_1	Light Green . .	5270.52
B	Red	6884.11	E_2	. . . .	5269.84
C	Orange	6563.07	F	Blue	4861.51
D_1	Yellow	5896.18	G	Indigo	4293.
D_2	. . . .	5890.22	H_1	Violet.	3968.

In white light the number of colors is infinite, and they pass into one another by imperceptible gradations of shade and refrangibility. Color stands related to light in the same way that pitch does to sound. In artificial lights certain colors are either feeble or wanting, as can be proved by an examination of their spectra. Hence, artificial lights are not white, but each one is characterized by the color that predominates in its spectrum.

288. Color of Opaque Bodies. — **Experiment.** — Paste a small rectangular strip of white paper on a sheet of black cardboard. View this strip through a glass prism, holding its edges parallel to the length of the strip. The image is a spectrum, colored like that produced by sunlight, but less bright. If a red strip of paper, similarly mounted, is examined in the same way, the spectral image is red at one end, while the colors belonging to the other end are dim or absent. In like manner if a blue strip is examined, the spectral image is blue, the other colors being mostly wanting.

Experiment. — Project the solar spectrum on a white screen. (Why white?) Hold pieces of colored paper or cloth successively in different parts of the spectrum. A strip of red flannel appears brilliantly red in the red part of the spectrum, and black elsewhere; a blue ribbon is blue only in the blue part of the spectrum, and a piece of black paper is black in every part of the spectrum.

The experiments show that the color of a body is due both to the light that it receives and the light that it reflects; that a body is red because it reflects chiefly, if not wholly, the red rays of the light incident upon it, the others being absorbed wholly or partly at its surface, and that it cannot be red if there is no red light incident upon it. In like manner a body is white if it reflects all the rays in about equal proportions, provided white light is incident upon it. It therefore appears that bodies have no color of their own, since they can exhibit no color not already present in the light which illuminates them. This truth is illustrated by the difficulty experienced in matching colors by artificial lights, and by the changes in shade some fabrics undergo when taken from sunlight into gaslight. Artificial lights are largely deficient in blue and violet rays; and hence all complex colors, into which blue or violet enters, as purple and pink, change their shade when viewed by them.

289. Color of Transparent Bodies. — **Experiment.** — Project the spectrum of the sun or of the arc light on the screen. Hold across the slit a flat bottle or cell filled with a solution of ammoniated oxide of copper. The spectrum below the green will be cut off. Substitute a solution of picric acid and the spectrum above the green will be cut off. Place both solutions across the slit and the green alone remains. It is the only color transmitted by both solutions. In like manner blue glass cuts off the less refrangible part of the spectrum, ruby glass cuts off the more refrangible, and the two together cut off the whole.

It thus appears that the color of a transparent body is determined by the colors that it absorbs. It is colorless if it absorbs all colors in like proportion, or absorbs none; but if it absorbs some colors more than others, its color is due to the mixed impression produced by the transmitted radiations.

Fig. 174.

290. Mixing Colors. — **Experiment.** — Cut out of colored paper several colored disks, about 15 cm. in diameter, with a hole at the centre for mounting them on the spindle of a whirling machine (Fig. 174), or for slipping them over the handle of a heavy spinning top. Slit them along a radius from the circumference to the centre, so that two or more of them can be placed together, exposing any proportional part of each one as desired (Fig. 175). Select seven disks, whose colors most nearly represent those of the solar spectrum; put them together so that equal portions of the colors are exposed. Clamp on the spindle of the whirling machine and rotate them rapidly. When viewed in a strong light the color is a dull white or uniform gray.

This method of mixing colors is based on the physiological fact that a sensation lasts longer than the impression producing it. Before the sensation caused by one impression has ceased, the disk has moved, so that a different impression is produced. The effect is equivalent to superposing the several colors on one another, as was done in the recomposition of white light (§ 278).

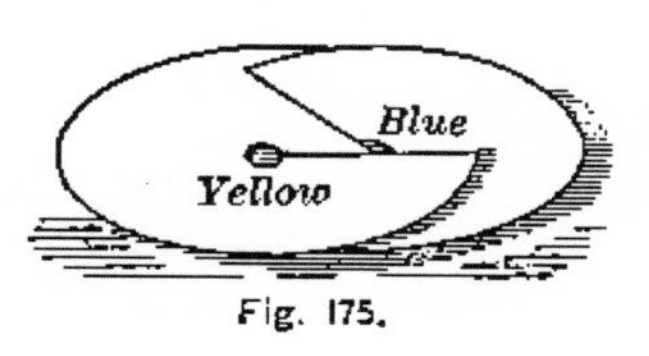

Fig. 175.

If red, green, and violet disks, or red, green, and blue ones, are used, exposing equal portions, gray or white is obtained on rotating them rapidly. If any two colors standing opposite each other in Fig. 176 are used, the result is white; and if any two alternate ones are used, the result is the intermediate one. By using the red, the green, and the violet disks, and exposing different proportions, it has been found possible to produce any color of the spectrum. This fact suggested to Dr. Young the theory that there are only three primary color sensations, and that our recognition of different colors is due to the excitation of these three in varying degrees. It is thus evident that a mixture of colored lights is very different from a mixture of pigments (§ 292).

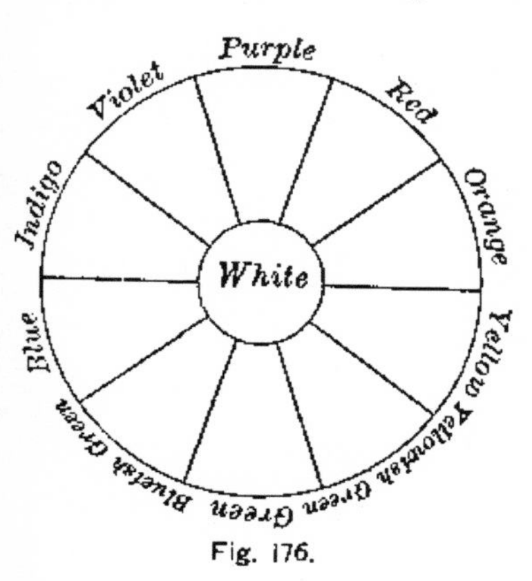

Fig. 176.

291. Complementary Colors. — Any two colors whose mixture produces on the eye the impression of white light are said to be *complementary*. Thus, red and bluish green are complementary; also orange and light blue. When complementary colors are viewed next to each other, the effect is a mutual heightening of color impressions.

Experiment. — Complementary colors may be seen by retinal fatigue. Cut some design out of paper, and paste it on red glass. Project it on a screen in a dark room. Look steadily at the screen for several seconds, and then turn up the lights. The design will appear on a pale green ground.

The explanation is that the portion of the retina on which the red light falls becomes tired of red, and refuses to convey as vivid a sensation of red as of the other colors when less intense white light is thrown on it. But it retains its sensitiveness in full for the rest of white light, and therefore conveys to the brain the stimulus of white light with the red cut out; that is, of the complementary color, green.

292. Mixing Pigments. — **Experiment.** — Draw a broad line on the blackboard with a yellow crayon. Over this draw a similar band with a blue crayon. The result will be a band distinctly green.

The yellow crayon reflects green light as well as yellow, and absorbs all the other colors. The blue crayon reflects green light along with the blue, absorbing all the others. Hence, in superposing the two chalk marks, the mixture absorbs all but the green. The mark on the board is green, because that is the only color that survives the double absorption. In mixing pigments, the resulting color is the residue of a process of successive absorptions. If the spectral colors, blue and yellow, are mixed, the product is white instead of green.

VIII. INTERFERENCE AND DIFFRACTION.

293. Newton's Rings. — **Experiment.** — Press together at their centre two small pieces of heavy plate glass, using a small iron clamp for the purpose. Then look obliquely at the glass; curved bands of color may be seen surrounding the point of greatest pressure.

This experiment is a modification of one performed by Hooke. It was afterwards repeated by Newton while attempting to determine the relation between the colors in the soap bubble and the thickness of the film. Each used a plano-convex lens of long focus resting on a plate of plane glass. Figure 177 shows a section of the apparatus. Between the lens and the plate there is a wedge-shaped film of air, very thin, and quite similar to that formed between the glass plates in the above experiment. If the glasses are viewed by reflected light, there is a dark spot at the point of contact, surrounded by several colored rings (Fig. 178); but if viewed by transmitted light, the colors are complementary to those seen by reflection (§ 291). The explanation of the phenomenon is to be found in the interference of two sets of waves, one reflected internally from the curved surface, *ACB*, and the other from the surface *DCE*, on which it presses. If light of one color is incident on *AB*, a portion will be reflected from *ACB*, and another portion from *DCE*. Since the light reflected from *DCE* has travelled farther by twice the thickness of the air-film than that from *ACB*, and the

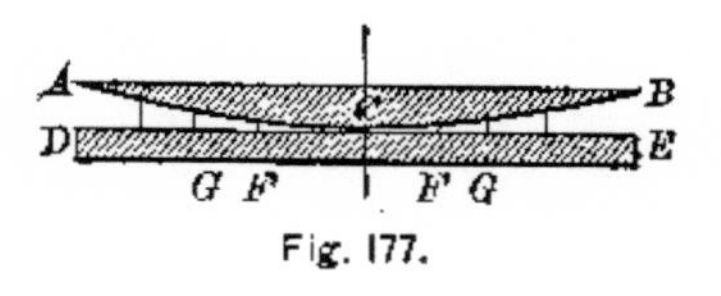

Fig. 177.

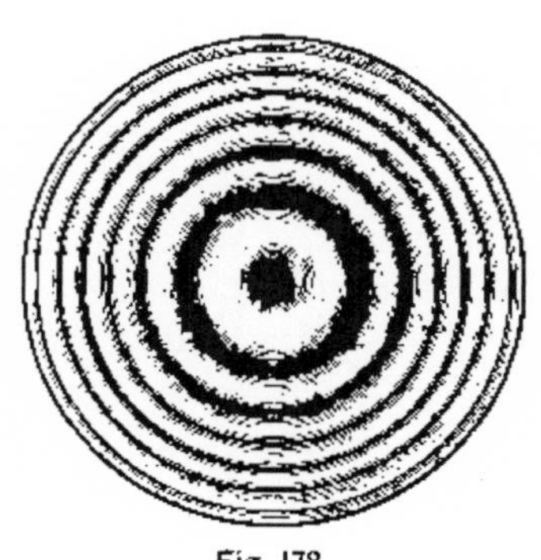
Fig. 178.

film gradually increases in thickness from *C* outward, it follows that at some places the two reflected portions will meet in like phase, and at others in opposite phase, causing a strengthening of the light at the former, and extinction of it at the latter. If red light be used, the appearance will be that of a series of concentric circular red bands separated by dark ones, each shading off into the other. If violet light be employed, the colored bands will be closer together on account of the shorter wave length. Other colors will give bands intermediate in diameter between the red and violet. From this it follows that if the glasses be illuminated by white light, at every point some one color will be destroyed, and the others will be either weakened or strengthened, depending on the thickness of the air-film at the point under consideration, the color at each point being the result of mixing a large number of colors in unequal proportions. Hence, the point *C* will be surrounded by a series of colored bands.[1]

The colors of the soap bubble, of oil on water, of heated metals which easily oxidize, of a thin film of varnish, and of the surface of very old glass, are all caused by the interference of light reflected from the two surfaces of a very thin film.

294. Diffraction. — **Experiment.** — Place across the opening of the porte-lumière two superposed pieces of perforated cardboard. The projected images of the very small holes, as one piece is moved across the other, are fringed with the spectral colors.

---

[1] The light from *ACB* differs in phase half a wave length from that reflected from *DE*, because the former is reflected in an optically dense medium next to a rare one, and the latter in an optically rare medium next to a dense one. This phase difference is additional to the one above described.

Experiment. — Transparent *diffraction gratings* are made by ruling with a fine diamond point a number of equidistant parallel lines very close together on glass. Substitute this for the prism in projecting the spectrum of sunlight or of the arc light on the screen (§ 284). There will be seen on the screen a central image of the slit, and on either side of it a series of spectra. Cover half of the length of the slit with red glass and the other half with blue. There will now be a series of red images and also a series of blue ones, the red ones being farther apart than the blue. Lines ruled close together on smoked glass may be used instead of a "grating."

These experiments illustrate a phenomenon known as *diffraction*. The colored bands are caused by the interference of the waves of light which are propagated in all directions from the fine openings, the effects being visible because the transparent spaces are so small that the intensity of the direct light from the source is largely reduced. Diffraction gratings are also made to operate by reflected light. Striated surfaces, like mother-of-pearl, changeable silk, and the plumage of many birds, owe their beautiful changing colors to interference of light by diffraction.

IX. OPTICAL INSTRUMENTS.

295. The Simple Magnifier, or *simple microscope*, is a double-convex lens, usually of short focal length. The

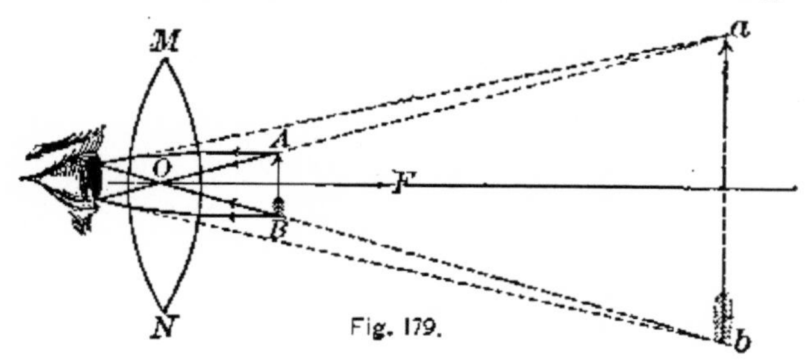

Fig. 179.

object must be placed nearer the lens than its principal focal length. The image is then virtual, erect, and enlarged. If AB is the object in Fig. 179, the virtual image is ab;

and if the eye be placed near the lens on the side opposite the object the impression received will be projected outward in the direction in which the light enters the eye, and the virtual image will be seen in the position of the intersection of the rays produced, as at *ab*.

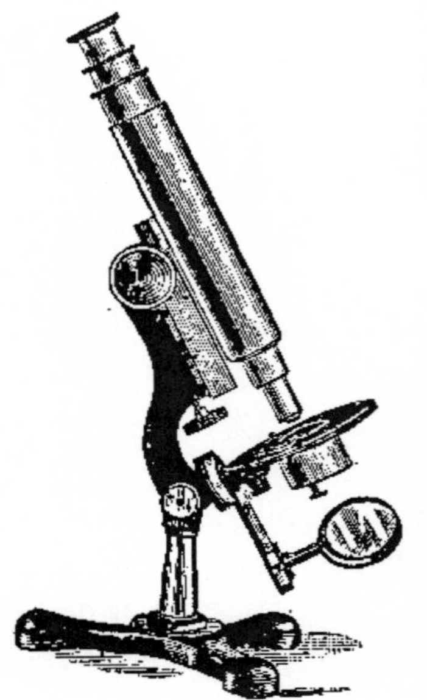

Fig. 180.

296. The Compound Microscope (Fig. 180) is an instrument designed to obtain a greatly enlarged image of very small objects. In its simplest form it consists of a converging lens *MN* (Fig. 181), called the *object glass*, and another converging lens *RS*, called the *eyepiece*. The two lenses are mounted in the ends of the tube of Fig. 180. The object is placed on the stage just under the objective, and a little beyond its principal focus. A real image *ab* (Fig. 181) is formed slightly nearer the eyepiece than its focal length. This

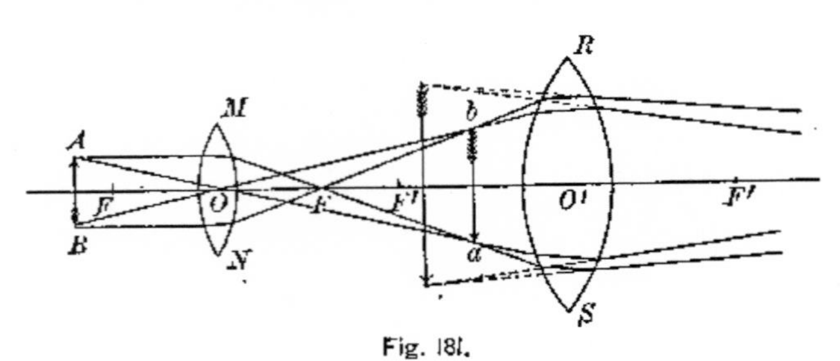

Fig. 181.

image formed by the objective is viewed by the eyepiece, and the latter gives an enlarged virtual image. (Why?) Both the objective and the eyepiece produce magnification.

297. The Astronomical Telescope. — The system of lenses in the refracting astronomical telescope (Fig. 182) is similar to that of the compound microscope. Since it is intended to view distant objects, the objective *MN* is of large aperture and long focal length. The real image given by it is the object for the eyepiece, which again forms a virtual image for the eye of the observer. The

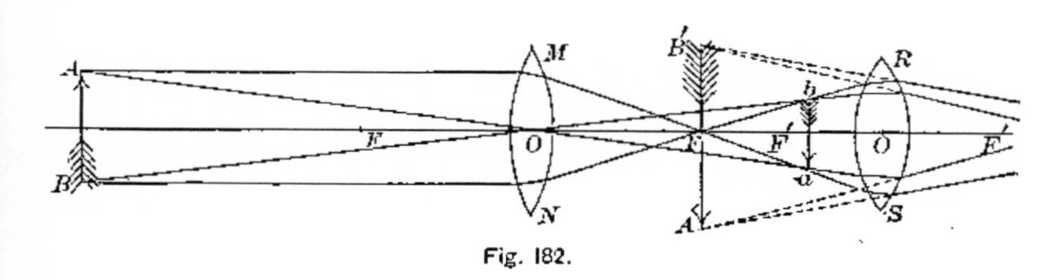

Fig. 182.

magnification is the ratio of the focal lengths of the objective and the eyepiece. The objective must be large, for the purpose of collecting enough light to permit large magnification of the image without too great loss in brightness.

The figure shows that the image in the astronomical telescope is inverted. In a terrestrial telescope the image is made erect by introducing near the eyepiece two double-convex lenses, in such relation to each other and to the first image that a second real image is formed like the first but erect.

298. Galileo's Telescope. — This earliest form of telescope produces an erect image by the use of a diverging lens for the eyepiece (Fig. 183). This lens is placed between the objective and the real image, *ab*, which would be formed by the objective if the eyepiece were not interposed. Its focus is practically at the image *ab*, and the rays of light issue from it slightly divergent for distant objects.

The image is therefore at $A'B'$ instead of at ab, and it is erect and enlarged. This telescope is much shorter than the astronomical telescope, for the distance between the

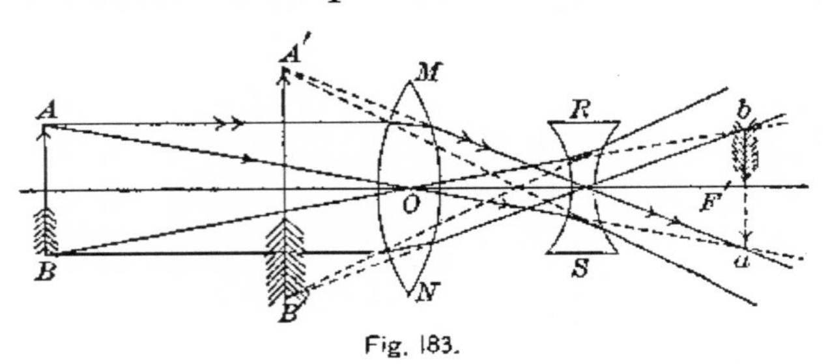

Fig. 183.

lenses is the difference of their focal lengths instead of their sum. In the *opera glass* two of Galileo's telescopes are attached together with their axes parallel.

299. The Projection Lantern is an apparatus by which a greatly enlarged image of an object may be projected on a screen. Its three essentials are a strong light, a con-

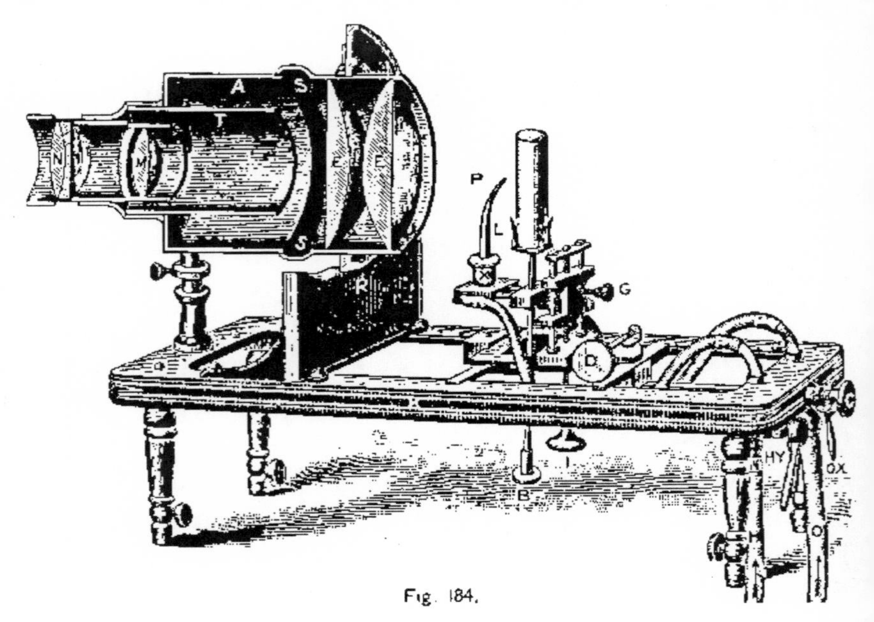

Fig. 184.

denser, and an objective. The light may be the calcium light, as shown in Fig. 184, the electric arc light, or a large oil burner. The condenser *E* is composed of a pair of plano-convex lenses, with their convex surfaces turned toward each other. It has for its chief purpose the collection of the light on the object by refraction, so as to bring as much as possible on the screen. The object, commonly a drawing or a photograph on glass, is placed near the condenser at *SS*, where it is strongly illuminated. The objective, *MN*, is an achromatic combination, acting as a converging lens to project on the screen a real, inverted, and enlarged image of the object.

300. The Photographer's Camera, a section of which is shown in Fig. 185, consists of a box *BC*, adjustable in length, blackened inside, and provided at one end with a double achromatic lens *LL'*, and at the other with a holder for the sensitive plate. If by means of the rack and pinion *D*, the lens be properly focussed for an object in front of it, an inverted image will be formed on the plate *E*. The light acts on the salts contained in the sensitized film on the plate, producing in them a modification which, by the processes of "developing" and "fixing," becomes a permanent negative picture of the object.

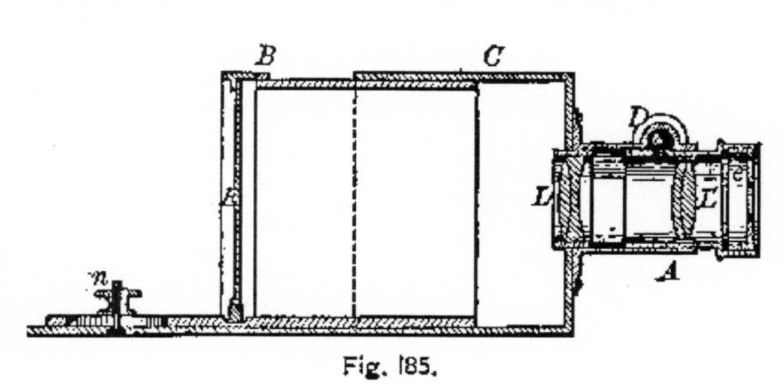

Fig. 185.

301. The Eye. — The eye is like a small photographic camera, with a converging lens, a dark chamber, and a sensitive screen. Fig. 186 is a vertical section through the axis. The outer covering, or *sclerotic* coat *H*, is a thick opaque substance, except in front, where it is extended as a transparent coat, called the *cornea A*. Behind the cornea is a diaphragm *D*, constituting the colored part of the eye, or the *iris*. The circular opening in the iris is the *pupil*, the size of which changes with the intensity of light. The pupil of the cat's eye is not round, but elongated. Between the iris and the crystalline lens *E* is a transparent fluid called the *aqueous humor*, while the large chamber behind the lens is filled with a jellylike substance called the *vitreous humor*. The *choroid coat* lines the walls of this posterior chamber, and on it is spread the *retina*, a membrane traversed by a network of nerves, branching from the *optic nerve M*. The choroid coat is filled with a black pigment, which serves to darken the cavity of the eye, and to absorb the light reflected internally.

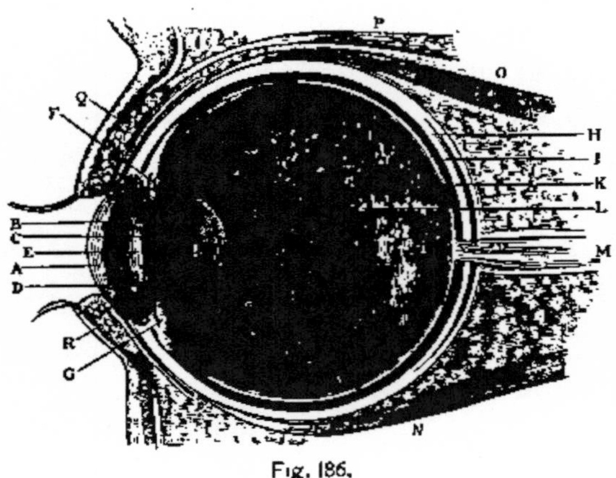

Fig. 186.

302. Production of Sight. — Rays of light diverging from the object enter the pupil and form an inverted image on the retina, precisely as in the photographic camera. In place of the sensitized plate is the sensitive retina, from which the sensation of sight is carried to the brain along the optic nerve.

In the camera the distance between the lens and the screen or plate must be adjusted for objects at different distances. In the eye the corresponding distance is fixed, and the adjustment for distinct vision is made by changing the curvature of the front surface of the crystalline lens. The curvature is increased by relaxing the ligament which connects it with the choroid, and the focal length is thus diminished. This capability of the lens of the eye to change its focal length for objects at different distances is called *accommodation.*

303. The Blind Spot. — There is a small depression where the optic nerve enters the eye. The rest of the retina is covered with microscopic rods and cones, but there are none in this depression, and it is insensible to light. It is accordingly called the *blind spot.* Its existence can be readily proved by the help of Fig. 187. Hold the book

Fig. 187.

with the circle opposite the right eye. Now close the left eye and turn the right to look at the cross. Move the book toward the eye from a distance of about a foot, and a position will readily be found where the black circle will disappear. Its image then falls on the blind spot. It may be brought into view again by moving the book either nearer the eye or farther away.

304. Defects of the Eye. — A normal eye is one which, in its passive condition, focuses parallel rays on the retina. If such rays focus in front of the retina, because the eyeball from front to back is too long, the eye is *myopic,* giving rise to *near-sightedness.* By intercepting the rays with a diverging lens of suitable focal length, the image may be made to fall on the retina and the vision becomes distinct.

If, on the other hand, the eyeball from front to back is too short, the focus for parallel rays will be back of the retina, and the eye is *hypermetropic.* One is then unable to see near objects distinctly, but can see distant ones. This kind of eye must not be confused with a *presbyopic* one, where through some functional defect the crystalline lens loses the power of adjusting itself to rays of considerable divergence. This defect is known as *far-sightedness*, and is corrected by means of a converging lens. Since the normal eye adapts itself without painful effort to the distinct vision of small objects, like the letters on this page, at a distance of 25 cm. (10 in.), it is customary to speak of this distance as that of normal distinct vision. Eyes are classified as *near-sighted* when the distance of distinct vision is less than 25 cm., and *far-sighted* when greater.

CHAPTER VI.

HEAT.

I. HEAT AND TEMPERATURE.

305. Sensations of Heat and Cold. — If one takes hold of an iron rod that has just been removed from the fire, it feels *hot;* on the other hand, if one touches a piece of ice, it feels *cold*. The cause of these sensations is said to be *heat*. The iron feels hot because it imparts heat to the hand, and the ice feels cold because the hand loses heat to the ice.

306. Nature of Heat. — For a long time it was believed that heat was a subtle and weightless fluid that caused all thermal phenomena by entering bodies and possibly combining with them. This fluid was called *caloric*. About the beginning of the last century certain experiments of Count Rumford and Sir Humphry Davy demonstrated that the materialistic view of heat was no longer tenable; and finally about the middle of the century, when Joule proved that a definite amount of mechanical work is equivalent to a definite amount of heat, it became evident that *heat is a form of energy*. The modern *kinetic* theory, briefly stated, is as follows: The molecules of a body have a certain amount of independent motion, generally very irregular. Any increase in the energy of this motion manifests itself by the body becoming warmer, and any decrease by its becoming cooler. The heating or the cooling of a body, by whatever process, is but the transference or the transformation of energy.

307. Temperature. — It is a matter of common observation that if we place a mass of hot iron in contact with a cold one, the latter becomes warmer and the former cooler, the heat flowing from the hot body to the cold one. The two bodies are said to differ in *temperature* or "heat level," and when they are brought in contact there is a flow of heat from the one of higher temperature to the one of lower till thermal equilibrium is established. *Temperature* may be defined as the thermal condition of a body which determines the transfer of heat between it and any body in contact with it. This transfer is always from the body of higher temperature to the one of lower. Temperature may be considered as a measure of the degree of hotness; it depends solely on the kinetic energy of the molecules of the body. Temperature must be distinguished from quantity of heat. A pint of water in a vessel may be at a much higher temperature than the water in a lake, yet the latter contains a vastly greater quantity of heat, owing to the greater quantity of water.

308. Measuring Temperature. — **Experiment.** — Select three basins *A*, *B*, and *C*. Fill *A* with hot water, *B* with cold water, and *C* with tepid water. Hold one hand in *A*, and the other in *B* for a few seconds; then transfer both to *C*. The water of *C* will feel cold to the hand from *A* and warm to the hand from *B*.

Experiment. — Hold the hand successively against a number of the various objects in the room, at about the same height from the floor. Metal, slate, or stone objects will feel colder than those of wood, even when side by side and of the same temperature.

It is therefore evident that the sense of touch cannot be depended on to give accurate information regarding the relative temperatures of bodies, and some method independent of bodily sensations must be resorted to for their reliable measurement. The one most extensively used is based on the regular increase in the length or volume of a body attending a rise in its temperature. This method is illustrated by the common mercurial thermometer.

II. THE THERMOMETER.

309. A Thermometer is an instrument for measuring temperatures. The common *mercurial thermometer* consists of a capillary glass tube of uniform bore, on one end of which is blown a bulb, either spherical or cylindrical (Fig. 188). Part of the air is expelled by heating, and while in this condition the open end of the tube is dipped into a vessel of pure mercury. As the tube cools, mercury is forced into the tube by atmospheric pressure. Enough mercury is introduced to fill the bulb and part of the tube at the lowest temperature which the thermometer is designed to measure. Heat is now applied to the bulb till the expanded mercury fills the tube; the end is then closed in the blowpipe flame. The mercury contracts as it cools, leaving a vacuum at the top of the tube.

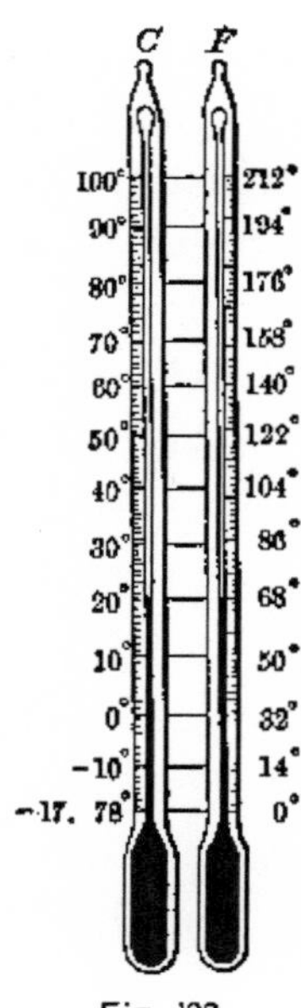

Fig. 188.

310. Necessity of Fixed Points. — It is evident that no two thermometers are likely to have bulbs and stems of the same capacity. Consequently, equal increments of temperature will not produce equal changes in the height of the mercury. If, then, the same scale were attached to all thermometers, their indications would differ so widely that the results would be worthless. Hence, if thermometers are to be compared, corresponding divisions on the scale of different instruments must indicate the same temperature. This may be done by graduating every thermometer by comparison with a standard, an expensive proceeding and for many purposes unnecessary, since mercury has a nearly uniform rate of expansion. If two points are marked on the stem, the others can be obtained by dividing the space between them into the proper number of equal parts. Careful investigations have made it certain that under a constant pressure the temperature of *melting ice* and that of *steam* are invariable. Hence, the temperature of melting ice and that of steam under a pressure of 76 cm. of mercury have been chosen as the fixed points on a thermometer.

311. Marking the Fixed Points. — The thermometer is packed in finely broken ice, as far up the stem as the mercury extends. The containing vessel (Fig. 189) has an opening at the bottom to let the water run out. After standing in the ice for several minutes the top of the thread of mercury is marked on the stem. This is called the *freezing point*.

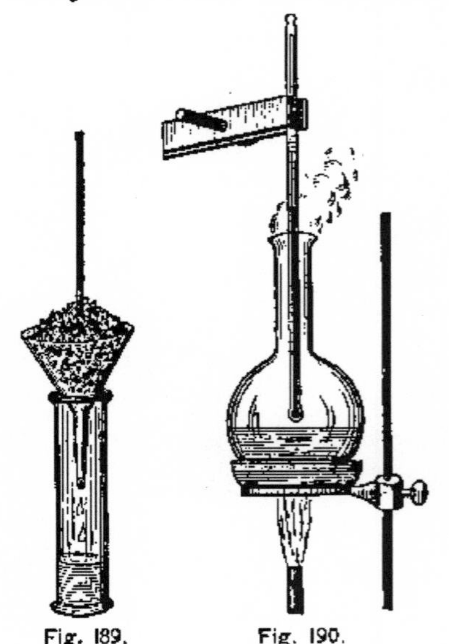

Fig. 189. Fig. 190.

The *boiling point* is marked by observing the top of the mercurial column when the bulb and stem are enveloped in steam (Fig. 190) under an atmospheric pressure of 76 cm. (29.922 in.). If the pressure at the time is not 76 cm., then a correction must be applied, the amount being determined by the approximate rule that the temperature of steam rises 0°.1 C. for every increase of 2.71 mm. in the barometric reading, near 100° C.

312. Thermometer Scales. — The distance between the fixed points is divided into equal parts called *degrees*. The number of such parts is wholly arbitrary, and several different scales have been introduced. Three of these are in use at the present time: the *Fahrenheit*, the *Centigrade*, and the *Réaumur*. The Fahrenheit scale was introduced by Fahrenheit about 1714, and is the one in common use in all English-speaking countries. For some unknown reason he marked the freezing point at 32° above the zero of the scale, and the boiling point at 212°, dividing the space between into 180 equal parts. The Centigrade scale was designed by Celsius about 1742. It differs from the Fahrenheit in making the freezing point 0° and the boiling point 100°, the space between being divided into 100 equal parts. This is the one in general use among scientific men. The Réaumur scale marks the freezing point 0° and the boiling point 80°. This is the household scale on the continent of Europe; in this country its use is restricted to breweries. Each of these scales is extended beyond the fixed points as far as desired. The divisions below 0° are read as negative; for example, – 10° signifies 10 degrees below zero. The reading according to any particular scale is indicated by affixing the initial letter of the name; for example, 5° F., 5° C., and 5° R. signify 5 degrees above zero on the Fahrenheit, Centigrade, and Réaumur scales respectively.

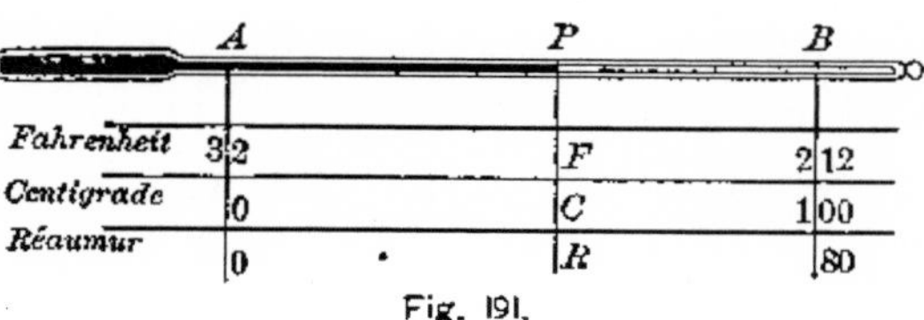

Fig. 191.

313. Comparison of Scales. — In Fig. 191 AB is a thermometer with three scales attached, P is the head of the mercury column, and F, C, and R are the readings on the scales respectively. On the Fahrenheit scale $AB = 180$ and $AP = F - 32$, since the zero is 32 spaces below A; on the Centigrade $AB = 100$ and $AP = C$; on the Réaumur $AB = 80$ and $AP = R$. Then the ratio of AP to AB is $\frac{F-32}{180} = \frac{C}{100} = \frac{R}{80}$. By substituting the reading on any one scale in this equation the equivalent on either of the other scales is easily obtained. For example, if it is required to express 68° F. on the Centigrade scale, then $\frac{68-32}{180} = \frac{C}{100}$ and $C = 20°$.

314. Limitations of the Mercurial Thermometer. — Since mercury freezes at $-38°.8$ C., it is evident that it cannot be used as the thermometric substance below this temperature. For temperatures below $-38°$ C. alcohol is substituted for mercury. Under a pressure of one atmosphere mercury boils at about 350° C. For temperatures approaching this value and up to about 550° C. the thermometer stem is filled with pure nitrogen under pressure. The pressure of the gas keeps the mercury from boiling (§ 337). For high temperatures the mercury thermometer must be calibrated by comparison with an air thermometer, or by reference to the temperature at which water boils under known high pressures.

315. The Air Thermometer was invented by Galileo about 1593 for the use of physicians. In its early form it consisted of a glass bulb on the end of a tube of small bore, supported vertically in front of a scale. By warming the bulb, part of the air is expelled, and then the stem is inserted in a liquid, as colored water, alcohol, or mercury. When the air cools it contracts and the liquid rises in the stem from atmospheric pressure. In Fig. 192 the bottle containing the liquid serves as a support. If the temperature rises, the liquid column is depressed; and if the temperature falls, the column rises. The instrument is remarkable for its sensitiveness, that is, for the large movements of the index for small changes of temperature; but unless greatly modified in construction, making it quite complex in plan, it is only a thermoscope, because its readings change with every change in barometric pressure.

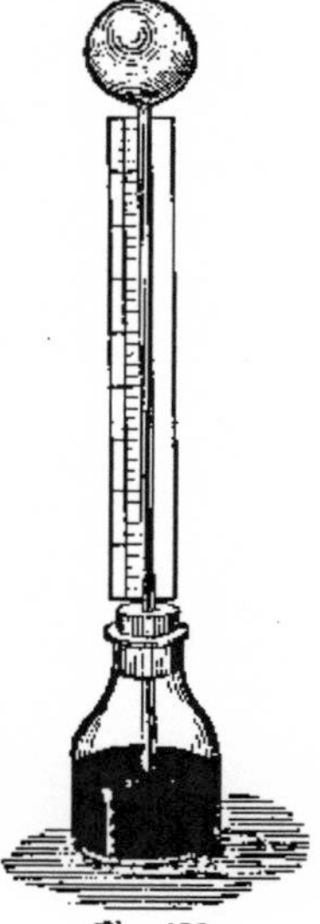

Fig. 192.

Questions and Problems.

1. What is the best way to graduate a water thermometer?

2. If the bulb of a thermometer be plunged into hot water, the mercury at first falls. Why?

3. What effect will it have on the distance between the fixed points to use a tube with a very small bore? To use a large bore?

4. Why is a thermometer with a cylindrical bulb preferable to one having a spherical bulb?

5. How would it affect the readings of a liquid thermometer if, after graduation, the bulb should contract?

6. What would be the nature of the error in a thermometer if the capillary tube tapered outward from the bulb?

7. In the air thermometer should the cork fit the bottle air-tight? Why?

8. Give several reasons why mercury is the most suitable of all known liquids for use in thermometers.

9. Express in the Centigrade scale the following: 30° F., – 4° F., 25° F., – 10° F.

10. Express in the Fahrenheit scale the following: 200° C., 20° C., – 4° C.

11. Find the difference in Centigrade degrees between 50° C. and 50° F.

12. Mercury freezes at – 38°.8 C. and boils at 350° C.; express these facts in the Fahrenheit scale.

13. Lead melts at 326° C. and in melting absorbs as much heat as will raise 5.37 times its mass of water 1° C. What numbers will take the place of 326 and 5.37 when the Fahrenheit scale is employed?

Ans. 618°.8 F.; 9.666.

14. By evaporating solid hydrogen Dewar obtained a temperature of –264° C., what would this be on the Fahrenheit scale?

Ans. – 443°.2 F.

15. In taking the temperature of a vessel of water with both a Centigrade and a Fahrenheit thermometer, it was noticed that the Fahrenheit reading was three times the Centigrade; what was the temperature?

16. A thermometer provided with the three scales was used in taking the temperature of some water. The sum of the readings was 140; what was the temperature?

17. At what temperature are the readings on the Centigrade and Fahrenheit scales identical?

18. Calculate the temperature of steam from water boiling under a barometric pressure of 74 cm.

19. The boiling point of water according to a certain thermometer was found to be 98°.5 C. when the barometric pressure was 745 mm.; what is the error in the boiling point, and what is the correct temperature when this thermometer reads 30° C., assuming the zero point correct?

Ans. 0°.95 too low; 30°.29 C.

20. In testing the accuracy of the fixed points on a thermometer the freezing point was found to be 0°.5 C.; what is the correct temperature when this thermometer reads 25° C., assuming the boiling point correct?

Ans. 24°.62 C.

21. A certain thermometer was found to read 0°.5 C. in melting ice and 99°.5 C. in steam when the barometric pressure was 74 cm. What is the correct temperature when the reading is 45° C.?

Ans. 44°.62 C.

22. If a thermometer is so graduated that it reads 95° C. for the boiling point of water, what would it read when placed in a bath whose true temperature is 20° C., assuming the zero point correct?
Ans. 19° C.

III. EXPANSION.

316. Expansion of Solids. — Experiment. — A metallic rod *S* (Fig. 193) is supported horizontally in such a manner that the end *A* rests firmly against a support, while the end *B* rests against

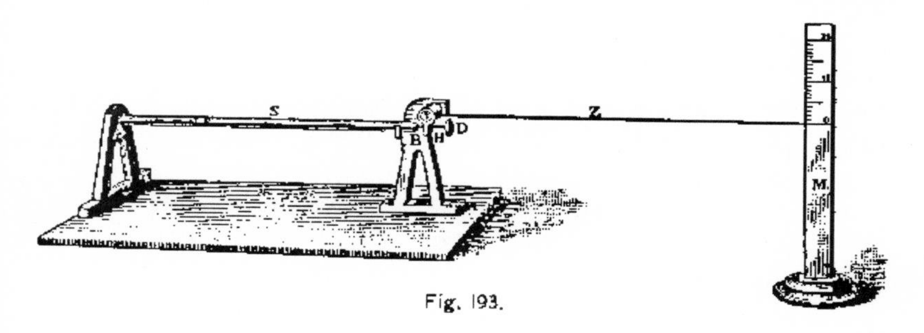

Fig. 193.

H, the short arm of a bent lever *Z*, the long arm moving over a scale *M*. Place a spirit lamp under *S*. The pointer *Z* will move upward on *M*, showing that the rod is increasing in length.

Experiment. — Rivet together at short intervals a strip of sheet iron and one of copper (Fig. 194). Support the ends and place a spirit lamp under the middle. This composite bar will bend into an arc with the copper on the convex side, showing that the two metals expand unequally and that the copper expands more than the iron.

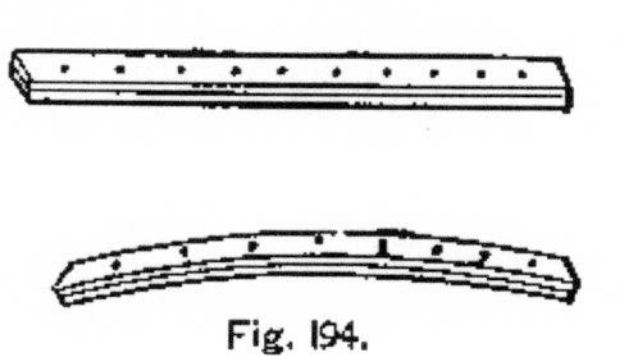
Fig. 194.

Fig. 195.

Experiment. — Fig. 195 illustrates a piece of apparatus known as Gravesande's ring. It consists of a metallic ball that at ordinary temperatures will just pass through the ring. Heat the ball in boiling water. It will now rest on the ring and will not fall through until it has cooled.

These experiments show that solids expand in every direction when heated and contract when cooled, the amount varying with the substance. Stretched india-rubber and iodide of silver are exceptions to this law; for within a certain range they contract when heated and expand when cooled.

317. Expansion of Liquids. — **Experiment.** — Prepare several glass tubes of the same bore, closing them at one end with a blowpipe. Fill each of them to the height of about 15 cm., but with different liquids, as water, alcohol, glycerine, etc., each colored with an aniline dye. Support the tubes in a vessel of hot water. The liquids will rise in the tubes, but not equally. If placed in ice-water they contract unequally.

The experiment shows that liquids, like solids, expand when heated and contract when cooled, the amount depending on the nature of the substance. It also shows that the expansion of liquids is greater than that of glass, otherwise there would have been no apparent increase in their volume. Some liquids do not expand when heated at certain points on the thermometric scale. Water, for example, on heating from 0° C. to 4° C. contracts, but above 4° C. it expands.

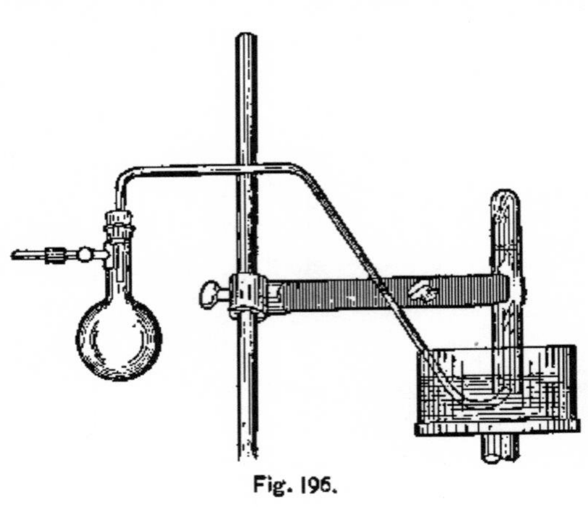

Fig. 196.

318. Expansion of Gases. — **Experiment.** — Fit a bent delivery-tube to a small Florence flask (Fig. 196). Fill the flask with air and place the upturned end of a delivery-tube under an inverted graduated glass cylinder filled with water. Heat the flask by immersing it in a vessel of hot water,

The air will expand and escape through the delivery-tube into the cylinder; note the amount. Now refill the flask with some other gas, as coal gas, and repeat the experiment. If the work is carefully done, it will be seen that the amount of gas collected each time is constant, showing that the expansion is the same.

The investigation of gases by Gay-Lussac, Charles, Regnault and many others has proved that all gases expand very nearly alike at atmospheric pressure, approaching equality as the pressure is diminished. Gases that are easily liquefied, as carbon dioxide, show the largest variation in their coefficient of expansion.

319. Coefficient of Expansion. — It appears from the preceding experiments that substances when heated expand in every direction. This expansion in volume is called *cubical expansion*, in distinction from *linear expansion*, or expansion in length, and *superficial expansion*, or expansion in area. The *coefficient of linear expansion* is the fraction of its length which a body expands when heated from 0° C. to 1° C.; the *coefficient of superficial expansion* is the fraction of its area which a body expands when heated from 0° C. to 1° C.; and the *coefficient of cubical expansion* is the fraction of its volume which a body expands when heated from 0° C. to 1° C. Since the expansion of most substances is found to be nearly constant for each degree of temperature, it is customary to determine the mean coefficient for a change of several degrees. If l_1 and l_2 represent the lengths of a metallic rod at the temperatures t_1 and t_2 respectively, then $\frac{l_2 - l_1}{t_2 - t_1} = \frac{l_2 - l_1}{t}$ is the expansion for 1°, in which t is the difference of temperatures. If a represents the mean coefficient of expansion, then $a = \frac{l_2 - l_1}{l_1 t}$; whence $l_2 = l_1(1 + at)$. In like manner for volumes, if k is the coefficient of cubical expansion, v_1 and v_2 the volumes at the temperatures t_1 and t_2 respectively, then $k = \frac{v_2 - v_1}{v_1(t_2 - t_1)} = \frac{v_2 - v_1}{v_1 t}$; whence $v_2 = v_1(1 + kt)$.

In the case of solids, superficial and cubical expansion are obtained by computation from the linear expansion, the coefficient of the former being twice the linear, and that of the latter three times.

320. Law of Charles.—It was shown by Charles, in 1787, that the volume of a given mass of any gas under constant pressure increases by a constant fraction of its volume at zero for each rise of temperature of 1° C. The investigations of Regnault and others show that the law is not rigorously true, and that the accuracy of Charles's law is about the same as that of Boyle's law. The coefficient of expansion of dry air is 0.003665, or about $\frac{1}{273}$. This fraction may be considered as the coefficient of any true gas.

321. The Absolute Scale. — The law of Charles leads to a fourth scale of temperature called the *absolute scale*. By this law the volumes of any mass of gas, under constant pressure, at 0° C., and at any other temperature t° C., are connected by the following relations: —

$$v = v_0(1 + \tfrac{1}{273}t) = \frac{v_0(273 + t)}{273}.$$

At any other temperature, t', the volume becomes

$$v' = \frac{v_0(273 + t')}{273}.$$

Then
$$\frac{v}{v'} = \frac{273 + t}{273 + t'}.$$

Suppose now a new scale is taken, whose zero is 273 Centigrade divisions below the freezing point of water, and that temperatures on this scale are denoted by T. Then $273+t$ will be represented by T, and $273+t'$ by T', and

$$\frac{v}{v'}=\frac{273+t}{273+t'}=\frac{T}{T'},$$

or *the volumes of the same mass of gas under constant pressure are proportional to the temperatures on this new scale.* The point 273° below 0° C. is called the *absolute zero*, and the temperatures on this scale, *absolute temperatures.* Up to the present it has not been found possible to cool a body to the absolute zero; but by evaporating liquid hydrogen under very low pressure, a temperature estimated to be within 9° of the absolute zero has been obtained by Professor Dewar.

322. The Laws of Boyle and Charles combined. — If v, p, and T represent the volume, pressure, and absolute temperature of a given mass of gas, then by Boyle's law $v \propto \frac{1}{p}$, when T is constant; and by the law of Charles, $v \propto T$, when p is constant. Therefore, by the principle of variation in algebra, $v \propto \frac{T}{p}$, or $\frac{vp}{T}$ is constant. This relation may be written as an equation,

$$vp = RT, \qquad (26)$$

that is, the product of the volume of a given mass of gas by its pressure is proportional to its absolute temperature. R is a constant.

For example, if 20 cm³. of gas, under a pressure of

76 cm. at 20° C., change to 30 cm³. at 50° C., the new pressure, determined by the equation

$$\frac{20 \times 76}{273 + 20} = \frac{30 \times p}{273 + 50},$$

is 55.9 cm. Again, if 1 cm³. of air at 0° C. and 760 mm. weighs 0.001293 gm., the weight of 200 cm³. of air at 27° C. and 500 mm. is found by the equation

$$\frac{1 \times 760}{0.001293 \times 273} = \frac{200 \times 500}{m \times 300}$$

to be 0.1548 gm.

323. Force of Contraction and Expansion.—**Experiment.**—Fill a small test-tube about one quarter full of water, and close the end by fusion. Lay it in an empty sand bath on the ring of an iron stand. Apply heat, and stand at a safe distance. In a few minutes there will be a loud report, caused by the bursting of the tube.

The force of expansion or of contraction of a substance is evidently equal to the force necessary to compress or expand it to the same extent by mechanical means, and hence can be computed by proceeding in the manner illustrated in the following example: A bar of malleable iron, one square inch in cross-sectional area, if placed under the tension of a ton, increases in length 0.0001 of itself. The coefficient of linear expansion of iron is 0.0000122. Since $0.0001 \div 0.0000122 = 8+$, a change of temperature of about 8° C. will produce the same change in the length of the bar as a force of one ton.

324. Applications of Expansion. — Many familiar phenomena are accounted for by expansion or contraction attending changes of temperature. If hot water is poured into a thick glass tumbler, the glass will probably break because of the strain produced by the sudden expansion of its inner surface. The principle of unequal expansion is employed in thermometers, in the compensated clock pendulum (§ 77), and in the balance wheel of a watch. Glass and platinum have nearly the same coefficient of expansion. For that reason platinum is in great demand in the manufacture of incandescent electric lamps, since it does not crack the glass when it cools. Iron tires are fitted to wheels and then expanded by heating so that they slip on easily; on cooling, they contract and compress the wheel. The rivets which hold together the plates of steam boilers are inserted red-hot, and hammered down. The contracting rivets press the plates together with great force. In all heavy iron structures, such as railroad bridges, a certain freedom of motion of the parts must be provided for; otherwise, the changes in length attending variations in temperature would have a disastrous effect. Sidewalks of artificial stone should have spaces left for expansion to prevent "buckling." Crystalline rocks, on account of unequal expansion in different directions, are slowly disintegrated by changes of temperature; and for the same reason quartz crystals, when strongly heated, fly in pieces.

Problems.

1. Why will heating the neck of a bottle frequently loosen the glass stopper?

2. Why does temperature affect density?

3. A metal rod was 2.3 m. long at 0° C. and 2.3027554 m. long at 100° C. What is the mean coefficient of linear expansion?
Ans. 0.00001198.

4. The volume of a gramme of water at 4° C. is 1 cm³. and at 60° C. it is 1.0169 cm³. What is the mean coefficient of cubical expansion?

5. Find the length of a bar of cast iron at 100° C. whose length at 18° C. is 5 ft., the coefficient of linear expansion being 0.00001125. *Ans.* 5.0046 ft.

6. A steel metre scale at 10° C. is 99.981 cm. long, and at 40° C. 100.015 cm. long. At what temperature will the scale be just one metre long?

7. An iron steam pipe is 40 ft. long at 0° C. What will be its length when steam at 100° C. passes through it, the linear coefficient being 0.0000122?

8. An iron yard measure is correct at 0° C. What is the error made in measuring 50 ft. with it at a temperature of 22° C., the linear coefficient being 0.0000122? *Ans.* 0.01342 ft.

9. The coefficient of linear expansion of glass is 0.0000087, and a certain glass flask holds exactly 100 cm³. of water at 0° C. What will be the volume of the water contained when the flask and its contents are heated to 100° C.?

10. A specific gravity bottle when empty weighs 6.8 gm., and when full of alcohol at 0°, 37.2 gm. It is then warmed to 20° C., wiped, and allowed to cool, and its weight is now found to be 36.51 gm. Calculate the coefficient of apparent expansion of alcohol, making no allowance for the expansion of the bottle. *Ans.* 0.001161.

11. The coefficient of linear expansion of copper is 0.000017. Find the increase in area of a copper plate which is 2 ft. square at 10° C., after heating it to 200° C.

12. If iron rails are 30 ft. long, and if the range of temperature throughout the year is 50° C., what space must be left between two consecutive rails to allow for expansion, the coefficient for iron being 0.00001125?

13. What will be the volume at 50° C. of a mass of air which has a volume of 200 cm³. at 10° C.?

14. At 50° C. the volume of a gramme of hydrogen is 13.2 l. What will be its volume at -10° C.?

15. The density of air at 0° C. is 0.001293, and at 200° C. is 0.0007457. Find the mean coefficient of expansion. *Ans.* 0.0036697.

16. If the volume of a quantity of air at 10° C. is 230 cm^3., at what temperature will its volume be increased to 300 cm^3.?

Ans. 96°.13 C.

17. A litre of dry air weighs 1.293 gm. at 0° C. and 76 cm. pressure. At what temperature will a litre of air weigh one gramme if the pressure is 74 cm.?

18. A flask containing air is corked at 20° C. Find the pressure inside the flask after it has stood for some time in a steam bath at 98° C., the original pressure being the standard atmospheric pressure of 76 cm. of mercury.

19. A quantity of gas is collected in a graduated tube over mercury. The volume of the gas at 10° C. is 50 cm^3., and the level of the mercury in the tube is 10 cm. above the level outside; the barometer stands at 75 cm. Find the volume which the gas would occupy at 0° C. and 76 cm. pressure. *Ans.* 41.25 cm^3.

20. A quantity of air is collected at 20° C. and 74 cm. pressure. The temperature is now increased to 100° C. What change must be made in the pressure to maintain the volume at its original value?

Ans. 94.2 cm.

21. A gas bag contains 14 cu. ft. of coal gas at 0° C. and under a pressure of 26 lb. per square foot. If the pressure is increased to 28 lb. per square foot, what will the volume become? To what temperature would the gas have to be warmed so as to recover its original volume?

Ans. 13 cu. ft.; 21° C.

22. A litre of a certain gas at 0° C. and 760 mm. pressure weighs 1.562 gm. What will be the weight of 500 cm^3. of the same gas at 25° C. and 780 mm.?

23. A certain quantity of gas is heated from 27° C. to 177° C., its volume being constant. Find in what proportion its elastic pressure is increased? *Ans.* 2 to 3.

24. A litre of air at 0° C. and 760 mm. pressure weighs 1.293 gm. What will a litre of air at 15° C. and 763 mm. weigh?

IV. MEASUREMENT OF HEAT.

325. Unit Quantity of Heat. — For the purpose of measuring the quantity of heat gained or lost by a body when its temperature or its state changes, it is necessary to adopt a unit of heat. The one most commonly used in connection with the metric system is the quantity of heat that will raise the temperature of one gramme of water one degree Centigrade. It is called a *calorie*. There is no agreement as to where the one degree shall be on the scale between the freezing and the boiling points. An exact definition requires the degree to be specified, because it is known that the heat required to raise a gramme of water from 4° to 5° is not the same as the quantity required to raise it from 14° to 15°, for example. The difference, however, is small; and in this book the heat necessary to raise a gramme of water through one degree at different temperatures will be assumed to be the same.

326. Thermal Capacity. — The number of calories required to raise the temperature of a body through one degree Centigrade is the *thermal capacity* of the body. The thermal capacities of equal masses of different substances differ widely. Thus, if 100 gm. of water at 0° be mixed with 100 gm. at 100°, the temperature of the whole mass will be very nearly 50°. But if 100 gm. of copper at 100° be cooled in 100 gm. of water at 0°, the final temperature will be about 9°.1. The heat lost by the copper in cooling through 90°.9 is sufficient to raise the same mass of water through only 9°.1.

327. Specific Heat. — The thermal capacity of a unit mass of a substance is its *specific heat*. In the metric system the number of calories necessary to raise the temperature of one gramme of the substance through one degree Centigrade, at any temperature, is its specific heat at that temperature. Specific heat varies a little with the temperature, but for most purposes it may be assumed to be constant. The specific heat of mercury is 0.033; this means that the heat which will raise 1 gm. of mercury through 1° C. will raise 1 gm. of water through only 0°.033 C.

The following table gives the specific heat of several subtances at the mean temperatures of the second column, and in terms of water at 15° C.

Substance	Temperature	Specific heat
Water	5°	1.0041
Water	15°	1.0000
Water	20°	0.9987
Ice	− 10°	0.502
Paraffin	10°	0.694
Copper	50°	0.092
Zinc	50°	0.093
Iron	15°	0.109
Platinum	50°	0.032
Mercury	20°	0.033

328. Specific Heat by Method of Mixtures. — **Experiment.** — Place a known number of grammes of lead shot in a test-tube, closing the end loosely with a plug of cotton. Suspend the test-tube for several minutes in boiling water. The temperature of the shot will then be that of the water. Now pour the shot quickly into a beaker containing a known quantity of water at the temperature of the room or a little lower. Stir gently with a thermometer for a few seconds, and record the temperature. The mass of the water in grammes, multiplied by the gain in temperature, will be the number of calories of heat gained by the water, and this same quantity is lost by the shot in cooling from the temperature of boiling water to that of the final temperature of the beaker and its contents. This number, divided by the product of the number of grammes of shot and its fall in temperature, will be the specific heat of lead.

The experiment illustrates the process of obtaining specific heat by the "method of mixtures." In practice it is necessary to take into account the thermal capacity of the vessel containing the known mass of water. Let a mass of m_1 gm. of a substance whose specific heat is s_1, and temperature t_1, be mixed with a mass of m_2 gm. of water, at a temperature t_2 (lower than t_1), in a vessel weighing m_3 gm., and of specific heat s_3. Let the common temperature acquired after the interchange of heat be t. Then the heat lost by the first body will be

$$H_1 = m_1 s_1 (t_1 - t),$$

and that gained by the water and the containing vessel will be

$$H_2 = m_2 s_2 (t - t_2) + m_3 s_3 (t - t_2).$$

If there is no other interchange of heat during the operation, H_1 equals H_2, and

$$m_1 s_1 (t_1 - t) = m_2 s_2 (t - t_2) + m_3 s_3 (t - t_2).$$

Solving for s_1,

$$s_1 = \frac{(m_2 s_2 + m_3 s_3)(t - t_2)}{m_1 (t_1 - t)}.$$

But the specific heat s_2 of water is one; and if m is put equal to $m_3 s_3$, then m is the "water equivalent," or the mass of water having the same thermal capacity as the vessel. Substituting

$$s_1 = \frac{(m_2 + m)(t - t_2)}{m_1 (t_1 - t)}. \qquad (27)$$

For example, let it be required to find the specific heat of iron from the following data: —

Iron, mass 20 gm. at 98° C.; water, 75 gm. at 10° C.; copper vessel, mass 15 gm., and specific heat 0.095; resulting temperature 12°.5 C.

Then $m = 15 \times 0.095 = 1.425$ gm. Substituting in (27),

$$s_1 = \frac{(75 + 1.425)(12.5 - 10)}{20(98 - 12.5)} = 0.112.$$

Problems.

1. If 30 gm. of iron nails at 100° C. are dropped into 60 gm. of water at 13°.4 C., and the final temperature is 18° C., what is the specific heat of the nails?

2. How much heat is required to raise 150 gm. of copper (specific heat 0.095) from 10° to 150° C.?

3. What is the thermal capacity of a copper beaker (specific heat 0.095) weighing 125 gm.?

4. A leaden bullet, mass 41.9 gm., at a temperature of 150° C., is plunged into 30 gm. of water at 10° C., contained in a copper vessel of mass 20 gm., and the resulting temperature is 15°.7 C. The specific heat of copper is 0.095. Find the specific heat of lead.

Ans. 0.032.

5. 10 gm. of alcohol, at 10° C., are contained in a copper calorimeter of mass 4 gm. A coppor ball, of mass 20 gm., at 65° C., is then dropped into the instrument, and the temperature rises to 23°.5 C. Find the specific heat of alcohol. *Ans.* 0.546.

6. The temperature of a copper vessel is 12° C.; 90 gm. of water, at 60° C., are poured in, and the temperature, after stirring, is found to be 52° C. Find the thermal capacity, or water equivalent, of the vessel.

7. 35 gm. of iron, at 98° C. (specific heat 0.112), are immersed in 70 gm. of water at 20° C., contained in a copper vessel whose mass is 25 gm. Calculate the resulting temperature.

8. A platinum ball, weighing 80 gm., is introduced into a furnace. When it has attained the temperature of tho furnace it is quickly transferred to a vessel containing 400 gm. of water at 15° C. The temperature of the water rises to 20° C. What was the temperature of the furnace, the specific heat of platinum being 0.032?

Ans. 801° C.

9. A copper vessel, of mass 30 gm., contains 80 gm. of water and an iron ball of mass 20 gm.; the whole is at 10° C. It is then set over a Bunsen burner and brought to 90° C. The specific heat of copper is 0.095, and of iron 0.112. How many units of heat are absorbed?

Ans. 6807.2 cal.

10. Calculate the specific heat of copper from the following data :—

Weight of copper	16.65 gm.
Weight of water in calorimeter	50 gm.
Initial temperature of copper	98° C.
Initial temperature of water and calorimeter .	12° C.
Final temperature of the mixture . . .	14°.5 C.
Water equivalent of calorimeter . . .	2.8 gm.

11. The specific heat of antimony is 0.0507. What mass of water can be raised from 0° C. to 15° C. by plunging into it 1 kgm. of antimony at 100° C.?

V. CHANGE OF STATE.

329. The Melting Point. — When a body changes from the solid to the liquid state by the application of heat, it is said to *melt*, or *fuse*, and the change is called *melting*, *fusion*, or *liquefaction*. The temperature at which fusion takes place is called the *melting point*. Solidification or freezing is the converse of fusion, and the temperature of solidification is usually the same as the melting point of the same substance. Water, if undisturbed, may be cooled a number of degrees below 0° C., but if it is disturbed it usually freezes at once, and its temperature rises to the freezing point.

The melting point of crystalline bodies is well marked. A mixture of ice and water will remain without change if the temperature is 0° C.; but if the temperature is above zero, some of the ice will melt; if it is below zero, some of the water will freeze. Some substances, like wax, glass, and wrought iron, have no sharply defined melting point. They first soften and then pass more or less slowly into the condition of a viscous liquid. It is this property which permits of the bending and moulding of glass, and the welding and forging of iron.

330. Change in Volume accompanying Fusion. — Most substances occupy a larger volume in the liquid state than in the solid; that is, they expand on liquefying. A few substances, like water and bismuth, expand on solidifying. When water freezes, its volume increases nine per cent. If this expansion is resisted, water in freezing is capable of exerting an enormous force.

Experiment. — Fit to a small bottle a perforated stopper through which passes a fine glass tube. Fill with water freed from air by boiling, the water extending halfway up the tube, and then pack in a mixture of salt and finely broken ice. The water column at first will fall slowly, but in a few minutes it will begin to rise, and will continue to do so till water flows out of the top of the tube. The water in the bottle freezes, and the attending expansion causes the overflow.

331. Laws of Fusion. — The following laws have been established by experiment: —

I. *Every crystalline substance begins to melt at a definite temperature, which is invariable for each substance if the pressure is constant.*

II. *The temperature of a body, when slowly melting, remains constant till the whole mass is melted.*

III. *Substances that expand on solidifying have their melting points lowered by pressure, and vice versa.*

The following interesting experiment illustrates the last law: —

Experiment. — Support a rectangular block or prism of ice on a stout bar of wood. Pass a small iron wire around the ice and the bar of wood, and suspend on it a weight of about 25 kgm. The pressure of the wire lowers the melting point of the ice, and the ice melts; the water, after passing around the wire, where it is relieved of pressure, again freezes. In this way the wire passes slowly through the ice, leaving the block solidly frozen.

332. Latent Heat. — When a body passes slowly from one state to another, as from the solid to the liquid, there is no rise of temperature, notwithstanding the constant application of heat. When this fact was first observed, it was generally believed that heat was a kind of matter, called *caloric.* This view led to the introduction of two terms, *sensible heat* and *latent heat;* the former denoting heat which changes the temperature of a body, and the latter heat which changes its state without affecting its temperature. The advocates of the caloric theory of heat thought that heat became hidden or concealed in the process of fusion, and they therefore called it "latent heat." We now know that this view is incorrect, and that the heat which disappears during a change of state ceases to be heat, and is energy converted into the potential form in the work of giving mobility to the molecules. The term *latent* should therefore no longer be applied to heat.

333. Heat of Fusion. — When a solid fuses, a quantity of heat disappears; and, conversely, when a liquid solidifies, the amount of heat generated is the same as disappears during liquefaction. The *heat of fusion* of a substance is the number of calories required to melt a gramme of it without change of temperature. The heat of fusion of ice is 80 calories. The manner of measuring it is illustrated by the following example: — Place 200 gm. of clean ice in 500 gm. of water at 60° C. The ice melts and reduces the temperature of the whole to 20° C. Then the heat lost by the 500 gm. of water equals the heat required to melt the ice plus the quantity required to raise the water formed from the ice from 0° C. to 20° C., or

$$500\,(60 - 20) = 200 \times L + 200 \times 20.$$

Whence the heat of fusion L equals 80.

334. Heat Lost in Solution. — **Experiment.** — Fill a test-glass part full of water at the temperature of the room, and add some finely divided ammonium nitrate. A thermometer will show a sensible fall of temperature.

Experiment. — Make a saturated solution of sodium sulphate at a temperature of 30° C. in a small flask or a large test-tube. Pass a thermometer through a stopper so that its bulb is in the solution. Cool slowly without disturbing the solution to about 20° C. The solution is then undercooled, but no crystals should form. Now remove the thermometer very carefully and allow the liquid on the bulb to evaporate till some crystals of the sodium sulphate have formed. Replace the thermometer in the solutions. Rapid crystallization will set in and extend through the whole solution. At the same time the temperature will rise to about 30° C.

The first experiment illustrates the fact that heat is absorbed when a body passes from the solid to the liquid state even by solution. It sometimes happens that this absorption of heat is masked by the heat evolved by chemical action between the dissolved body and the solvent. Freezing mixtures are based on the principle of the absorption of heat during the passage of bodies from the solid to the liquid state. When salt and pounded ice are mixed, both solids become liquid and absorb heat in the transition from the one state to the other.

The second experiment is the converse of the first, and shows that heat is evolved when a substance becomes a solid by crystallization from solution.

335. Vaporization. — **Experiment.** — Pour a few drops of ether into a beaker and cover loosely with a plate of glass. After a few seconds bring a lighted taper to the mouth of the beaker. A sudden flash will show that the vapor of ether was mixed with the air.

Experiment. — Support on an iron stand a beaker two-thirds full of water and apply heat. In a short time bubbles of steam will form at the bottom of the beaker, rise through the water, and burst at the top, producing violent agitation throughout the mass.

Vaporization is the conversion of a substance into the gaseous form. If the change takes place slowly, as in the first experiment, and from the surface of a liquid, it is called *evaporation;* but if the liquid is visibly agitated by rapid internal evaporation, the process is called *ebullition* or *boiling.*

There are two other varieties of vaporization, namely, the *spheroidal state* and *sublimation.* When a small quantity of liquid is placed on hot metal, as water on a red-hot stove, it assumes a globular or spheroidal form, and evaporates at a rate between ordinary evaporation and boiling. The vapor acts like a cushion and prevents actual contact between the liquid and the metal. The globular form is due to surface tension. Liquid oxygen at a very low temperature assumes the spheroidal form when placed on water. The temperature of the water is relatively high compared with that of the liquid oxygen. When a substance passes directly from the solid to the gaseous form without passing through the intermediate state of a liquid, it is said to *sublime.* Arsenic, camphor, and iodine sublime at atmospheric pressure, but if the pressure be sufficiently increased, they may be fused. Ice also evaporates slowly at a temperature below freezing.

336. Laws of Evaporation. — The laws of evaporation established by experiment are as follows: —

I. *The rate of evaporation increases with rise of temperature.*

II. *The rate of evaporation increases with the free surface of the liquid.*

This principle is utilized in the manufacture of salt by using large shallow pans for the brine, or by allowing the brine to trickle over bundles of twigs.

III. *The rate of evaporation is increased by a continual change of air in contact with the liquid.*

If the surrounding air is at rest, it soon becomes saturated with vapor from the liquid, and the rate of evaporation is checked. The drying action of the wind on roads after a rain and on wet cloth hanging in the air are illustrations in point.

IV. *The rate of evaporation is increased by diminishing the vapor pressure.*

In order that syrups may be concentrated at a low temperature to avoid burning, the operation is carried on in large covered pans from which the air and vapor are exhausted by air-pumps.

337. Laws of Ebullition. — The following laws express the results of experiment: —

I. *Each liquid has its own boiling point, which is invariable for that liquid under the same conditions.*

II. *The boiling point is dependent upon the character of the inner surface of the containing vessel.*

The temperature of boiling water is slightly higher if the inner surface of the containing vessel is smooth than if it is rough. But the temperature of the vapor given off is independent of the nature of the vessel. Hence, in fixing the boiling point on a thermometer, the thermometer is immersed in the steam and not in the water itself.

III. *The boiling point is raised by salts and lowered by gases dissolved in the liquid.*

When the air has been boiled out of water, the temperature may rise several degrees before ebullition sets in; and in case the inner surface of the vessel is very smooth, the boiling proceeds intermittently and explosively. The phenomenon is called "*bumping.*"

IV. *The boiling point rises with increase of pressure and falls with decrease of pressure.*

The effect of pressure on the boiling point is seen in the low temperature of boiling water at high elevations, and in the high temperature of the water under pressure in digesters used for extracting gelatine from bones. The change in the boiling point of water near 100° C is 0°.1 for an increase of pressure of 2.71 mm. of mercury. The following experiments show the effect of reduced pressure: —

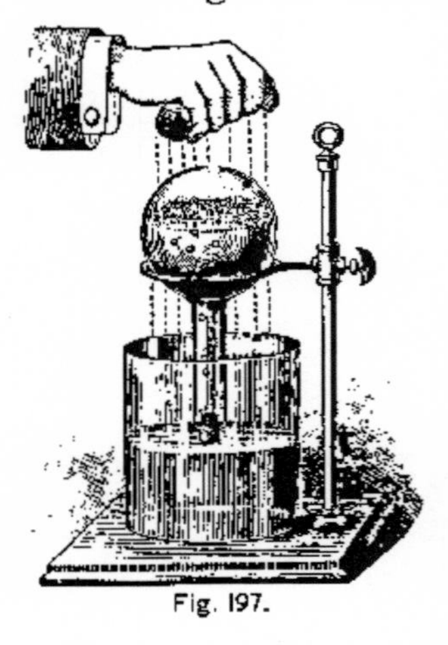
Fig. 197.

1. Place a flask of warm water under the receiver of an air-pump. It will boil violently when the receiver is exhausted.

2. Fill a round-bottomed Florence flask half full of water and heat till it boils vigorously. Cork the flask, invert, and support it on a ring stand (Fig. 197). The boiling ceases, but is renewed by applying cold water to the flask. The cold water condenses the vapor, and reduces the pressure within the flask so that the boiling begins again.

338. Relation of Altitude to the Boiling Point. — It has already been stated that, since atmospheric pressure decreases with the elevation, the boiling point of a liquid also decreases. Hence, the boiling point of water may be used as an indicator of the height of a place above the level of the sea. A change of elevation of about 295 m. makes a difference of 1° C. in the boiling point. Thus, at Quito, the highest city in the world, the average boiling point is 90°.1 C. Hence, the height above sea level is $295 \times (100 - 90.1) = 2920.5$ metres, a quantity greater than the true height by 34.4 metres.

339. Cold by Evaporation. — **Experiment.** — Put a few drops of ether on the bulb of an air thermometer (§ 315). The index at once begins to rise, showing that the bulb has been cooled.

In the evaporation of the ether, some of the heat of the thermometer bulb has been used to do work on the liquid. The rapid evaporation of liquid ammonia is utilized in the artificial production of ice. Sprinkling the floor of a room cools the air, because of the heat expended in evaporating the water. Porous water vessels keep the water cool by the evaporation of the water from the outside surface. Liquid carbon dioxide is readily frozen by its own rapid evaporation. Dewar liquefied oxygen by means of the low temperature obtained through the successive evaporation of liquid nitrous oxide and ethylene. In like manner, by the evaporation of liquid air he has liquefied hydrogen. The evaporation of liquid hydrogen under reduced pressure has enabled him to maintain a temperature within less than 16° of the absolute zero (§ 321).

340. Condensation and Distillation. — When a vapor is liquefied, all the heat that has disappeared during vaporization is generated again. This fact is applied in steam heating. Some gases may be made to assume a liquid form through their affinity for a liquid. Thus, for instance, when ammonia gas is brought in contact with water, it is rapidly absorbed with a marked rise of temperature.

Distillation involves both evaporation and condensation. Pure water, free from foreign substances such as vegetable and mineral matter, is obtained by distillation. If two liquids are mixed together, the more volatile will be vaporized by heat first, and it may be condensed and collected by itself. In this way alcohol is separated from fermented liquors. The apparatus used for evaporating the liquid is called the *still*, and that for liquefying, the *condenser*. The latter is usually a coiled tube, called the *worm*, surrounded by water. Fig. 198 illustrates one of the forms used in laboratories.

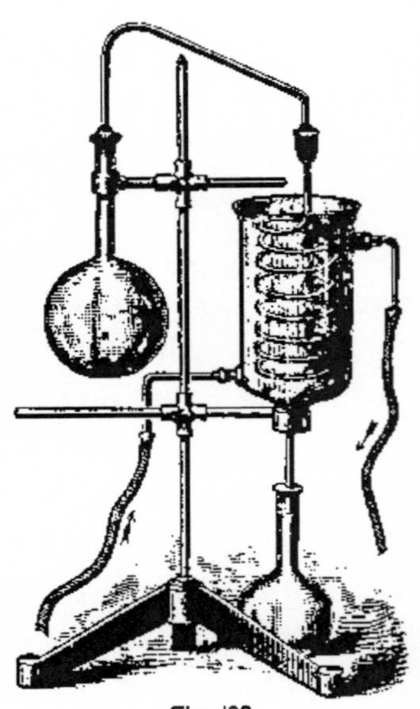

Fig. 198.

341. The Dew Point.—**Experiment.** In a polished nickel-plated copper beaker pour some tepid water. Stir gently with a thermometer, reducing the temperature by the introduction from time to time of small pieces of ice. In time a mist will form on the outside of the vessel. Note the temperature of the water on the first appearance of this mist.

The dew point is the temperature at which the aqueous vapor of the atmosphere begins to condense. It may be determined as outlined in the preceding experiment. In the formation of clouds, the precipitation of dew, and in the "sweating" of pitchers of ice-water, we have evidence of the existence of water vapor in the atmosphere. The amount of moisture that the air can retain depends on the temperature. The terms *dryness* and *moistness*, applied to the air, are purely relative, and indicate the proportion of vapor actually present, in comparison with what the air could contain at the existing temperature. At the dew point the air is saturated. *Relative humidity*, or wetness, is expressed by the number of parts by weight of aqueous vapor contained by the air to every 100 that it could contain. Saturation is represented by 100, and absolute dryness by 0. A humidity of 60 to 70 is demanded for health.

342. The Heat of Vaporization is the number of calories required to change a unit mass of a liquid at its boiling point into vapor at the same temperature. Water has the greatest heat of vaporization of all liquids. The following process of obtaining the heat of vaporization of water will make clear the principles that underlie the general problem : —

Set up apparatus like that shown in Fig. 199. The steam from the boiling water is conveyed into a beaker containing a known quantity of water at a known temperature. The increase in the mass of the water gives the amount of steam condensed. The "trap" in the delivery-tube catches the water that condenses before it reaches the beaker. Suppose that the experiment gave the following data: Amount of water in the beaker, 400 gm. at the beginning, 414.1 gm. at the end; temperature at the beginning, 20° C., and at the end, 41° C.; observed boiling point, 99° C.; there were 14.1 gm. of steam condensed. Now, by the principle that the heat lost or given off by the steam equals that gained by the water, we have

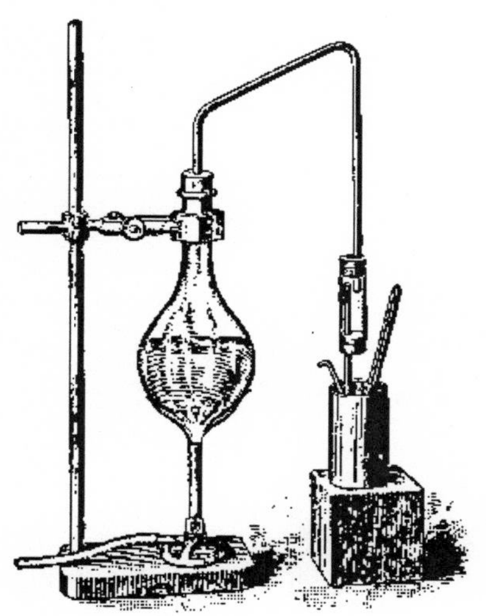

Fig. 199.

$$400 \times (41 - 20) = 14.1 \times l + 14.1 \times (99 - 41);$$

whence $l = 537.7$ cal. The most carefully conducted experiments show that the heat of vaporization of water is 535.9.

Questions and Problems.

1. How can pure water be obtained from sea water?

2. How can water be heated above the ordinary boiling point?

3. Account for the fog which forms in the receiver of an air-pump when the air is exhausted.

4. Why does a current of air cause one to feel cold?

5. Why is an iceberg frequently enveloped in fog?

6. Why is it necessary to take the barometric pressure into account in fixing the boiling point on a thermometer scale?

7. Why does a wet cloth feel cold to the hand?

8. Why does warming a room make it feel drier?

9. Why do morning fogs and mists frequently disappear when the sun gets well up?

10. What mass of water at 50° C. will melt 5 kgm. of ice at 0° C.?
Ans. 8 kgm.

11. Calculate the height of a mountain where water boils at 85° C.
Ans. 4425 m.

12. Mt. Washington is 6288 ft. above sea level; at what temperature will water boil on its top?

13. What mass of water at 74° C. will convert 100 gm. of ice at – 14° C. into water at 4° C.? (Specific heat of ice 0.5.)

14. Water boils in the city of Mexico at 92°.3 C. What is its height above sea level?

15. If 1 kgm. of water at 100° C., mixed with 1 kgm. of ice, yields 2 kgm. of water at 10°.36 C., what is the heat of fusion of ice?
Ans. 79.28.

16. An aluminum ball, of mass 20 gm., is heated to 100° C., and placed on a block of ice. It is found to melt 5.35 gm. Taking the heat of fusion of ice to be 80, find the specific heat of aluminum.
Ans. 0.214.

17. How much heat would it require to raise 250 gm. of ice from 0° to 100° C., and to convert it into steam?

18. A copper calorimeter, of mass 10 gm., contains 30 gm. of water at 5° C.; 20 gm. of cadmium are melted, and poured at the melting point, which is 315° C., into the water. The final temperature is 24° C. The specific heat of solid cadmium being 0.055, calculate its heat of liquefaction, taking the specific heat of copper as 0.095.

Ans. 13.4.

19. 10 gm. of steam at 100° C., condensed in 1 kgm. of water at 0° C., raised the temperature to 6°.3 C. Calculate the heat of vaporization of water.

20. How much steam at 100° C. is required to raise 150 gm. of water from 0° to 100° C.?

21. Steam at 100° C. was passed into 500 gm. of water at 12° C. until the temperature rose to 26° C. The weight of the contents of the calorimeter was now 511.4 gm. Calculate the heat of vaporization of water?

22. How much steam will it take to melt two pounds of ice.

23. How much ice must be dissolved in a litre of water at 20° C., in order to reduce its temperature to 5° C.?

24. Find the result of mixing 2 lbs. of ice at 0° C. with 3 lbs. of water at 45° C.

25. A glass beaker contains 1 lb. of water and some ice. When 1 oz. of steam at 100° C. has been supplied to the vessel the contents are at a temperature of 3° C. How much ice was there to begin with?

Ans. 0.44 lb.

VI. TRANSMISSION OF HEAT.

343. Three Modes of Transmitting Heat. — Experiment. — Place one end of a metal rod in a Bunsen flame and the other in melting ice. It will be found that heat passes along the rod and melts the ice. Hold the hand high above the flame; it will be warmed by a rising current of hot air. Hold the hand by the side of the flame; again a sensation of heat will be perceived.

This simple experiment illustrates the three ways in which heat may be transmitted from one point to another. They are: —

1. *Conduction*, in which heat is conveyed by matter without any visible motion of the matter itself. It is passed on from the hotter to the colder particles by some invisible molecular motion.

2. *Convection*, in which heat is transferred by the visible motion of heated matter, as by a current of hot air or the flow of hot water through pipes.

3. *Radiation*, in which heat is propagated like light, by a wave motion in the ether, without the aid of matter. It is by this method that radiant energy (heat and light) reaches us from the sun.

344. Conduction.—Experiment.—Twist together two stout wires, iron and copper, of the same diameter, forming a fork with long paral-

Fig. 200.

lel prongs and a short stem. Support them on a wire stand (Fig. 200), and heat the twisted ends. After several minutes find the point on each wire, farthest from the flame, where a sulphur match ignites when held against the wire. This point will be found farther along on the copper than on the iron, showing that the former has led the heat farther from its source.

Fig. 201.

Experiment. — Prepare a cylinder of uniform diameter, half of which is made of brass and half of wood. Hold a piece of writing paper firmly around the junction like a loop (Fig. 201). By applying a Bunsen flame the paper in contact with the wood is soon scorched, while the part in contact with the brass is scarcely injured. The metal conducts the heat away and keeps the temperature of the paper below the point of ignition.

These experiments show that solids differ in their conductivity for heat. The metals are the best conductors; wood, leather, flannel, and organic substances in general are poor conductors; so also are all bodies in a powdered state, owing doubtless to a lack of continuity in the material.

It is a common mistake to assume that the rate at which the temperature rises is a measure of conductivity. For example, if equal bars of iron and lead are arranged so that one end of each is heated alike, pieces of phosphorus at the same distance from the heated ends will be found to take fire first on the lead, although iron is the better conductor. This is due to the fact that iron has about four times the specific heat of lead, and hence requires four times as much heat to produce the same change of temperature. The lead therefore acquires the necessary temperature to ignite the phosphorus long before the iron.

Fig. 202.

345. Conductivity of Liquids.— Experiment. — Pass the tube of a simple air thermometer through a cork fitted to the neck of a large funnel. Support the apparatus as shown in Fig. 202. Fill the funnel with water, covering the bulb to the depth of about one centimetre. Pour a spoonful of ether on the water and set it on fire. The steadiness of the index shows that little if any of the heat due to the burning ether is conducted to the bulb.

It appears from this experiment that water is a poor conductor of heat. This is equally true of all liquids except molten metals.

346. Conductivity of Gases. — The conductivity of gases is very small, and its determination is very difficult because of radiation and convection. The conductivity of hydrogen is about 7.1 times that of air, while the conductivity of water is 25 times as great.

347. Applications. — Some articles in a room feel cold to the touch while others feel warm. An explanation will be found in the fact that those which feel cold are good conductors of heat, and those which feel warm are bad conductors. The former conduct away the heat from the hand faster than the body supplies it, causing the sensation of cold; the latter do not carry off the heat, and consequently they do not feel cold.

The handles on metal instruments that are to be heated are usually made of some poor conductor, as wood, bone, etc.; or else they are insulated by the insertion of some non-conductor, as in the case of the handles to silver teapots, where pieces of ivory are inserted to keep them from becoming too hot.

The non-conducting character of air is utilized in houses with hollow walls, in double doors and double windows, and in clothing of loose texture. The warmth of woollen articles and of fur is due mainly to the fact that much air is enclosed within them on account of their loose structure.

348. Convection. — **Experiment.** — Remove the bottom from a wide-mouthed bottle. Fit a double-perforated stopper to the mouth and pass through it two glass tubes; one of these *CD* should be straight, and the other *EF* should have some such form as shown in Fig. 203. The other ends of these tubes pass through a stopper fitted to the Florence flask, *B*. The tube *CD* should extend from the top of the flask nearly to the top of the open vessel; and the tube *EF* should reach from the bottom of the flask to the bottom of the open vessel. Support the apparatus on a heavy ring stand. Fill the flask with water colored with red aniline, and the open vessel with water colored with blue aniline. Blow through the straight tube till all air-bubbles are removed. Now place a Bunsen burner beneath the flask. In a short time the red liquid will be seen gathering on the top of vessel *A*, and the blue liquid at the bottom of vessel *B*.

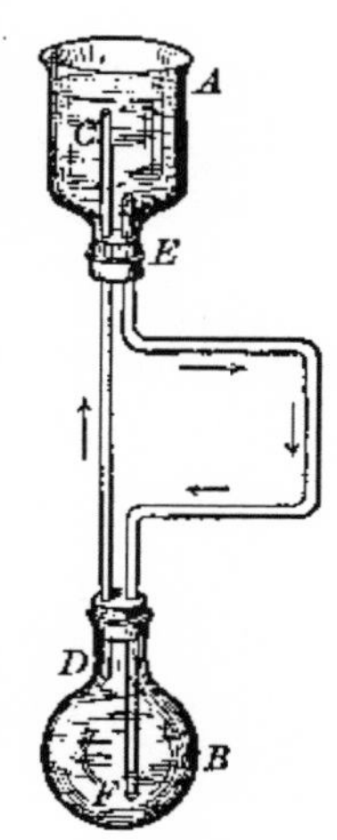

Fig. 203.

The water in *B*, on being heated, is expanded and rendered less dense. Hence, it is forced to rise by the downward pressure of the colder water in *A*. This circulation down *EF* and up *DC* will continue so long as *A* is colder than *B*.

Experiment. — Fill a large glass beaker about three-fourths full of cold water, and pour on it carefully enough warm water, colored with an aniline dye, to form a layer about 2 cm. in thickness. Fill a large test-tube full of a freezing mixture of salt and ice, and hold it in the colored water. Streams of the colored water will soon begin to descend through the uncolored part. The freezing mixture reduces the temperature of the colored water below that of the uncolored, making it heavier.

These experiments show that by raising the temperature of the lower part of a liquid in a vessel, or by lowering the temperature of the upper part, currents can be set up within it. The same is true of gases, as may be seen in the rising currents of air above a hot stove. The currents set up in fluids through differences of temperature are called *convection currents.* The heating of buildings by hot water circulating through pipes, or by hot-air furnaces, is a familiar application of convection currents. Land and sea breezes, the trade winds, and, in fact, winds in general, are convection currents on a gigantic scale.

349. Ventilation. — **Experiment.** — Support in a shallow dish a short piece of candle, and place over it a lamp chimney. Pour enough water into the dish to close the lower end of the chimney. The flame is soon extinguished. Why? Relight the candle, and insert a cardboard partition in the chimney, as in Fig. 204. The candle will now burn, and if a piece of lighted touch-paper[1] is held over the top of the chimney, it will show that there is a current of air down one side and up the other.

Fig. 204.

The office of a lamp chimney is to increase the supply of oxygen to the flame. The air within it is heated by the flame and rises, and cold air flows in through the bottom to restore the equilibrium, becomes heated in passing over the flame, and thus keeps up the high temperature of the chimney. This principle underlies the ordinary methods of ventilating rooms. A flue carries off the impure air, and fresh air flows in to take its place, after passing over heated pipes or through a heated furnace.

That the existence of flues opening into a room does

---

[1] Made by soaking porous paper in a strong solution of saltpetre, and drying. It burns without flame and gives off smoke.

not insure ventilation, unless means are adopted to make certain the upward movement of the air in such flues, is illustrated in the following experiment: —

Experiment. — Fit to a wide-mouthed bottle of about 2 l. capacity a cork, through which pass two glass tubes, each at least 2 cm. in diameter, and 20 cm. in length, the corresponding ends of the tubes being at the same level (Fig. 205). A wire also passes through the cork, carrying a candle at the lower end. The wire is bent so that the candle can be brought directly under either tube, or can be turned away from both of them. First, set the candle in the last position, light it, and insert the cork with its tubes in the bottle. The flame will soon go out, no air entering through either tube, although both are open. Second, blow out the foul gas, relight the candle, turn the wire till the flame is directly under one of the tubes, and insert the cork in the bottle. The candle will continue to burn brightly. If lighted touch-paper is applied to the top of the tubes in succession, it will be found that there is a downward current in one, and an upward current in the other.

Fig. 205.

350. Radiation. — The heat perceived when one stands in the vicinity of a hot stove is not received by conduction, nor is it conveyed by the air. The heat energy of a hot body is constantly passing into space as radiant energy in the luminiferous ether. Radiant energy becomes heat again only when it is absorbed by bodies upon which it falls. Energy transmitted in this way is, for convenience, referred to as *radiant heat*, although it is transmitted as radiant energy, and is transformed into heat by absorption. Radiant heat and light are physically identical, but are perceived through different avenues of sensation. Radiations that produce sight when received through the eye give a sensation of warmth through the nerves of touch, or heat a thermometer when incident upon it. The long ether waves do not affect the eye, but they heat a body which absorbs them.

351. Laws of Heat Radiation. — The following laws have been established experimentally: —

I. *Radiation proceeds in straight lines.* This law is illustrated in the use of fire screens and sun shades.

II. *The amount of radiant energy received by a body from any small area varies inversely as the square of its distance from this area as a source.*

III. *Radiant energy is reflected from a polished surface so that the angles of incidence and reflection are equal.* Archimedes is said to have set fire to the Roman ships during the siege of Syracuse in 212 B.C., by concentrating on them the heat of the sun by the aid of a large concave reflector.

IV. *The capacity of a surface to reflect radiant energy depends both on the polish of the surface and the nature of the material.* Polished brass is the best reflector, and lampblack is the poorest.

V. *The rate at which the temperature of a cooling body falls by radiation is proportional to the excess of its temperature over that of the surrounding medium.* This is known as Newton's *Law of Cooling*, and holds approximately for small differences of temperature but fails when the excess is large. According to this law a body at a temperature of 30° C. cools twice as fast as one of 25° C. in air at 20° C., for the excess 10°, in the first case is twice 5°, the excess in the second.

352. Absorption of Heat. — **Experiment.** — In slots cut 10 cm. apart in a narrow board, support two pieces of bright tin plate, each

10 cm. square. Coat the inner face of one of these squares of tin with lampblack. Stick balls of equal size, one at the centre of each outside face, with shoemaker's wax, using as little as possible. Hold a heated plate of iron midway between the two tin squares. The ball will soon fall from the blackened plate, showing that lampblack is a ready absorber of heat.

By using plates coated with different substances, it will be found that these substances differ in their capacity of absorbing heat. Leslie discovered that the best absorbers, as lampblack, ashes, and rough surfaces, are bad reflectors; while good reflectors, as polished metals, are bad absorbers.

353. The Radiometer. — This instrument was invented by Sir William Crookes in 1873 while investigating the properties of highly attenuated gases. It consists of a glass bulb from which the air has been exhausted till the pressure does not exceed 7 mm. of mercury (Fig. 206). Within the bulb is a light cross of aluminum wire carrying small diamond-shaped vanes of mica, one face of each being coated with lampblack; the whole is mounted to revolve on a vertical pivot. When the instrument is placed in the sunshine or in the radiation from any heated body, the cross revolves with the blackened faces of the vanes moving away from the source of heat.

Fig. 206.

The explanation of this interesting phenomenon is to be found in the kinetic theory that the mean free path of the molecules between collisions with other molecules becomes, at this low pressure, at least equal to the distance between the vanes and the wall of the bulb. The infrequent collisions among the molecules in such a vacuum prevents the equalization of pressure throughout the tube. Now the blackened sides of the vanes absorb more heat than the bright ones, and the gas molecules rebound from the warmer surfaces with a greater velocity than from the others, thus giving the vanes an impulse in the opposite direction. This impulse is the equivalent of a pressure, and the residual gas has lost the power of rapid adjustment of pressure throughout its mass. When the vacuum is not so good, no difference of pressure on the two sides of the vanes can exist, and there is no motion of the vanes.

354. Selective Absorption. — **Experiment.** — Fill a large flat bottle with clear water, and place it between a lamp and the radiometer. The rate of rotation of the vanes will be much reduced. Repeat the experiment, using a similar bottle filled with a solution of iodine in carbon disulphide. There will be no perceptible effect on the rate of rotation.

These experiments show that water cuts off the greater part of the radiant heat, while the solution of iodine does not perceptibly affect the intensity. Substances which transmit radiant heat are called *diathermanous*, and those which do not, *athermanous*. Rock salt is the most highly diathermanous substance known. On the other hand, alum, sugar, glass, water, and ice are extremely athermanous. The diathermanous character of a substance varies with the temperature of the radiant. Such substances as alum, water, etc., transmit little or none of the radiation from a surface of low temperature. The radiant energy from the sun passes readily through the atmosphere to the earth, warming its surface; but the radiations from the earth are stopped by the enveloping atmosphere. So also the radiant heat from the sun passes readily through the glass of the greenhouse, but that from within is unable to pass outward.

Questions.

1. When snow and ice melt why do they not liquefy all at once?

2. Why does a metal liquefy so rapidly when it begins to melt?

3. Why does snow protect from cold?

4. Why will a current of air extinguish a candle?

5. Why will ice packed in sawdust or straw keep from melting?

6. Why do men working about smelting-furnaces wear flannel clothing?

7. Why is paper so effective in protecting plants from frost?

8. Why is it difficult to boil water in a "furred" kettle?

9. Why is the direct radiation of the sun on the top of a mountain more intense than at the base?

10. Why is there little or no dew on a windy night?

11. Should the surfaces of stoves and heat radiators be rough or polished?

12. If a pond is freezing over, what is the temperature of the water at the bottom?

13. Why does increasing the height of a chimney increase the draught?

VII. HEAT AND WORK.

355. Heat and Mechanical Action. — **Experiment.** — Strike the edge of a piece of flint a glancing blow with a piece of hardened steel. Sparks will fly at each blow.

Experiment. — Pound a bar of lead vigorously with a hammer. The temperature of the bar will rise.

Experiment. — Place a small piece of tinder, such as is employed in cigar lighters, in the cavity at the end of the piston of a fire syringe (Fig. 207). Force the piston quickly into the barrel. If the piston is quickly withdrawn the tinder will probably be on fire.

These experiments illustrate the transformation of mechanical energy into heat. Some of the energy of the descending flint, the hammer, and the piston have in each case been transferred to the molecules of the bodies themselves, increasing their kinetic energy; that is, raising their temperature. Savages kindle fire by rapidly twirling a dry stick, one end of which rests in a notch cut in a second dry piece. The axles of carriages and the bearings in machinery are heated to a high temperature when not properly lubricated. The heating of drills and bits in boring, the heating of saws in cutting timber, the burning of the hands by a rope slipping rapidly through them, the stream of sparks flying from an emery wheel, are instances of the same kind of transformation; the work done against friction produces kinetic energy in the form of heat.

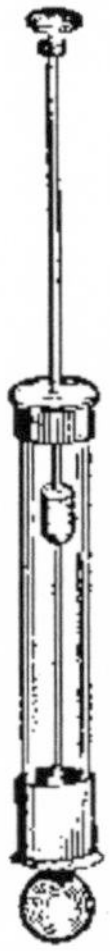

Fig. 207.

356. Numerical Relation between Heat and Work. — In 1840 Joule of Manchester began a series of experiments to determine the numerical relation between the unit of heat and the foot-pound. Joule's experiments by a number of methods extended over a period of nearly forty years. His most successful method consisted in determining the heat produced when a known amount of work was expended in heating water by stirring it with paddles driven by weights falling through a known height. He concluded that 772 foot-pounds of work, when converted into heat, will raise the temperature of one pound of water 1° F. The equivalent for 1° C. is 1390 foot-pounds.

The investigations of Rowland in 1879, and of Griffiths in 1893 have shown that 778 foot-pounds for 1° F., or 427 kilogramme-metres for 1° C., are the nearest whole number values; that is, 778 foot-pounds of work when converted into heat will raise the temperature of one pound of water 1° F., or 427 kilogramme-metres of work when converted into heat will raise the temperature of one kilogramme of water 1° C. This numerical relation between heat and work is known as the *mechanical equivalent of heat*. Its value expressed in absolute units is 4.19×10^7 ergs. This is the energy value of one calorie.

The following problem illustrates one of the uses which may be made of this relation: A mass of iron weighing 10 kgm. (specific heat 0.112) falls through a height of 100 m. Find the heat generated when it strikes the ground.

The work done by gravity is $10 \times 100 = 1000$ kgm.-m.; $1000 \div 427 = 2.342$ kgm.-degrees of heat, that is, the heat that would raise 2.342 kgm. of water through 1° C. Then $2.342 \div 0.112 = 20.9$, the number of kgm. of iron that 2.342 kgm.-degrees of heat will raise through 1° C. $20.9 \div 10 = 2.09$ degrees. If all the heat were confined to the iron, its temperature would rise 2.°09 C.

357. The Steam Engine, in its most essential features, is the invention of James Watt. It is a device for transforming the energy stored in steam into that of mechanical motion. In the more common of its many modern forms it consists of a strong cylinder in which a piston is made to move to and fro by applying the pressure of steam to its two faces alternately.

Figure 208 shows a longitudinal section of a simple engine divested of many of the more complicated accessories designed to improve its efficiency. The piston M moves to and fro in the cylinder D by virtue of the pressure of the steam supplied by the boiler through the tube

F. In the steam chest *E* works the slide valve *R*, which admits the steam alternately to the ends of the cylinder through *N* and *O*. When the valve is situated as shown, the steam passes into the upper end of the cylinder and drives the piston down. At the same time the other end is connected with an exhaust pipe, shown at *P*, through which the steam either escapes into the air, as in *high pressure* or *non-condensing engines*, or into a large chamber, as in *low pressure* or *condensing engines*, where it is condensed to water, reducing the pressure on that face of the piston. The slide valve is moved by the rod *H*, connected to an eccentric *C*, a wheel pivoted a little to one side of its

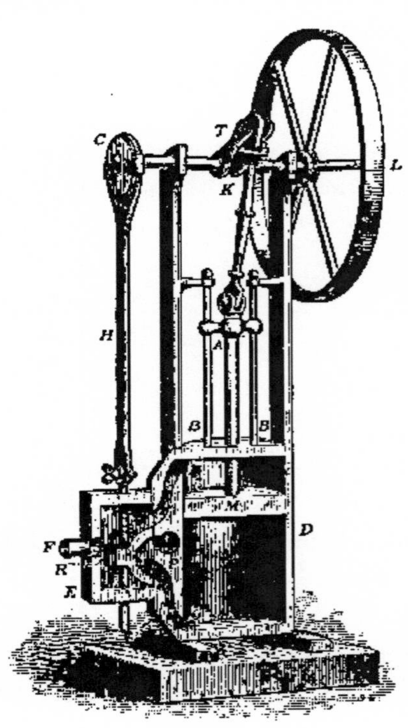

Fig. 208.

centre, on the horizontal shaft *K*. This shaft receives its motion from the piston by means of the jointed rod *A* and the crank *T*. The flywheel *L* serves the double office of belt pulley and reservoir of energy. It is made with a heavy rim in order that when the piston is at the end of the cylinder, and the direction of motion must change, the energy stored in it may be sufficient to carry the shaft beyond these *dead points* to a position where the piston can again turn the shaft. It also serves to give uniformity of motion to the shaft, which would otherwise vary because the effective part of the force exerted on the crank is not constant, being greatest when the crank is at right angles to the connecting rod, and diminishing to nothing when parallel to it.

In order that the piston rod may always move in a straight line, and the piston maintain a steam-tight fit in the cylinder, the former is attached to a transverse bar, or cross head *A*, which slides on two guide-bars *B B*, firmly bolted to the framework of the engine, and adjusted accurately parallel to each other and to the piston rod. In many large engines the cylinder is given a horizontal position.

Problems.

1. A mass of 500 gm. moving with a velocity of 40 m. per second is stopped. If all this energy is converted into heat, how much will it yield? (Kinetic energy $= \frac{1}{2}mv^2$.) *Ans.* 95.46 cal.

2. Through what distance must 10 kgm. of water fall to raise its temperature 10° C., if all the energy goes to heating the water? *Ans.* 4270 m.

3. How much heat is produced in stopping a train of 100 metric tons mass, running at 60 km. an hour? *Ans.* 3314×10^3 cal.

4. A mass of rock weighing 500 kgm. falls from a cliff 80 m. in height. If the specific heat of rock be taken at 0.2, how much will the temperature of the rock be raised, supposing all of the heat generated to be concentrated in the rock? *Ans.* 0°.93 C.

5. A leaden bullet weighing 30 gm. strikes a target with a velocity of 400 m. per second. Assuming 25 per cent of the kinetic energy of the bullet to remain in the bullet as heat, find how much its temperature will be raised, the specific heat of lead being 0.03. *Ans.* 159°.1 C.

CHAPTER VII.

MAGNETISM AND ELECTRICITY.

I. MAGNETS AND MAGNETIC ACTION.

358. The Natural Magnet or Lodestone. — Certain ores, consisting of iron and oxygen, sometimes possess the property of attracting and holding small particles of iron. This property was known to the ancients and was exhibited in a marked degree by iron ores from Magnesia in Asia Minor; they were therefore called *magnetic stones.* They are now known as *natural magnets*, and the properties exhibited by them are called magnetic properties.

Experiment. — Sprinkle iron filings over a piece of natural magnet. The filings will adhere to it in tufts, not uniformly over the surface, but chiefly at the ends and on projecting edges.

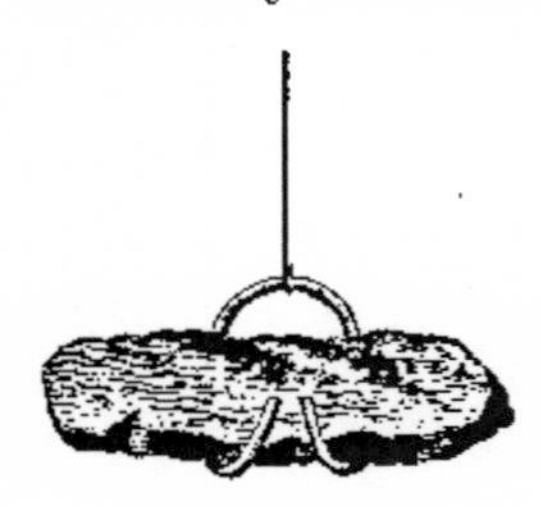
Fig. 209.

Experiment. — Make a stirrup out of wire, place in it the piece of natural magnet, and suspend it by an untwisted thread (Fig. 209). Carefully exclude all air-currents and allow the magnet to come to rest. Note its position, then disturb it slightly, and again let it come to rest. It will be found that it invariably returns to the same position, the line connecting the two ends to which the filings chiefly adhered in the preceding experiment lying north and south.

This property of the natural magnet was early turned to account in navigation, and secured for it the name of *lodestone* (leading-stone).

359. Artificial Magnets.— **Experiment.** — Stroke a large darning-needle from end to end, and always in the same direction, with one end of the lodestone. Roll it in iron filings and they will cling to its ends as they did to the lodestone. The needle has become a magnet.

Experiment.— Use the needle of the last experiment to stroke another needle. This second needle also acquires magnetic properties, and the first one has suffered no loss.

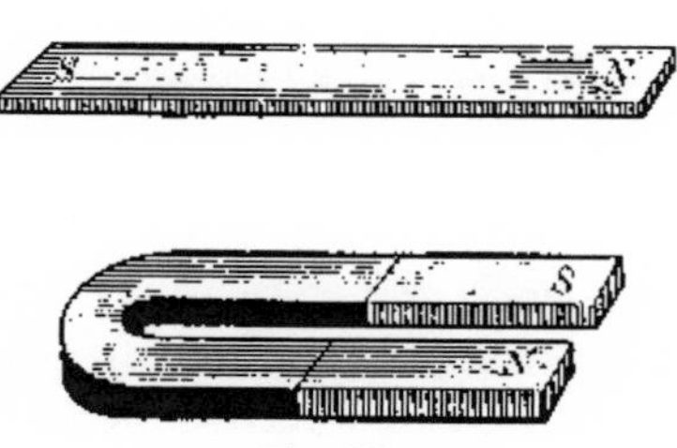

Fig. 210.

Bars of hard steel, that have been made magnetic by the application of a lodestone or of some other magnetizing force, are called *artificial magnets*. The form of artificial magnets, or simply magnets, most commonly met with are the *bar* and the *horseshoe* (Fig. 210), so called from their shape.

360. Polarity. — **Experiment.** — Roll a bar magnet in iron filings. It will become thickly covered with the filings near its ends. Few, if any, will adhere at the middle (Fig. 211).

Fig. 211.

The experiment shows that the greater part of the magnetic attraction is concentrated at the ends of the magnet. These are called its *poles*, and the magnet is said to have *polarity*. The line joining the poles of a long slender magnet is its *magnetic axis*.

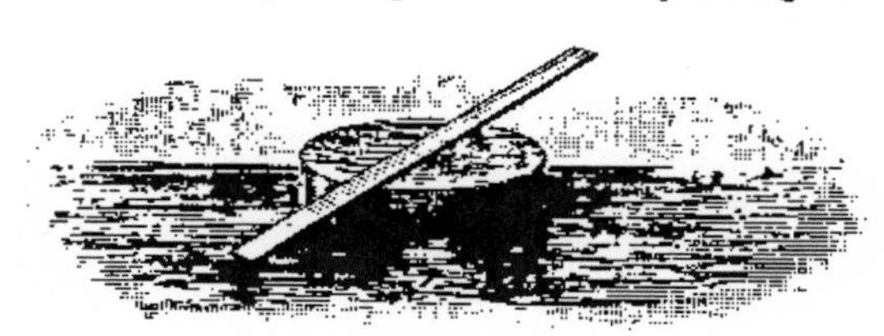
Fig. 212.

361. Experiment. — Straighten a piece of steel clock spring and float it on a piece of cork in a glass vessel filled with water. Note its position after it comes to rest (Fig. 212). Change its position several times. It will be found that there is no uniformity in the direction it takes when at rest. Now stroke the spring from end to end with one pole of a magnet and repeat the tests. The floating spring will now always come to rest in a north-and-south line and with the same end to the north.

The end pointing toward the north is called the *north-seeking pole*, and the other the *south-seeking pole* of the magnet. They are commonly called simply the *north pole* and the *south pole*.

362. Consequent Poles. — **Experiment.** — Draw the temper of a knitting-needle slightly at two or three points, and then stroke it from end to end with one pole of a strong magnet. Dip it in iron filings. They will adhere in tufts at the points where the temper was drawn as well as at the ends, showing that there are several poles.

Magnetic poles intermediate between those at the ends are called *consequent poles*.

363. A Magnetic Substance is one which is attracted by a magnet, or which can be magnetized. Faraday showed that most substances are influenced by magnetism. Ordinary magnets, however, produce a noticeable effect on but few substances besides iron and its compounds. Cobalt and nickel stand next to iron in respect to magnetism. Some substances, like antimony and bismuth, are slightly repelled by powerful magnets. They are said to be *diamagnetic*.

364. Magnetic Transparency. — **Experiment.** — Cover the pole of a strong bar magnet with a thin plate of glass. Bring the face of the plate opposite the pole in contact with a pile of iron tacks. A number will be found to adhere, showing that the attraction takes place through glass. In like manner, try thin plates of mica, wood, paper, zinc, copper, and iron. No perceptible difference will be seen except in the case of the iron, where the number of tacks lifted will be much less.

Magnetic force acts freely through all substances except those classed as *magnetic*. Soft iron serves as a more or less perfect screen to magnetism. Watches may be protected from magnetic force that is not too strong by means of an inside case of soft sheet iron.

365. Magnetic Needle. — A slender magnetized bar, suspended by an untwisted fibre or pivoted on a point, like a compass needle, is a *magnetic needle*. The direction in which it comes to rest without torsion or friction is the *magnetic meridian*.

Experiment. — Magnetize a piece of watch spring about 2 cm. long. Fasten a fibre of unspun silk to the bit of magnetized steel so that it will hang horizontally. Suspend it inside a wide-mouthed bottle by attaching the fibre to a cork fitting the mouth of the bottle. The little magnetic needle will then be protected from currents of air. It may be made visible at a distance by sticking fast to it a piece of thin white paper.

366. Mutual Action between Magnets. — **Experiment.** — Magnetize a piece of large knitting-needle, about four inches long, by stroking it from the middle to one end with the north pole of a bar magnet, and then from the middle to the other end with the south pole. Repeat the operation several times. Suspend the needle in a small stirrup like that of Fig. 209 and mark the north pole with red paint.

Present the north pole of the magnetized knitting-needle to the north pole of the needle suspended in the bottle. The latter will be repelled. Present the same pole to the south pole of the little magnetic needle; it will be attracted. Repeat with the south pole of the knitting-needle and note the deflections.

The results may be expressed by the following law of magnetic attraction and repulsion:—

Like poles repel and unlike poles attract each other.

The suspended magnet affords a ready means of ascertaining which pole of another magnet is the north pole, for the north pole of one will repel the north pole of the other. Repulsion is always a more reliable indication of polarity than attraction. The reason will be obvious from the experiments which follow.

367. Induced Magnetism.—Experiment.—Hold one end of a short rod of soft iron near one pole of a strong bar magnet, and while in this position dip the other end into iron filings. They adhere to it as to a magnet, but fall off when the magnet is removed.

Magnetism produced in magnetic substances by the influence of a magnet is said to be *induced.*

368. Polarity of the Iron Bar.—Experiment.—Support a strong horseshoe magnet in a vertical plane, with its poles uppermost, and the line joining them horizontal (Fig. 213). Suspend by a thread a short rod of soft iron so that it hangs horizontally above and near the poles of the magnet. Now bring near one end of this rod a bar magnet, so that its pole is opposite in name to that of the vertical magnet. The repulsion of the rod indicates that its polarity is the same as that of the bar magnet, and hence the reverse of that of the horseshoe magnet.

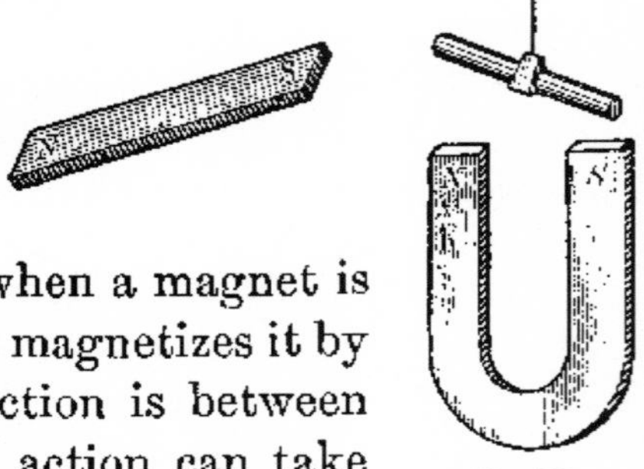

Fig. 213.

It appears, therefore, that when a magnet is brought near a piece of iron it magnetizes it by induction, and that the attraction is between unlike poles. The inductive action can take place through a series of iron rods, as exemplified by the attraction of a bunch of filings or tacks.

369. Permanent and Temporary Magnetism. — In the last two experiments the soft iron ceases to be a magnet when removed to a distance from the bar or horseshoe magnet. When a piece of hardened steel is brought near a magnet, it acquires magnetism as the piece of soft iron does under the same conditions; but the steel retains its magnetism when the magnetizing force is withdrawn, while the soft iron does not. In addition, therefore, to the *permanent magnetism* exhibited by the magnetized steel, we have *temporary magnetism* induced in a bar of soft iron when it is brought near a magnet or in contact with it.

II. NATURE OF MAGNETISM.

370. Magnetism a Molecular Phenomenon. — **Experiment.** — Magnetize a piece of watch spring, then heat it red hot and test it for magnetism. It will be found to have lost the power of attracting iron filings.

Experiment. — Magnetize a knitting-needle and find by averaging several trials how many tacks can be lifted by it. Now hold one end firmly against the edge of the table or in a vise and, by plucking the free end, cause the needle to vibrate vigorously for a few seconds. The power of the magnet to pick up tacks will be found to be appreciably lessened by the vibration.

Experiment. — Take a piece of iron wire, about 30 cm. long and 1.5 mm. diameter, and carefully anneal it. Bend it to the form shown in Fig. 214. Stroke it carefully several times with a strong magnet. It will be a weak magnet. (How shown?) Now hold it by the turned-up ends and give the wire a sudden twist. If the wire is again tested for magnetism it will be found to have lost nearly all.

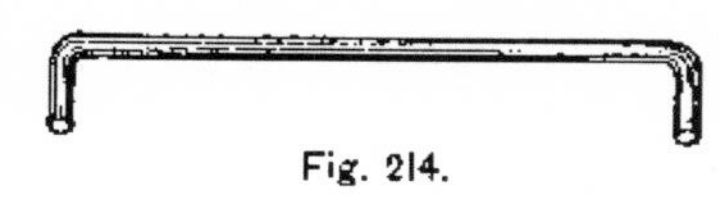

Fig. 214.

In each of the preceding experiments the molecular arrangement has been disturbed; and it is interesting to note that in each the magnetism has been weakened.

The conclusion is that magnetism is connected in some way with the molecular arrangement of the iron or steel.

371. Further Evidence. — **Experiment.** — Magnetize a piece of straightened watch spring; notice that it has two poles, one at each end, the centre being neutral. Break it at the neutral point; each piece will be found to have two poles, two new ones having been formed at the point that was formerly neutral. If these pieces in turn be broken, their parts will be magnets with two poles like the original.

There is no apparent limit to the extent to which this process may be carried, indicating that possibly if carried as far as the molecules, they too would prove to be magnets.

Experiment. — Fill a slender glass tube nearly full of steel filings, closing the ends with cork. Stroke the tube from end to end with one pole of a strong magnet; the filings acquire magnetic properties. Shake up the filings thoroughly; all polarity is lost.

A minute examination of each steel particle will show that it is a magnet. The loss of polarity is evidently due to the neutralization of the actions of many little magnets by disturbing their arrangement. Undoubtedly the polarity would be restored if the particles could be restored to their original positions. The experiment strongly supports the theory that each molecule of a magnetic substance is a magnet.

It is worthy of notice that magnetization is facilitated by jarring the substance, or by heating it and then cooling it, while under the magnetizing influence.

372. Retentivity. — **Experiment.** — Prepare three bars of the same size, one each of soft iron, soft steel, and hard steel. Dip one end of each in succession into iron filings, and bring a strong magnet in contact with the other end. When the bars are withdrawn and the magnet is removed, most of the filings drop from the iron, the hard steel retaining the largest number.

The difference exhibited by these substances is due to what is called *retentivity*, or the ability to retain magnetism.

373. Theory of Magnetism. — Experiments like the preceding ones lead to the conclusion that magnetism is a molecular phenomenon. The most probable theory is that the individual molecules of a magnetic substance are always magnetized. In an unmagnetized bar the poles of these molecular magnets are turned in all directions, or else the little magnets form stable combinations or closed chains, so that no magnetism external to the bar is exhibited. When they are turned by an external magnetizing force, so that a certain portion of the molecules have their poles pointing in the same direction, then the bar is magnetized. The larger the proportion of the molecules which have their molecular axes turned in the same direction, the stronger the magnet.

The molecules of soft iron are readily turned by a magnetizing force; but when this force is withdrawn, they revert to the unmagnetized state. Hardened steel, on the other hand, requires a greater magnetizing force to shift the magnetic axes of the molecules, but once shifted they remain in the new position, and the bar is permanently magnetized.

III. THE MAGNETIC FIELD.

374. Lines of Magnetic Force. — **Experiment.** — Place a sheet of paper or glass over a small bar magnet and sift iron filings evenly over it from a muslin bag, tapping the paper or glass gently to aid the filings in arranging themselves under the influence of the magnet. They will cling together in curved lines, which diverge from one pole of the magnet and meet again at the opposite pole.

These lines are called *lines of magnetic force* or of *magnetic induction.* Each particle of iron becomes a magnet by induction ; hence the lines mapped out by the filings are the lines along which magnetic induction takes place. Figure 215 was made from a photograph which was taken by sifting iron filings on the sensitized side of a photographic plate with a piece of magnetized watch spring under it. The plate was then exposed for about a second to the light of an incandescent lamp, and was developed in the usual way. These lines of force spring from the north pole, curve round through the air to the south pole, and complete their circuit through the magnet itself.

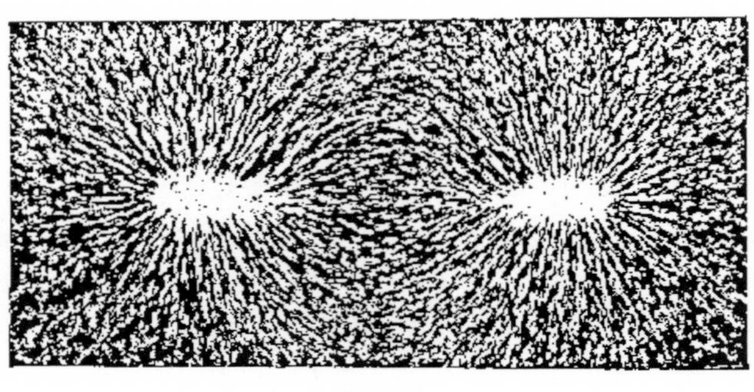

Fig. 215.

Figure 216 was made from two magnets with their unlike poles turned toward each other. The lines of force from the north pole of one extend across to the south pole of the other. Lines of magnetic force are always to be considered as under tension and as possessing elasticity. Figure 216 is therefore a picture of attraction.

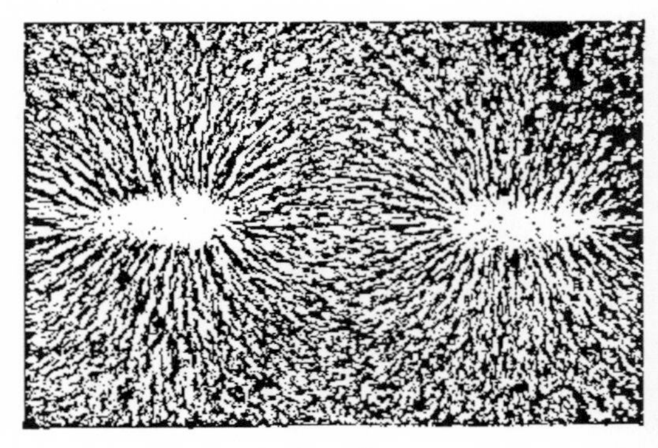

Fig. 216.

Figure 217 was made from two magnets with their like poles turned toward each other. None of the lines springing from one of them enter the other. This figure is a picture of magnetic repulsion.

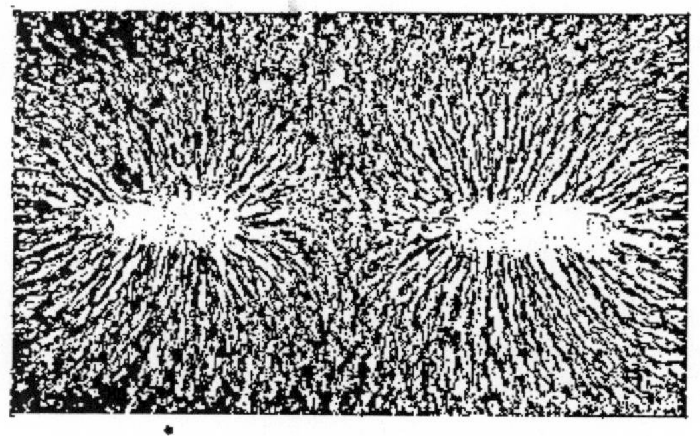

Fig. 217.

375. Magnetic Field. — The region around a magnet, or a space within which there are lines of magnetic force, is called a *magnetic field of force*. Faraday introduced the method of studying magnetic fields by means of the lines of force illustrated above.

376. Direction of Lines of Force. — The direction of a line of force at any point is that of a line drawn tangent to the curve at the point ; the direction along it is the same as that in which a north pole is urged. The north pole of a magnetic needle is repelled from the north pole of a bar magnet. Hence, if an observer stands with his back to the north pole of a magnet, he is looking in the direction of the lines of force coming from that pole.

377. Permeability. — Experiment shows that when iron is placed in a magnetic field, the lines of force are concentrated by it. This property possessed by iron, when placed in a magnetic field, of concentrating the lines of force and increasing their number is known as *permeability*. The superior permeability of soft iron explains the action of magnetic screens (§ 364). In the case of the watch shield, the lines of force pass through the iron and not across it; the watch is thus protected from magnetism because the lines of force do not enter it.

IV. TERRESTRIAL MAGNETISM.

378. The Earth a Magnet. — **Experiment.** — Place a long bar magnet on the table and suspend over it a magnetic needle mounted so as to turn readily about a horizontal axis. When over the north-seeking pole, the needle will be vertical, with its south-seeking pole down; when over the middle, it is horizontal; and when over the south-seeking pole, the needle is again vertical, but with its north pole pointing downward.

The earth acts in a similar manner toward such a needle when moved over its surface from pole to pole. At a point in Boothia Felix, west of Baffin's Bay, the needle is nearly vertical, with its north pole down; at points successively farther south, it dips less and less toward the vertical, becoming horizontal near the earth's equator, and again gradually inclining toward the vertical, with its south pole down as it nears the south magnetic pole of the earth. If a bar magnet, about half the length of the earth's diameter, were thrust through the earth's centre, making an angle of about 20° with its axis, it would account for many of the phenomena of terrestrial magnetism.

379. Earth's Induction. — **Experiment.** — Procure a thoroughly annealed iron bar about 75 cm. long, showing little or no polarity when tested with a magnetic needle while the bar is supported horizontally in an east-and-west line. Hold this bar in a meridian plane, but with its north end dipping down some 70° below the horizontal. Tap the end of the bar with a hammer and then test for polarity. The lower end will be found strongly north-seeking, and the upper end south-seeking. If the bar is turned end for end, and again tapped with a hammer, the lower end again becomes north-seeking.

The experiment illustrates the inductive action of the earth. If we examine any iron object which has remained undisturbed for some time, as a stove or a supporting column of a building, we shall find that it is polarized like the bar of the above experiment. The inductive action of the earth probably accounts for the existence of natural magnets.

380. Magnetic Dip. — Experiment. — Thrust through a cork an unmagnetized knitting-needle, and at right angles to this two short pieces (Fig. 218). Support the apparatus on the edges of two wine glasses, with the axis in an east-and-west line, and the needle adjusted so as to rest horizontally. Now magnetize the needle, being careful not to displace the cork. It will no longer assume a horizontal position, the north pole dipping down as if it had become heavier.

Fig. 218.

The angle made by this needle with a horizontal plane is called the *inclination* or *dip* of the needle. A magnetic needle mounted so as to move freely in a vertical plane, and provided with a graduated arc for measuring the inclination, is called a *dipping needle* (Fig. 219).

Fig. 219.

The magnetic poles of the earth are points where the dip is 90°; the dip at the magnetic equator is 0°. Lines on the earth's surface, passing through points of equal dip, are called *isoclinic lines;* they are irregular in direction, though resembling somewhat parallels of latitude.

381. Declination. — The magnetic poles do not coincide with the terrestrial poles, and consequently in most places the direction of the magnetic needle is not that of the meridian of the place. The direction of the magnetic needle at any place is that of the magnetic meridian of the place. The *declination* of the needle is the angle between the magnetic and the geographical meridian.

382. The Line of no Declination passes through those places where the needle points true north. Such a line, in 1900, ran from the north magnetic pole across the eastern end of Lake Superior, through Lansing, Mich., Columbus, Ohio, through West Virginia and South Carolina, and left the mainland at Charleston, on its way to the south magnetic pole. The returning line through the eastern hemisphere is quite irregular in direction. At places east of this line the needle points west of north, and west of the line it points east of north. Lines passing through points of equal declination are called *isogonic lines.*

V. ELECTRIFICATION.

383. Electrical Attraction. — **Experiment.** — Cut a number of small balls out of the pith of common elder. Place them in a pile on the table and touch them with a rod of sealing-wax. Notice that the rod does not affect the balls in the least. Now rub the rod with dry flannel and again bring it up to the pile of balls. They will be alternately attracted and repelled.

Rods of glass, shellac, sulphur, very dry wood, ebonite, etc., may be substituted for the sealing-wax, and a collection of any light objects, bits of tissue paper for example (Fig. 220), for the pith-balls.

Fig. 220.

Bodies which exhibit the power of attracting light bodies after being rubbed are said to be *electrified*. *Electrification* may be brought about in a variety of ways in addition to friction, as will appear in the course of this chapter.

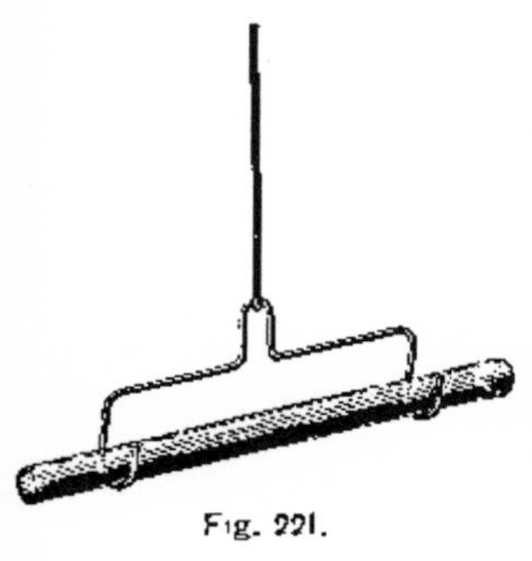
Fig. 221.

384. Attraction Mutual. — Experiment. — Prepare a glass tube about 2 cm. in diameter and 40 cm. long. Remove all sharp corners by fusion in the flame of a blowpipe. Electrify the tube by friction with a piece of silk, and hold it near the end of a long wooden rod resting in a wire stirrup suspended by a silk thread. The suspended rod is attracted. Now, replace the rod by the electrified tube (Fig. 221). When the rod is held near the rubbed end of the glass tube, the latter moves as if attracted by the former.

The experiment teaches that each body attracts the other; that is, that *the action is mutual.*

385. Electrical Repulsion. — Experiment. — Suspend several pith-balls by fine linen threads from a glass rod, and touch them with an electrified glass tube (Fig. 222). At first they are

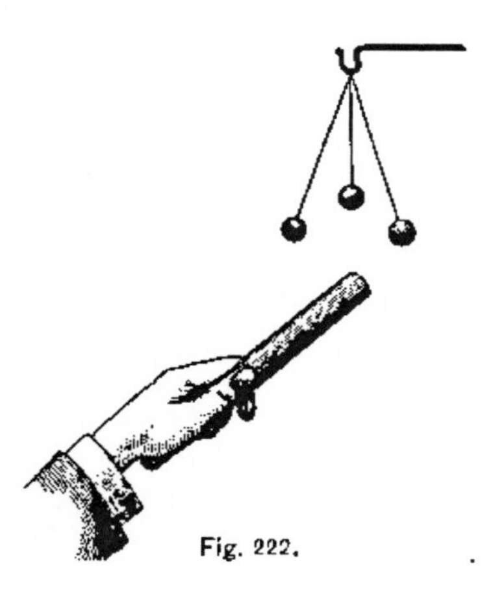
Fig. 222.

attracted, but they soon fly away from the tube and from one another. When the tube is removed to a distance, the balls no longer hang side by side, but keep apart for some little time. If we bring the hand near the balls they will move toward it as if attracted, showing that the balls are electrified.

It thus appears that bodies become electrified by coming in contact with electrified bodies, and that electrification may show itself by repulsion as well as by attraction.

386. Two Kinds of Electrification. — **Experiment.** — Rub a glass tube with silk and suspend it as in Fig. 221. Excite a second glass tube and hold it near one end of the suspended one. The suspended tube will be repelled. Bring near the suspended tube a rod of sealing-wax excited by friction with flannel. The suspended tube is now attracted. Repeat these tests with an electrified rod of sealing-wax in the stirrup instead of the glass tube. The electrified sealing-wax will repel the electrified sealing-wax, but there will be attraction between the sealing-wax and the glass tube.

The experiment shows that there are *two kinds of electrification:* one developed by rubbing glass with silk, and the other by rubbing sealing-wax with flannel. In the former case the body is said to be *positively* electrified; in the latter case *negatively* electrified.

387. First Law of Electrostatic Action. — It was seen in the last experiment that there was repulsion between the electrified glass tubes, and that the electrified sealing-wax attracted the electrified glass. These facts are expressed by the following law: —

Electrical charges of like sign repel each other; electrical charges of unlike sign attract.

388. The Electroscope, as the name implies, is an instrument for detecting electrical charges. The most common form is the *gold-leaf electroscope.* It consists of a glass flask, through the top of which passes a brass rod terminating in a ball on the outside (Fig. 223), and two strips of thin gold or aluminum foil on the inside, hanging parallel and close together. If an electrified object is brought in contact with the rod, the metal strips become similarly charged, and are repelled from each other.

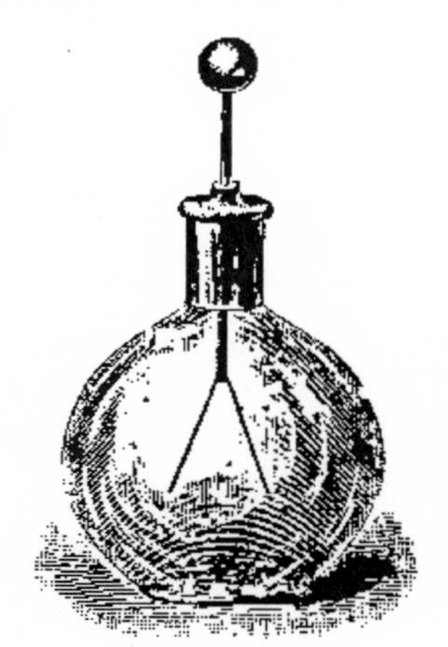
Fig. 223.

389. Use of the Electroscope. — In order to determine the kind of electrification on a body, a *proof-plane* may be used. It is a small metal disk cemented to one end of an ebonite or shellac rod (Fig. 224). Charge the electroscope by touching its knob with a glass rod excited by friction with silk. Slide the metal disk of the proof-plane along the surface to be tested, and then bring it near the ball of the electroscope. If the leaves diverge further, the body in question is positively charged (§ 394); if they diverge less, the charge is probably negative. An increase of divergence is a more reliable indication than a decrease, because the divergence will decrease when the proof-plane is not charged (§ 406). The test may be varied by touching the charged proof-plane to the knob of the electroscope. In either case, if there is a decrease in the divergence of the leaves, repeat the test by charging the electroscope by means of a stick of sealing-wax rubbed with flannel. If the leaves now diverge, the body in question is negatively electrified.

Fig. 224.

390. Simultaneous Development of the Two Electrifications.— Experiment.— Fit to the end of a rod of sealing-wax a cap of flannel, three or four inches long, with a silk cord attached to draw it off (Fig. 225). Electrify the rod by turning it around inside the cap, and then hold it near the knob of the electroscope. No divergence will be observed. By the aid of the cord remove the flannel cap, and present it to the positively charged electroscope. The increased divergence of the leaves shows that the cap is positively charged. If we test the rod of sealing-wax in the same way, it will be found to be negatively charged.

Fig. 225.

The experiment shows (1) *that one kind of electrification is not developed without the other; and* (2) *that the two kinds of electrification are produced in equal quantities*, as demonstrated by the fact that the quantity in the rubber exactly neutralizes that on the rod when the two are in contact. The two charges behave like equal positive and negative quantities.

391. Conductors and Non-conductors.— Experiment.—Fasten a smooth metallic button to a rod of sealing-wax. Connect it with the knob of the electroscope by a fine copper wire, 50 to 100 cm. long. Hold the sealing-wax in the hand and touch the button with an electrified glass tube. The divergence of the leaves indicates that they are electrified. If we repeat the experiment, using a silk thread in place of the wire, no effect will be produced on the leaves.

All substances may be roughly classed under two heads, *conductors* and *non-conductors.* In the former if one point of the body is electrified by any means, the electrification spreads over the whole body, but in a non-conductor the electrification is confined to the vicinity of the point where it is excited. Non-conductors are commonly called *insulators.* Substances differ greatly in their conductivity, so that it is not possible to divide them sharply into two classes. Metals, carbon, and the solution of some acids and salts are the best conductors. Among the best insulators are paraffin, turpentine, silk, sealing-wax, india-rubber, gutta-percha, dry glass, porcelain, mica, shellac, spun quartz fibres, and liquid oxygen. Some insulators, like glass, become good conductors when heated to a semi-fluid condition.

392. Probable Nature of Electrification. — It was suggested by Faraday, and a multitude of facts tend to confirm his view, that the electrification of a body is a strained condition of the ether which surrounds it and pervades it. Conductors differ from insulators in this: in the former, the molecular mobility is such that this state of strain is continually giving way, while in the latter considerable distortion is possible before the molecular structure yields to the strain. The phenomena of attraction and repulsion exhibited by electrified bodies are due to the attempt of the strained ether in and around the bodies to return to its normal condition. In producing electrification, work is done in distorting the medium; hence electrification is a form of potential energy.

VI. ELECTROSTATIC INDUCTION.

393. Electrification by Induction. — **Experiment.** — Excite a glass tube by friction with silk. Bring it gradually near the ball of an electroscope. The leaves begin to diverge when the tube is some distance from the knob, and the amount of divergence increases as the tube is brought nearer. When the tube is removed the leaves collapse.

It is evident, since the leaves do not remain apart, that there has been no transfer of electrification from the tube to the electroscope. The electrified condition, produced in the electroscope when the electrified body is brought near it, is due to what is called *electrostatic induction.* Why such an effect should occur is easily understood when we recall Faraday's views of electrification, that it is a distortion of the ether about the body. Evidently, then, any body placed within this *electrical field* would be electrified.

394. Charging a Body by Induction. — **Experiment.** — Support a smooth metallic ball on a dry plate of glass. Connect it with the knob of the electroscope by means of a metallic wire, the ends of which are bent into a loop and smoothly soldered. The ball and the electroscope now form one continuous conductor. Bring near the ball an electrified glass tube; the leaves of the electroscope diverge. Before removing the excited tube, remove the wire, handling it with some non-conductor. The electroscope remains charged, and it will be found to be positive. A similar test made of the ball will show that it is negatively charged. Repeat the experiment without removing the connecting wire. There are no signs of electrification after removing the excited tube.

Hence, we learn that *when an electrified body is brought near an object it induces the opposite kind of electrification on the side next it and the same kind on the remote side.*

395. The Inducing Charge equal to the Induced Charge. — **Experiment.** — Support a metallic vessel, like the one shown in Fig. 226, on a glass plate and connect it with the knob of an electroscope by a fine wire. Attach a silk thread to a metallic ball about an inch in diameter, and charge the ball, holding it by the silk thread. Lower the charged ball into the insulated vessel and observe that the leaves of the electroscope diverge as the ball enters the vessel. The divergence increases till the ball has been lowered perhaps two inches below the top, and then remains the same, even when the ball touches the bottom and communicates its charge to the insulated vessel. Suppose the ball charged positively; it induces a negative charge on the interior of the vessel and repels a positive charge to the outside. This positive charge is equal to the charge on the ball, for the divergence of the leaves does not change when the ball gives up its charge to the vessel. The charge on the ball neutralizes the equal negative charge on the interior, leaving the equal positive charge on the exterior.

Discharge the electroscope and charge the ball a second time. After it has been lowered into the insulated vessel without touching it, place the finger on the ball of the electroscope; the leaves will collapse. Remove the finger and lift the ball by the silk thread; the leaves will again diverge. Lower the ball again till it touches the vessel, and the leaves will again collapse. The charge induced on the inside is exactly neutralized by the inducing charge on the ball.

Hence, *the induced and the inducing charges are equal to each other.*

396. Charging an Electroscope by Induction.— Experiment.— Hold one finger on the ball of the electroscope and bring near it an electrified glass tube. Remove the finger before taking away the tube and the electroscope will be charged. Explain. What kind of electrification will then be on the electroscope? How can you modify the intensity of the charge?

VII. ELECTRICAL DISTRIBUTION.

397. The Charge on the Outside of a Conductor. — Experiment. — Place a cylindrical metallic vessel of about one litre capacity on an insulated support (Fig. 226). A vessel free from sharp edges should be selected. Electrify strongly and test in succession both the inner and the outer surface, using a proof-plane to convey the charge to the electroscope. It will be found that the inner surface gives no sign of electrification.

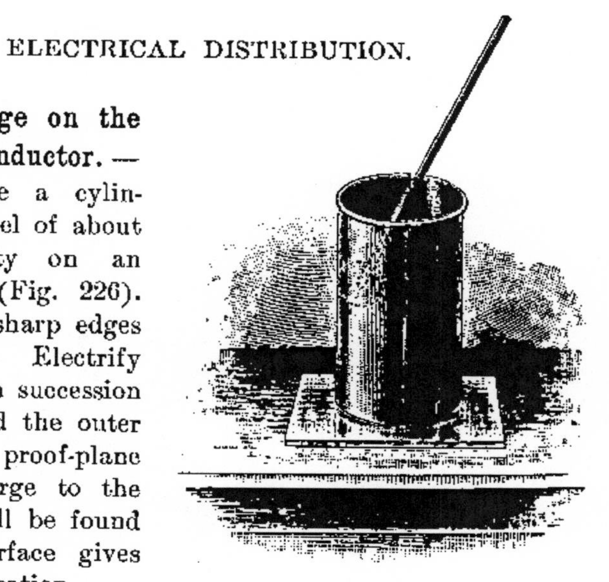

Fig. 226.

Hence, it appears that *the electrical charge of a conductor is confined to its outer surface.*

398. Effect of Shape. — **Experiment.** — Charge electrically an insulated egg-shaped conductor (Fig. 227). Touch the proof-plane to the large end, and convey the charge to the electroscope. Notice the amount of separation of the leaves. Test the small end of the conductor in the same way. A greater divergence of the leaves will be observed in the latter case.

Fig 227.

The distribution of the charge is, therefore, affected by the shape of the conductor, the surface density being greater the greater the curvature. By *surface density* is meant the quantity of electrification on a unit area of the surface of the conductor. The experiment shows that the surface density is greatest at the small end of the conductor.

399. Effect of Area. — **Experiment.** — Employing an electroscope provided with a disk instead of a ball, or with an insulated disk connected with the ball by a conductor, place on the disk a chain, and charge the electroscope by induction so that the leaves diverge widely. Now lift the chain by a dry glass rod, so as to increase the surface of the conductor. The leaves of the electroscope will slowly collapse. When the chain is lowered they will again diverge.

Hence, *with a given charge, the larger the surface the smaller the surface density.*

400. Action of Points. — **Experiment.** — Cement the middle of a sewing-needle to a stick of sealing-wax, as an insulator, so that the needle and the wax form a T. Place the eye end of the needle against the ball of the electroscope, and hold over and near its point an excited glass tube. When both are removed to a distance, the leaves of the electroscope remain separated. The approach of the excited tube will increase their divergence, showing that they are positively charged.

Discharge the electroscope and hold the eye end of the needle against the brass lower down where it joins the glass of the instrument. Now bring the excited glass tube near the top of the ball. When the tube is withdrawn the leaves will again diverge. The divergence decreases upon the approach of the excited tube, showing that the electroscope is now charged negatively. In the first case the attracted negative passed off from the electroscope by the point of the needle, leaving the leaves with a positive charge. In the second case the repelled positive was discharged into the air by the needle point, leaving the electroscope with a negative charge.

From these experiments it appears that electrification is discharged by pointed conductors. This conclusion might have been drawn from § 398. When the curvature becomes very great and the surface assumes the shape of a point, the surface density becomes very great. The particles of air near the point become highly charged and are then repelled. The charge is thus conveyed away, and the air current produced is called an *electric wind* (§ 417).

VIII. ELECTRIC POTENTIAL AND CAPACITY.

401. The Unit of Electrification or Charge. — The electrification of a conductor is capable of exact measurement, though it is known only by the phenomena it presents. The unit of electric quantity may be readily defined. Imagine two minute bodies similarly charged with equal quantities. They will repel each other. If the two equal and similar charges are one centimetre apart in air, and if they repel each other with a force of one dyne, then the charges are both unity. *The electrostatic unit of charge is that quantity which will repel an equal and similar quantity at a distance of one centimetre in air with a force of one dyne.* It is necessary to say "in air" because, as will be seen later, the force between two charged bodies depends on the nature of the medium between them (§ 407).

402. Coulomb's Law. — Coulomb demonstrated that the force exerted on each other by two charged conductors, the size of which is very small in comparison with the distance between them, is directly proportional to the product of the charges, and inversely proportional to the square of the distance between them.

Let q and q' be the number of units of charge on the two bodies respectively, and let d be the distance between them in centimetres. Then the force which they exert on each other in dynes is given by the equation

$$F = \frac{qq'}{d^2}. \qquad (28)$$

If the charges are of the same sign, the force will be a repulsion; if they are of opposite sign, it will be an attraction.

403. Difference of Potential. — Imagine two conductors similarly charged. Work will have to be done to bring one of them nearer the other. The force worked against is the force of repulsion between like charges. *The difference of potential between the two conductors is the work required to transfer a unit quantity of electrification from one of the conductors to the other.*

If two similar conductors, which are unequally charged, are brought into contact, or are connected by a wire, a part of the charge on the conductor with the higher charge will pass to the one with lower charge till equilibrium is established. The conductor which parts with some of its charge is said to be of the higher potential. Difference of electric potential is, therefore, like difference of temperature for heat, or difference of level for water. As heat flows from bodies of higher to bodies of lower temperature, and as water flows from a higher to a lower level, so positive electrical charges flow from conductors of higher to conductors of lower potential.

404. Zero Potential. — In measuring the potential difference between a conductor and the earth, the potential of the earth is assumed to be zero. The potential difference is then numerically the *potential of the conductor*. If a conductor of positive potential be connected with the earth by an electric conductor, the positive charge will flow to the earth. If the conductor has a negative potential, the flow of the positive quantity will be in the other direction.

405. Electrostatic Capacity. — **Experiment.** — Suspend a small, smooth metallic ball by a silk thread. Charge a gold-leaf electroscope till the leaves diverge widely. Bring the small ball in contact with the knob of the electroscope; the leaves will partly collapse, showing that the potential has been lowered.

If the charge on an insulated conductor is doubled, the force on a unit charge, anywhere in the neighborhood of the conductor, will also be doubled. The work required to bring a unit charge from the earth, whose potential is zero, to the conductor will then be doubled also. This means that the potential of the conductor is doubled by doubling the charge. There is, therefore, a constant ratio between the charge on a conductor and its potential. This ratio is its *electrostatic capacity*. In symbols, if a charge Q raises a conductor to the potential V, its capacity is

$$C = Q/V. \qquad (29)$$

An equivalent definition of *capacity is the charge required to raise the potential of a conductor from zero to unity.* From (29) $Q = CV$, and $V = Q/C$.

Electric potential is analogous to pressure in a tank containing a gas. The capacity of the tank is the quantity of gas which it holds when the pressure is one atmosphere. The whole quantity of gas is proportional to the pressure, or $Q = CP$, and $C = Q/P$, relations precisely like those expressed by equation (29).

The experiment shows that the capacity of a conductor is increased by increasing its surface, since the potential decreased when the surface was augmented, the charge remaining the same. Conversely, with a larger surface a greater charge is required to raise the conductor to the same potential.

406. Condensers. — **Experiment.** — Support a metal plate in a vertical position by means of two silk cords. Connect it with the knob of an electroscope by a fine copper wire. Charge the plate till the leaves of the electroscope show a wide divergence. Now bring an uninsulated conducting plate near the charged one and parallel to it. The divergence of the leaves will decrease; remove the uninsulated plate, and the divergence will increase again.

The capacity of an insulated conductor is not dependent on its dimensions alone, but it is increased by the presence of another conductor connected with the earth. The effect of this latter conductor is to decrease the potential to which a given charge will raise the insulated one (§ 405). Such an arrangement of parallel conductors separated by an insulator or *dielectric* is called a *condenser*.

A condenser is a device which greatly increases the charge on a conductor without increasing its potential. In other words, the plate connected with the earth greatly increases the capacity of the conductor.

407. Influence of the Dielectric. — **Experiment.** — With the apparatus of the last experiment, and with the uninsulated plate at a convenient distance from the charged plate and parallel to it, thrust suddenly between the two a cake of clean paraffin as large as the metal plates or larger, and from 2 to 4 cm. thick. Note that the leaves of the electroscope collapse slightly. Remove the paraffin quickly, and the divergence will increase again. A cake of sulphur from 5 to 10 cm. thick will produce a marked effect on the divergence of the leaves.

The presence of the paraffin or the sulphur increases the capacity of the condenser and, hence, decreases its potential, the charge remaining the same. Paraffin and sulphur, as examples of dielectrics, are said to have a larger *dielectric capacity* or *dielectric constant* than air. Glass has a dielectric capacity from four to ten times greater than air.

408. The Leyden Jar is a common and convenient form of condenser. It consists of a glass jar coated part way up, both inside and outside, with tin-foil (Fig. 228). Through the wooden or ebonite stopper passes a brass rod, terminating on the outside in a ball and on the inside in a metallic chain which reaches the bottom of the jar. The inner foil represents the collecting surface, and the glass the dielectric separating the two conductors.

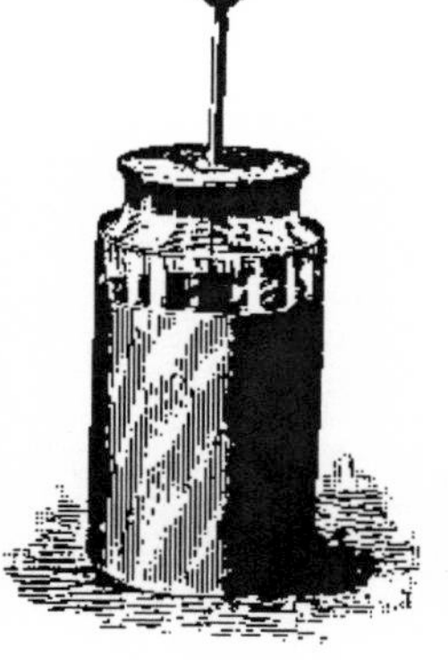

Fig. 228.

409. Charging and Discharging Jars. — To charge a Leyden jar connect the outer surface with the earth, either by a metallic conductor or by holding the jar in the hand. Place the ball in contact with the source of electrification, as, for example, the conductor of an electrical machine (§ 414). To discharge a Leyden jar bend a wire into the form of the letter V. With one end of the wire touching the *outer* surface of the jar (Fig. 229) bring the other around nea the ball, and the discharge wil take place.

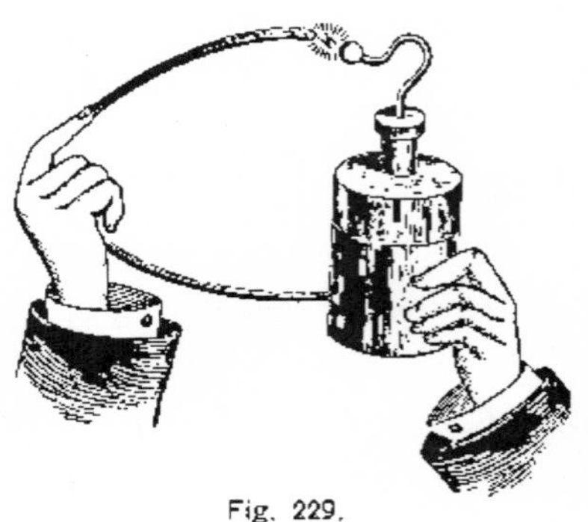

Fig. 229.

410. Action of Jar Explained — When a positive charge i communicated to the inne surface of a Leyden jar it act inductively through the glass attracting to the outer surfac an equal negative charge an repelling positive to the earth. These two charges ac inductively on each other through the glass and are sai to be "bound," in distinction from the condition wher an electrified conductor is at some distance from any othe conductor, in which case the whole charge is "free." When the outer surface of the jar is connected with th earth, the electrical capacity of the inner coating of the jar is largely increased. (Why?)

Fig. 230.

411. Seat of Charge. — **Experiment.** — Charge a Leyden jar made with movable metallic coatings (Fig. 230) and set it on an insulating stand. Lift out the inner coating, and then, taking the top of the glass vessel in one hand, remove the outer coating with the other. The coatings now exhibit no sign of electrification. Bring the glass vessel near a pile of pith-balls; they will be attracted to it, showing that the glass is electrified. Reach over the rim with the thumb and forefinger and touch the glass. A slight discharge ma be heard. Now build up the jar by putting the parts together; the ja will still be highly electrified and may be discharged in the usual way

This experiment, due to Franklin, shows that the electrification is a phenomenon of the glass. Faraday proved that during the act of charging the jar the glass is strained, the office of the conductors being to facilitate the release from strain. This view is supported by the facts that thin jars can be broken by overcharging; that a jar enlarges on charging; that on heating a jar its charge disappears; and that on charging a jar heavily and then discharging in the usual way, a second charge accumulates after a few minutes, the time being lessened by tapping the jar. The glass acts as if it were strained or distorted to so great a degree that, like a twisted glass fibre, it does not return at once to its normal state when released. This second charge is called the *residual charge.*

Questions and Problems.

1. Stand a charged Leyden jar on a cake of paraffin. Touch the ball; it is not discharged. Why?

2. Connect the inner surface of an uncharged Leyden jar to an electroscope, and insulate the outer surface from the earth. Notice that a very small quantity of electricity causes a violent separation of the leaves. Discharge and connect the outer coating to the earth. A much greater quantity is needed to affect the electroscope to the same extent as before. Explain.

3. Charges of 30 and 40 positive units are given respectively to two small spheres separated by a distance of 20 cm. in air. What is the force of repulsion between them?

4. What is the distance between two small bodies charged respectively with 60 and 73.5 units which repel each other in air with a force of 10 dynes?

5. Two small spheres, A and C, charged respectively with 64 and 144 positive units, are placed 20 cm. apart. Show where a third charge B must be placed, between A and C, to be in equilibrium.

Ans. 8 cm. from A.

IX. ELECTRICAL MACHINES.

412. Friction Machines. — We have thus far refrained from describing any device for obtaining electrification beyond such simple means as rubbing a glass tube with silk or sealing-wax with flannel. The explanation of modern electrical machines involves some knowledge of electrostatic induction; it has therefore been deferred till the subject of induction has been introduced.

The oldest forms of apparatus for producing electrification by mechanical means consisted of a glass cylinder or plate, mounted so as to be capable of rotating, and provided with silk-covered pads pressing against the glass. When the glass was revolved the friction electrified it positively. Opposite the rubbers was an insulated conductor with sharp points just grazing the surface of the glass. The attracted negative charge passed off these points and neutralized the positive on the glass, leaving the insulated conductor positively electrified.

These machines were very unsatisfactory, and they have been long displaced by machines which depend for their action on the inductive influence of an electrified body on an insulated conductor.

413. The Electrophorus. — The simplest induction (or influence) electrical machine is the *electrophorus*, invented by Volta. It is shown in Fig. 231. A cake of resin or disk of vulcanite rests in a metallic base. Another metallic disk or cover is provided with an insulating handle. The resin or vulcanite is electrified by rubbing with dry flannel or striking with a catskin, and the metal disk is then placed on it. Since the cover touches the nonconducting sole in a few points only, the negative charge due to the friction, is not removed. The two disks with the film of air between them form a condenser of great capacity. Touch the cover momentarily with the finger, and the repelled negative charge passes to the earth, leaving the cover at zero potential. Lift it by the insulating handle, the positive charge becomes free, and a spark may be drawn by presenting the finger. This operation may be repeated an indefinite number of times without reducing the charge on the vulcanite.

Fig. 231.

When the cover under inductive influence from the base is touched, it possesses no potential energy. (Why?) But when it is lifted by the insulating handle, work is done against the electrical attraction between the negative charge on the vulcanite and the positive on the cover. The energy of the charged cover represents this work. The electrophorus is, therefore, a device for the continuous transformation of mechanical work into the energy of electric charges.

414. Influence Electrical Machines. — The influence or induction electrical machine has undergone many modifications since its first introduction. It must suffice here to describe only one machine.

The Holtz machine, as modified by Toepler and Voss,

is illustrated in Fig. 232. There are two glass plates, e' and e, about 5 mm. apart, the former stationary and the latter turning about an insulated axle by means of the handle h and a belt. The stationary plate supports at the back two paper sectors, c and c', called *armatures*. Underneath them are disks of tin-foil connected by a narrow strip of the same material. The disks are electrically connected with two bent metal arms, a and a' (opposite a), which carry tinsel brushes long enough to rub against low

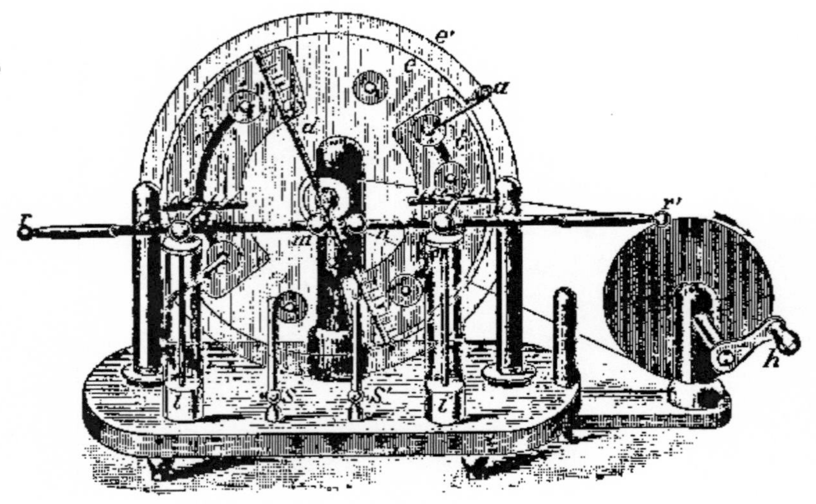

Fig. 232.

brass buttons cemented to small tin-foil disks, called *carriers*, on the front of the revolving plate. Opposite the paper sectors and facing them are two combs with several sharp-pointed teeth set close to the revolving plate, but not touching the metal buttons and carriers. The diagonal neutralizing rod d has tinsel brushes in addition to the combs. The two insulated conductors, terminating in the balls, m and n, have their capacity increased by connection with the inner coating of two small Leyden jars, i and i'; the outer coatings are connected under the base of the machine.

415. Action of the Machine. — The operation of all influence machines depends on the employment of a small initial charge to act inductively on the conductors or carriers. The attracted charges pass off by the sharp points of the combs or by the tinsel brushes, and are conveyed away by the carriers and the revolving plate. They go to increase the initial charges and to electrify the insulated conductors.

Suppose the two armatures slightly charged, the one on the left positively. (There is usually enough excitation due to friction, or to the contact of dissimilar substances, to furnish the very slight initial charges.) The brushes on the neutralizing rod d are set so as to connect two carriers at opposite ends of the rod just before they pass beyond the influence of the armatures, c and c'. They thus acquire by induction negative and positive charges respectively, which they carry forward till they are brought into momentary connection with the armatures by means of the brushes and the bent rods a and a'. They then deliver to them their small charges. This action is repeated by each pair of carriers twice during each revolution. In this way the armature c becomes more highly positive and c' more highly negative. When the armatures are highly electrified the carriers do not give up their entire charge to them; the collecting combs attached to the rods m and n receive the residue, in addition to the charges carried on the glass. The positive charge on m and the negative on n increase till a spark passes between the balls, or till further accumulation is prevented by leakage.

X. EXPERIMENTS WITH ELECTRICAL MACHINES.

416. Attraction and Repulsion. — **Experiment.** — Charge the electrophorus as directed. Before lifting the plate, place a handful of small bits of paper on it. Why do they fly off when the plate is lifted?

Experiment. — Support a metallic plate on a block of wood; on it place the glass cylinder of § 152, and let a second metallic plate rest on this glass. Connect the bottom plate to one conductor of the induction machine and the top one to the other. In the glass vessel put a handful of pith-balls. Work the machine, and account for the dancing of the pith-balls.

Experiment. — Support a small glass funnel, having an aperture of about one-eighth of an inch, in some suitable way, and fill it with fine dry sand. Notice that the sand runs out in a smooth stream. Pass one end of a wire into the sand in the funnel, and connect the other end with one of the conductors of an electrical machine. Set the machine in action, and observe that the sand of the escaping stream scatters. Explain.

417. Discharging Points. — **Experiment.** — Suspend two pith-balls side by side from one of the conductors of an electrical machine. When the plate is turned the balls separate widely. Why? Now hold a pointed rod near the conductor; the balls drop, showing that the conductor is discharged. Explain.

Experiment. — Fasten a cambric needle to one of the conductors of an electrical machine, so that the point projects. Try to charge the machine. Account for the failure to obtain sparks between the conductors. Hold the flame of a candle near the point. It is driven away as by a wind. Account for this air current.

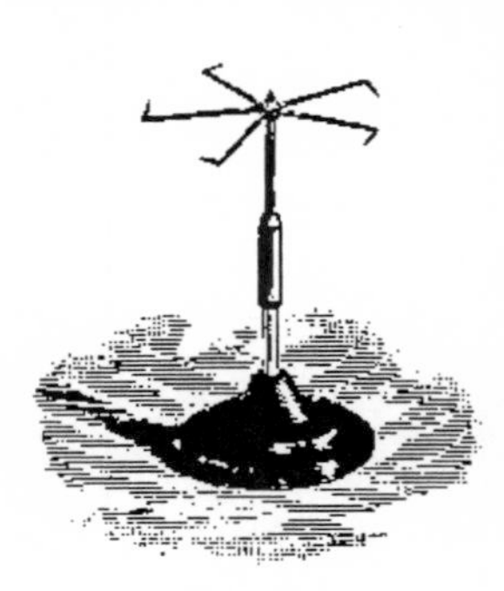

Fig. 233.

Experiment. — Connect an *electric tourniquet* (Fig. 233) to one of the conductors of an electrical machine, the other one being grounded. Notice the shape of its arms. Work the machine; the whirl rotates rapidly. Why?

Experiment. — Obtain two smooth round metal plates 10 or 15 cm. in diameter. Lay one of them on the top of a tumbler, and support the other a few centimetres above the first by means of a stick of sealing-wax cemented to its centre. Connect the two plates by wires or chains with the positive and negative conductors respectively of an electrical machine. Place a small bullet or a piece of metal shaped like a collar button on the lower plate. When the machine is worked, sparks will pass across between the elevated metal and the upper plate. They will be heavier and brighter if the plates are connected, the one with the outer and the other with the inner coating of a Leyden jar (Fig. 234).

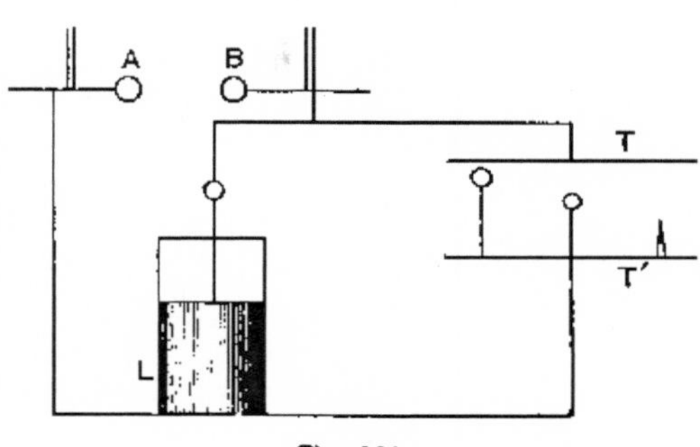

Fig. 234.

Now run a pin through a small cork so that it will stand on the lower plate with its point up. No sparks now pass across the space between the insulated plates. The sharp pin protects the projecting metal, even when it is higher than the pin.

418. Mechanical Effects of a Discharge. — **Experiment.** — Hold a piece of cardboard between the discharging balls of an electrical machine. It will be perforated by a spark and the hole will be burred outward on both sides. A thin dry plate of glass, or a thin test-tube over a sharp point, may be perforated by a very heavy discharge.

419. Heating Effects. — **Experiment.** — Charge a Leyden jar. Connect its outer surface with a gas-burner by a chain or wire. Turn on the gas and bring the ball of the Leyden jar near enough to the mouth of the burner for a spark to pass. The gas will be lighted by the discharge. Explain.

Experiment. — Ignite ether in an iron spoon by proceeding as in the last experiment.

Experiment. — Make a torpedo as follows: Fit corks to the ends of a paper tube, about 5 cm. long and 1 cm. diameter. Through one of them thrust two pieces of copper wire, the ends within the tube not quite touching. Fill the tube with fine gunpowder, close it, and place at a safe distance. Discharge a heavily charged Leyden jar through the two copper wires, a piece of wet string being included in the discharge circuit.

420. Magnetic Effect. — **Experiment.** — Make a helix of copper wire insulated with gutta-percha by winding it half a dozen times round a lead pencil. Place inside the helix an unmagnetized sewing-needle, and discharge a large Leyden jar through the helix. The needle will be magnetized. Compare the polarity of the needle with the direction of the discharge round it.

421. Brush Discharge. — When the balls of an electrical machine are separated too far for a spark to pass, the discharge then takes the form of an *electric brush*, accompanied by a characteristic hissing noise. In the dark the brush is seen to consist of innumerable branching streams of pale blue light diverging from a point not far from the metal. The brush appears to tear away metallic particles from the electrode, and to form more readily at the negative pole than at the positive. It is also brighter, smaller, and less finely divided at the negative pole.

XI. ATMOSPHERIC ELECTRICITY.

422. Lightning. — It was demonstrated by Franklin in 1752 that lightning is identical with the electric spark. He sent up a kite during a passing storm, and found that as soon as the hempen string became wet, long sparks could be drawn from a key attached to it, Leyden jars could be charged, and other effects characteristic of static electrification could be produced.

423. Lightning Flashes are discharges between oppositely charged conductors. They occur either between two clouds or between a cloud and the earth. The rise of potential in a cloud causes a charge to accumulate on the earth beneath it. If the stress in the air reaches a certain limiting value, the air breaks down, or is ruptured, like any other dielectric, and the two opposite charges unite in a long zigzag flash. This occurs when the electric tension reaches about 400 dynes per square centimetre. A lightning flash allows the strained medium to return to equilibrium.

424. Atmospheric Electrification. — No satisfactory explanation of the cause of atmospheric electrification has yet been given. It has been ascribed to evaporation and to friction between solid and liquid particles.

The potential of the air in clear weather is generally positive, and it is sometimes nearly as high as during a storm, but it shows smaller fluctuations. The observations at the Blue Hill Observatory near Boston show an average potential difference between the earth and the air of 540 volts (§ 465) for an elevation of 138 m., or nearly 4 volts per metre. This is equivalent to 0.00013 electrostatic unit (§ 403) per centimetre of elevation. During thunder-storms this potential difference may amount to 35 volts per metre. It is possible to obtain sparks from a cloudless sky when the exploring apparatus is at an elevation not exceeding 500 m.

425. Thunder. — A flash of lightning ruptures the air and heats it along the path of the discharge, producing a sudden expansion equivalent to a partial vacuum. Since the pressure on the walls of this opening in the air is 15 pounds per square inch, they come together with a violent crash. At a distance the direct report is mingled with its echoes from the clouds and the earth, producing the low reverberations of distant thunder.

426. Oscillatory Discharge.—When a Leyden jar is highly charged, the potential difference between its coatings increases till the dielectric between the discharge terminals suddenly breaks down and a spark passes. This discharge usually consists of several oscillations or to-and-fro discharges, like the vibrations of an elastic system, or the surges of a mass of water after sudden release from pressure. Imagine a tank with a partition across the middle and filled on one side with water. If a small hole be made in the partition near the bottom, the water will slowly reach the same level on both sides without agitation; but if the partition be suddenly removed, the first violent subsidence will be succeeded by a return surge, and the to-and-fro motion of the water will continue with decreasing violence till the energy is all expended.

A series of similar surges occurs when a condenser is suddenly discharged by the breaking down of the dielectric. The oscillatory character of such electric discharges was discovered by Joseph Henry in 1842. Its importance has been recognized only in recent times. Similar electric oscillations probably take place in some lightning flashes.

427. The Aurora.—The *aurora* is due to silent discharges in the upper regions of the atmosphere. Within the arctic circle it occurs almost nightly, and sometimes with indescribable splendor. The illumination of the aurora is due to positive discharges passing from the higher regions of the atmosphere to the earth. In our latitude these silent streamers in the atmosphere are infrequent. When they do occur they are accompanied by great disturbances of the earth's magnetism and by earth currents. Such magnetic disturbances sometimes occur at the same time in widely separated portions of the earth.

XII. ELECTRIC CURRENTS.

428. An Electric Current. — When a condenser is discharged through a wire, there is produced in and around the wire a state called an *electric current.* Electrification is a condition of strain in the dielectric; the electric current rapidly relieves this strain through the discharging conductor. If the state of strain is reproduced by the "generator" as fast as it is relieved by the conductor, the result is a continuous current. To accomplish this result work must be done, and therefore an electric current represents energy. The expression, "current of electricity," was introduced when electricity was regarded as a fluid which flowed from higher to lower potential through a wire, just as water flows through a pipe from a higher to a lower level. So far as we know, however, the only thing transferred is energy, and the belief is growing that the energy is not transmitted by the wire at all, but by the ether surrounding the wire. However that may be, a uniform electric current through a conductor requires the maintenance of a constant potential difference between its terminals. One of the simplest means of doing this is the primary cell or battery.

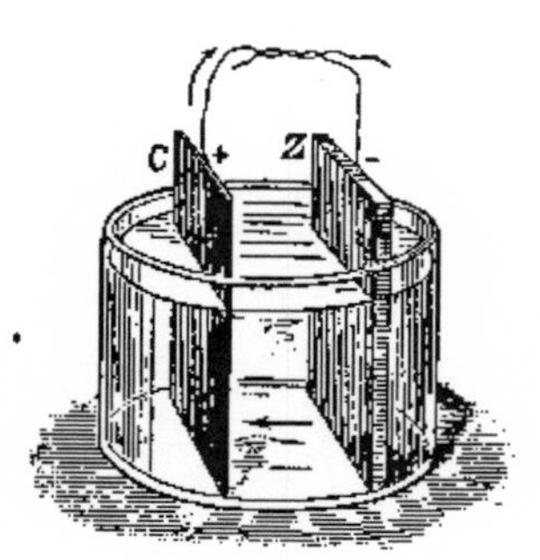

Fig. 235.

429. The Simple Voltaic Cell. — **Experiment.** — Cut a strip of heavy sheet zinc and one of sheet copper, each about 10 cm. long and 3 cm. wide. Scour the zinc with emery paper till it is bright. Support these strips side by side in a glass vessel nearly full of dilute sulphuric acid (one part acid to twenty of water). When the strips are brought together, innumerable bubbles of gas will rise from the copper strip, and some also from the zinc. This gas is hydrogen. Remove either strip, or do not allow the two to touch, and the chemical action is much diminished. If a little mercury be now rubbed on the zinc, no gas will be given off by it; but if the upper ends of the two strips be connected by any good conductor (Fig. 235), gas will again come off freely from the copper. This action will cease if the connection be made by any non-conductor. If the action is continued for some time, the zinc will be found to waste away, while the copper is unaffected.

Such a combination of two conductors, immersed in a compound liquid, called an *electrolyte*, which is capable of reacting chemically with one of the conductors, is called a *voltaic cell* or *element*. The name is derived from Volta of Padua, who first described such a cell in 1800.

A sufficiently sensitive test shows that the copper strip or plate is positive and the zinc negative. A potential difference is therefore established between the plates by immersion in the acid solution. The copper plate is called the *positive electrode* or the *cathode*, and the zinc the *negative electrode* or the *anode*. A current always leaves the solution by the *cathode* (way down or out).

430. The Circuit of a voltaic cell comprises the entire path traversed by the current, including the electrodes and the liquid in the cell as well as the external conductor. *Closing the circuit* means joining the two electrodes by a conductor; *breaking* or *opening the circuit* is disconnecting them. The flow of current in the external circuit is from the positive electrode (copper) to the negative (zinc), and in the internal part of the circuit from the negative electrode to the positive (Fig. 235).

431. Electrochemical Actions in a Voltaic Cell. — The modern theory of dissociation furnishes an explanation of the manner in which an electric current is conducted through a liquid. It is briefly as follows: When a salt or an acid, such as hydrochloric acid (HCl), for example, is dissolved in water, some of the molecules at least split into two parts ($\overset{+}{H}$ and $\overline{Cl}$, for example), one part having a positive electrical charge and the other a negative one. The two parts of the dissociated substance with their electrical charges are called *ions*. An electrolyte is a compound capable of such dissociation into ions. It conducts electricity only by means of the migration of the ions resulting from the splitting in two of the molecules. The separated ions convey their charges with a slow and measurable velocity through the liquid. Electropositive ions, such as zinc and hydrogen, carry positive charges in one direction, electronegative ions, such as chlorine and "sulphion" (SO_4), carry negative charges in the opposite direction, and the sum of the two kinds of charges carried through the liquid per second is the measure of the current.

The active components in a simple voltaic cell set up with hydrochloric acid may be represented as follows:—

$$\begin{array}{cccccc} & \overset{+}{H} & \overset{+}{H} & \overset{+}{H} & \overset{+}{H} & \\ Zn & & & & & Cu \\ & \overline{Cl} & \overline{Cl} & \overline{Cl} & \overline{Cl} & \end{array}$$

Immediately after the circuit has been closed this becomes

$$\begin{array}{ccccccc} & & & \longrightarrow & & & \\ & \overset{+}{Zn} & \overset{+}{H} & \overset{+}{H} & \overset{+}{H} & & \\ Zn \quad ZnCl_2 & & & & & H & Cu \\ & \overline{Cl} & \overline{Cl} & \overline{Cl} & \overline{Cl} & & \\ & & & \longleftarrow & & & \end{array}$$

Zinc goes into solution as zinc chloride ($ZnCl_2$), and hydrogen appears as free hydrogen gas at the copper plate. Zinc ions crowd out hydrogen ions, while the positive and negative charges brought to the copper and the zinc plate respectively reunite as a current through the external conductor.

432. Electromotive Force. — A voltaic cell is an electric generator. It is analogous to a rotary pump which produces a difference of pressure between its inlet and its outlet. Such a pump may cause water to circulate through a system of horizontal pipes against friction. In any portion of the pipe system the force producing the flow is the difference of water pressure between those points. But the force is all applied at the pump, and this produces a pressure throughout the whole circuit.

A voltaic cell generates electric pressure called *electromotive force.* It does not generate electricity, but it supplies the electric pressure to set electricity flowing. This electromotive force (E.M.F.) is numerically equal to the work which must be done to transport a unit quantity of electricity entirely round the circuit. Work is required to effect this transfer, because all conductors offer resistance to the passage of a current. The energy thus expended goes to heat the conductor.

The difference of potential between two points on the external conducting circuit is the work done in carrying a unit quantity of electricity from one point to the other. If E denotes this potential difference and Q the quantity conveyed, then the whole work done is the product EQ. But the quantity conveyed by a conductor per second is called the *strength of current*, I. The energy transformed, therefore, when a current I flows through a conductor, under an electric pressure or potential difference of E units between its ends, is EI ergs per second.

433. Detection of Current. — **Experiment.** — Solder a copper wire to each of the strips of a voltaic cell, and connect the wires with some form of key to close the circuit. Stretch a portion of the wire over a mounted magnetic needle (Fig. 236), holding it parallel to it and as near as possible without touching. Now close the circuit, and observe that the needle is deflected; after a few oscillations it comes to rest at an angle with the wire. Next form a rectangular loop of the wire, and place the needle within it. A greater deflection is now obtained. If a loop of several turns is formed, the deflection is still greater. A magnetic needle employed in this way becomes a *galvanoscope*, a detector of electric currents.

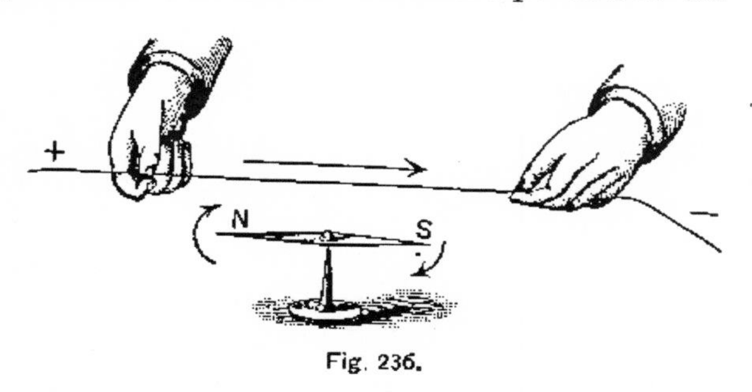

Fig. 236.

This experiment, first performed by Oersted in 1819, shows that the region round the wire has magnetic properties during the flow of electricity through it. In other words, it is a magnetic field (§ 375). Water flowing through a pipe produces no disturbance in the region round it, corresponding to the magnetic field round a conductor conveying a current of electricity. The analogy between a current of water and a current of electricity fails therefore in this respect.

434. Relation between the Direction of the Current and the Direction of Deflection. — **Experiment.** — Making use of the apparatus of § 433, compare the direction of the current through the wire with that in which the north pole of the needle turns. Cause the current to pass in the reverse direction over the needle; the deflection is reversed. Now hold the wire below the needle, and the direction of deflection is again reversed.

The direction of the deflection may always be predicted by the following rule: *Stretch out the right hand along the wire, with the palm turned toward the magnetic needle, and with the current flowing in the direction of the extended fingers. The outstretched thumb will then point in the direction of deflection of the north pole of the needle.* By the converse of this rule, the direction of the current may be inferred from the direction in which the needle is deflected.

435. Local Action. — **Experiment.** — Place a strip of commercial zinc in dilute sulphuric acid. Hydrogen is liberated during the chemical action, and after a few minutes the zinc becomes black from particles of carbon exposed to view on dissolving away the surface. If the experiment is repeated with zinc amalgamated with mercury, there will be little or no chemical action.

This experiment shows that the amalgamation of commercial zinc with mercury changes its properties. If in the experiment with the simple voltaic cell, a galvanoscope is inserted in the circuit both before the zinc has been amalgamated and afterward, it will be found that a larger deflection will be obtained in the second case.

The chemical action going on in a voltaic cell which contributes nothing to the current flowing through the circuit is known as *local action.* It is probably due to the presence of carbon, iron, etc., in the zinc; these with the zinc form miniature voltaic cells, the currents flowing round in short circuits from the zinc through the liquid to the foreign particles and back to the zinc again.

This local action is prevented by amalgamating the zinc; that is, by coating it with an alloy of mercury and zinc. The amalgam brings pure zinc to the surface, covers the foreign particles, and above all forms a smooth surface, so that a film of hydrogen clings to it and protects it from chemical action save when the circuit is closed.

436. Polarization. — **Experiment.** — Connect the poles of the voltaic cell with a galvanoscope and note the deflection. Let the cell remain in circuit with the galvanoscope for some time, the deflection will gradually become less and less. Now stir up the liquid vigorously with a glass rod, inserting the rod between the plates and brushing off the adhering gas bubbles; the deflection will increase to nearly its original amount.

The diminution in the intensity of the current is due to several causes, but the chief one is the film of hydrogen which gathers on the copper plate, causing what is known as the *polarization* of the cell. The hydrogen on the positive plate not only introduces more resistance to the flow of the current, by diminishing the available surface of copper, but it diminishes the electromotive force to which this flow is due. The presence of hydrogen on the copper plate sets up an inverse E.M.F., which either reduces or stops the flow of current.

437. Remedies for Polarization. — **Experiment.** — Place enough pure mercury in a quart jar to cover the bottom, and hang above it a piece of sheet zinc. Fill the jar with a nearly saturated solution of salt water, and place in the mercury the exposed end of a copper wire insulated with gutta percha, the upper end forming the positive pole of the battery.

If now the circuit is closed through a telegraph sounder (§ 510) of ten or fifteen ohms resistance, the armature will at first be attracted strongly; but in the course of a few minutes it will be released and will be drawn back by the spring. Polarization has then set in to the extent that the current is insufficient to operate the instrument.

Next take a small piece of mercuric chloride ($\dot{H}gCl_2$) no larger than the head of a pin, and drop it in on the surface of the mercury.

The armature of the sounder will instantly be drawn down, showin that the current has recovered its normal value. The hydrogen ha been removed by the chlorine of the mercuric chloride. In a fe minutes the chlorine will be exhausted, and polarization will agai set in. A little more of the chloride will again restore the activity o the cell.

This experiment illustrates a chemical method of re ducing polarization. Means are adopted to replace th hydrogen ions with others, such as copper or mercury which do not produce polarization when they are deposite on the positive electrode; or else the positive electrod is surrounded with a chemical which furnishes oxygen o chlorine to unite with the hydrogen before it reaches th electrode. In both cases the electrode is kept nearly fre from hydrogen.

438. The Daniell Cell illustrates the first of these chemi cal methods of preventing polari zation. In its most common for (Fig. 237) it consists of a glass ja containing a saturated solution o copper sulphate ($CuSO_4$), and in i a cylinder *C* of copper, which i usually cleft down one side. Withi the copper cylinder is a porous cu of unglazed earthenware containin dilute sulphuric acid, or preferabl a dilute solution of zinc sulphat ($ZnSO_4$). The porous cup contain also the zinc prism *Z*. The porou cup allows the ions to pass through its pores, but i prevents the rapid admixture of the two sulphates. Th copper sulphate must not be allowed to come in contac with the zinc electrode.

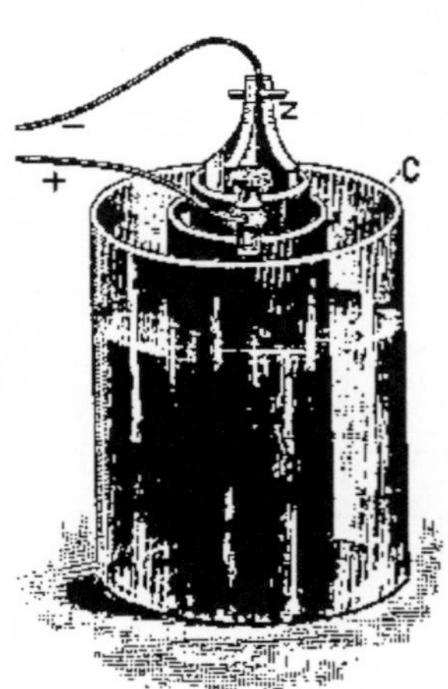

Fig. 237.

With sulphuric acid round the zinc, the hydrogen ions are intercepted at the porous cup by the copper sulphate. The positive copper ions then migrate toward the copper electrode and are there deposited as metallic copper. The $\overset{-}{SO}_4$ ions go to the zinc electrode with their negative charges, as in the case of the simple voltaic cell.

If the zinc is immersed in zinc sulphate, then both sulphates undergo partial ionization or dissociation, and there are no hydrogen ions. Only zinc and copper ions travel toward the copper electrode. Zinc ions never reach the copper, because zinc in copper sulphate invariably replaces the copper, forming $ZnSO_4$ in place of the $CuSO_4$. Graphically the operation may be represented as follows : —

$$\begin{array}{cccccc} & \overset{+}{Zn} & \overset{+}{Zn} & \overset{+}{Cu} & \overset{+}{Cu} & \\ Zn & & & & & Cu \\ & \overset{-}{SO}_4 & \overset{-}{SO}_4 & \overset{-}{SO}_4 & \overset{-}{SO}_4 & \end{array}$$

It is easy to see that as soon as the circuit is closed, $ZnSO_4$ will be formed at the zinc electrode, and copper will be precipitated on the copper electrode. At the same time there is a loss of copper sulphate corresponding exactly to the increase of zinc sulphate. Polarization is entirely obviated in the Daniell cell, and it is one of the most constant elements yet devised.

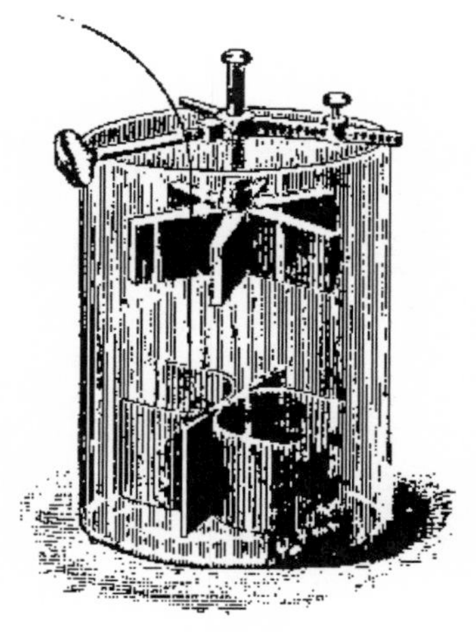

Fig. 238.

439. The Gravity Cell (Fig. 238) is a modified Daniell. The porous cup is omitted, the partial separation of the liquids being secured by difference in density. The copper electrode is placed at the bottom in saturated copper sulphate, while the zinc is suspended near the top in a weak solution of zinc sulphate, floating on top of the copper sulphate. The zinc should never be placed in the solution of copper sulphate. The saturated copper sulphate is more dense than the dilute zinc salt, and so remains at the bottom, except as it slowly diffuses upwards.

440. The Bunsen Cell (Fig. 239) consists of a glass jar containing dilute sulphuric acid, and a hollow cylinder of zinc immersed in it. Within the zinc cylinder is a porous cup containing strong nitric acid and a prism of compressed carbon.

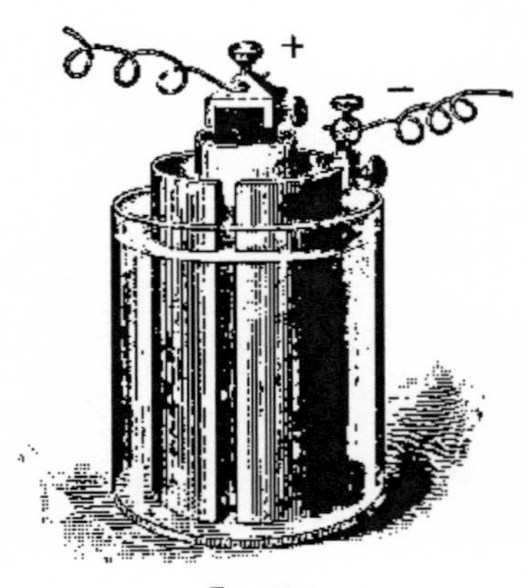

Fig. 239.

The hydrogen ions from the sulphuric acid are intercepted on their way to the positive electrode by oxygen from the nitric acid, and are oxidized with the production of water. The nitric acid molecule is broken up and yields a brownish red gas, which is very corrosive and poisonous.

The Grove cell differs from the Bunsen in no respect except that the positive electrode is a sheet of platinum instead of a prism of carbon.

441. The Chromic Acid Cell usually consists of a plate of zinc between two carbon plates dipping into a glass vessel containing dilute sulphuric acid, to which is added either chromic acid or the bichromate of potassium or of sodium. The sodium salt is much to be preferred to the potassium salt. With the bichromates an additional quantity of sulphuric acid is needed to liberate chromic acid.

Fig. 240 illustrates a form of this cell, called the *Grenet cell*, which is very convenient, but is open to the objection that, since the carbon plates are left standing in the solution, the liquid soon works up and attacks the connections at the top, making it difficult to keep the cell in good order. The zinc is attached to a sliding rod, *A*, so that it may be lifted out of the liquid when the cell is not in use.

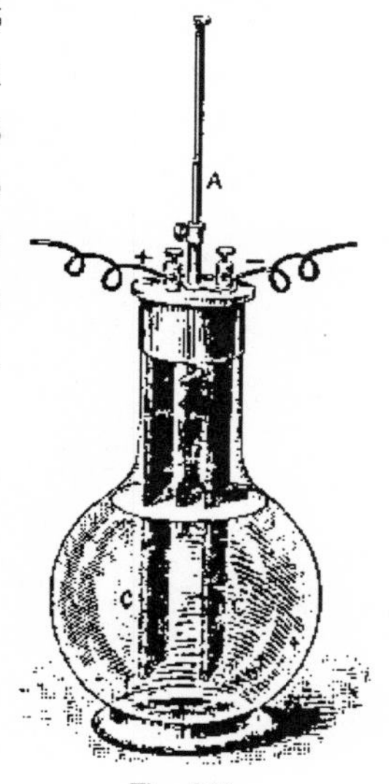

Fig. 240.

The hydrogen coming from the ionized acid is oxidized to water by the chromic acid, and polarization is prevented.

Fig. 241 illustrates a form of chromic acid battery, where the several cells composing it have their carbons and zincs suspended from a frame. It is known as a *plunge battery*, and is a very convenient form for experimental work.

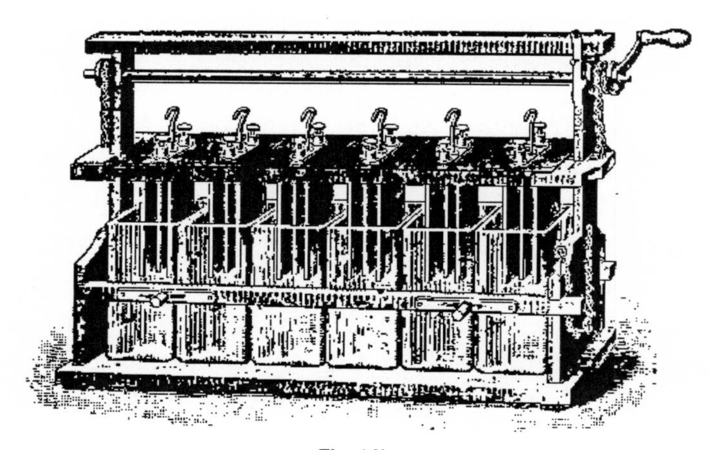

Fig. 241.

442. The Leclanché Cell (Fig. 242) consists of a glass vessel containing a saturated solution of ammonium chloride (sal ammoniac) in which stands a zinc rod and a porous cup. In this porous cup is a bar of carbon very tightly packed in a mixture of manganese dioxide and graphite, or granulated carbon.

Fig. 242.

The zinc is acted on by the chlorine of the ammonium chloride, liberating ammonia and hydrogen. The ammonia in part dissolves in the liquid, and in part escapes into the air. The hydrogen is slowly oxidized by the manganese dioxide. The cell is not adapted to continuous use, as the hydrogen is liberated at the positive electrode faster than the oxidation goes on, and the cell is polarized. If, however, it is allowed to rest, it recovers from polarization.

XIII. EFFECTS OF ELECTRIC CURRENTS.

a. HEATING EFFECTS.

443. Heating of a Conductor. — **Experiment.** — Close the circuit of a chromic acid battery through a piece of No. 30 platinum wire about 3 cm. long. The wire becomes red hot and possibly may fuse. If copper wire is substituted for the platinum, a smaller change of temperature will be observed.

In a battery, the potential energy of chemical separation is transformed into the energy of an electric current. When the current does no work this energy is all converted into heat in the circuit. The relative amounts of the heat generated in the external circuit and in the battery itself are strictly proportional to the external and internal resistances (§ 460). If in the experiment the circuit is closed with the fine wire omitted, more heat is generated in the liquid of the battery than with the fine wire in the circuit.

444. Laws of the Development of Heat by a Current. — The following laws were discovered experimentally by Joule in 1841: —

I. *The heat developed is proportional to the square of the current strength.*

II. *In any portion of a circuit it is proportional to the resistance of that portion.*

III. *The heat is also proportional to the time during which the current flows.*

The heating effects of an electric current are utilized in firing blasts and cannon, in exploding torpedoes, in cauterizing, in welding, in heating rooms, in cooking, in producing high temperatures for melting refractory substances, and for conducting chemical processes which require extreme heat.

b. CHEMICAL EFFECTS.

445. Electrolysis. — **Experiment.** — Bend a glass tube of about 1.5 cm. diameter and 15 cm. long into a V-form (Fig. 243). Close the ends with corks and thrust through them platinum wires, terminating within the tube in narrow strips of platinum foil. Support the tube in some convenient way after filling it two-thirds full of a solution of sodium sulphate, colored with the extract of purple cabbage. Connect the terminals to the poles of two or three cells joined in series (§ 468). After the circuit has been closed for a few minutes, the liquid around the anode will turn red, showing the presence of an acid, while that around the cathode will turn green, showing the presence of an alkali.

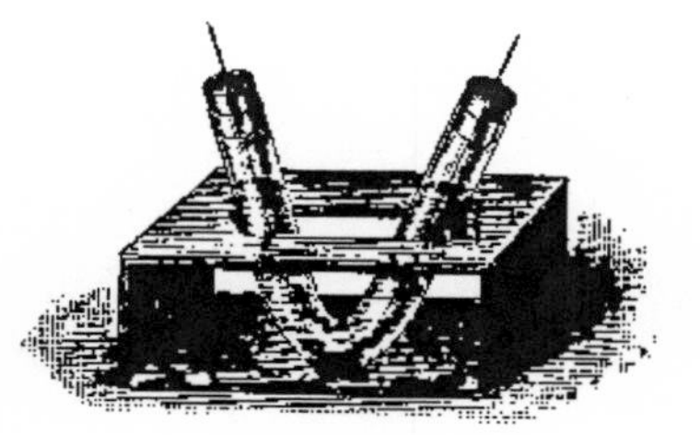

Fig. 243.

The experiment shows that the electric current in its passage through a liquid decomposes it. To this process of decomposing liquids by means of an electric current Faraday gave the name of *electrolysis;* to the substance decomposed, *electrolyte;* to the parts of the separated electrolyte, *ions.* The *anode* is the electrode by which the current enters the electrolytic cell, and the *cathode* the electrode by which it leaves.

446. Electrolysis of Copper Sulphate. — **Experiment.** — Fill the V-tube of the last experiment about two-thirds full of a solution of copper sulphate. After the circuit has been closed a few minutes, the cathode will be covered with a deposit of copper, and bubbles of gas will rise from the anode. These bubbles are oxygen.

When copper sulphate is dissolved in water it suffers dissociation to some extent. If, therefore, electric pressure is applied to the solution through the electrodes, the electropositive ions ($\overset{+}{Cu}$) are set moving from higher to lower potential, while the electronegative ions ($\overset{-}{SO_4}$) carry their negative charges in the opposite direction. The Cu ions are therefore driven against the cathode, and, giving up their charges, become metallic copper. The sulphions ($\overset{-}{SO_4}$) go to the anode; and, giving up their charges, they take hydrogen from the water present, forming sulphuric acid (H_2SO_4) and setting free oxygen, which comes off as bubbles of gas. If the anode were copper, the sulphion would unite with it, and copper would then be removed from the anode as fast as it is deposited on the cathode. The result of the passage of a current would then be the transfer of copper from the anode to the cathode. This actually takes place in the electrolytic refining of copper.

It will be seen that the passage of an electric current

through an electrolyte is accomplished in the same way, whether it is in a voltaic cell or in an electrolytic cell.

447. Electrolysis of Water. — Experiment. — Insert two burette tubes into a cork which has been boiled in paraffin, and which fits a wide-mouthed bottle. The glass stopcocks are at the top, as represented in Fig. 244. The thistle-tube is added for convenience in filling. Solder platinum wires to two copper wires, insulated with gutta percha, the platinum wires terminating in strips of platinum foil. The copper must be completely insulated, leaving only platinum exposed. Insert the wires and the foil as shown in the figure. Open the stopcocks and fill through the thistle-tube with water, acidulated with sulphuric acid, till the solution rises to the stopcocks, which should then be closed. Connect the apparatus by means of the connectors with two or more chromic acid cells joined in series (§ 468). Bubbles of gas will at once begin to rise from the platinum electrodes. After a few minutes the tube over the cathode will be found to contain about twice as much gas as the other one. The gas given off at the cathode is hydrogen and at the anode oxygen. The apparatus is called a *voltameter*.

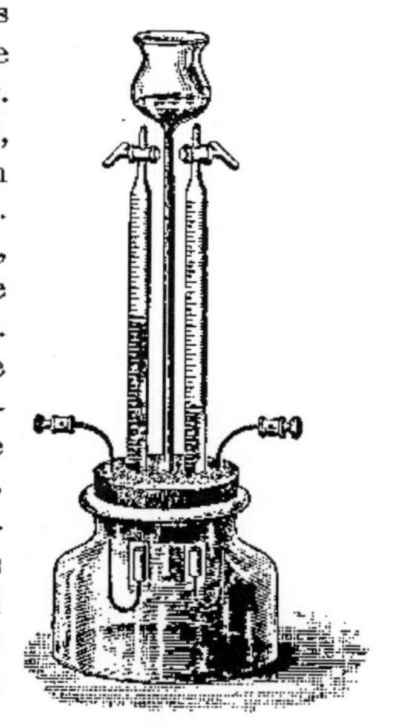

Fig. 244.

During the passage of the current the hydrogen ions, coming from the dissociated sulphuric acid, move with the current and the sulphions (SO_4) against it. The latter release oxygen from the water by taking away hydrogen, precisely as in the electrolysis of copper sulphate. If a copper or brass anode be used, the sulphion will attack it and no oxygen will appear.

448. Laws of Electrolysis. — The following fundamental laws of electrolysis were established by Faraday : —

I. *The mass of an ion liberated is proportional to the quantity of electricity which passes through the electrolyte.*

The mass of an ion liberated in one second is, therefore, proportional to the strength of current.

II. *The masses of different ions liberated per second by the same current are proportional to their chemical equivalents.*

By "chemical equivalents" are meant the relative quantities of the ions which are chemically equivalent to one another, or take part in equivalent chemical reactions. Thus, 32.5 gm. of zinc or 31.7 gm. of copper take the place of one gm. of hydrogen in sulphuric acid (H_2SO_4) to form zinc sulphate ($ZnSO_4$) and copper sulphate ($CuSO_4$) respectively.

The first law of electrolysis affords a valuable means of comparing the strength of two electric currents by determining the relative masses of any ion, such as silver or copper, deposited by the two currents in succession in the same time (§ 464).

449. Electroplating consists in covering bodies with a coating of any metal by means of the electric current. The process may be summarized as follows: Thoroughly clean the surface to remove all fatty matter. Attach the article to the negative electrode of a battery, and suspend it in a solution of some chemical salt of the metal to be deposited. If silver, cyanide of silver dissolved in cyanide of potassium is used; if copper, sulphate of copper. To maintain the strength of the solution a piece of the metal of the kind to be deposited is attached to the positive electrode of the battery. The action is similar to that heretofore given. Articles of iron, steel, zinc, tin, and lead cannot be silvered or gilded unless first covered with a thin coating of copper.

450. Electrotyping consists in copying medals, wood-cuts, type, and the like in metal, usually copper, by means of the electric current. A mould of the object is taken in wax or plaster of Paris. This is evenly covered with powdered graphite to make the surface a conductor, and treated very much as an object to be plated. When the deposit has become sufficiently thick it is removed from the mould and backed or filled with type-metal.

451. The Secondary or Storage Battery. — Experiment. — Connect the apparatus of § 447 to a suitable battery. After passing the current for a short time, causing an evolution of gas, disconnect the battery and put a galvanoscope in its place. The needle will be deflected, showing that a current is now passing through the apparatus in a direction opposite to the battery current.

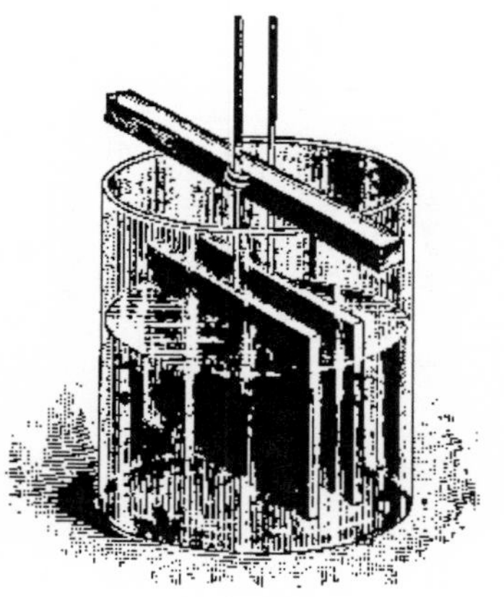
Fig. 245.

Experiment. — Support two lead plates with attached copper wires by a strip of wood (Fig. 245), and immerse the plates in dilute sulphuric acid, one part of acid by measure to five of water. Pass a current from a suitable battery through this electrolytic cell for a few minutes, and then disconnect the battery and connect the cell with an electric bell (§ 515). It will furnish for some time a current sufficient to ring the bell. As soon as it is discharged connect again to the battery as before and repeat the operation.

The last experiment illustrates the *lead storage battery*. The electrolysis of water liberates oxygen at the anode, which combines with the lead electrode to form a chocolate-colored deposit of lead peroxide (PbO_2). Hydrogen accumulates on the cathode. When the battery is disconnected and the lead plates are joined by a conductor, a current flows in the external circuit from the oxidized plate, which is called the positive electrode, to the other one, called the negative, the lead peroxide is reduced to spongy lead on the positive plate, while some lead sulphate is formed on the negative. During subsequent charging this lead sulphate is reduced by the hydrogen to spongy lead. Note that the charging current passes through the storage cell in the opposite direction to the discharge current furnished by the cell itself.

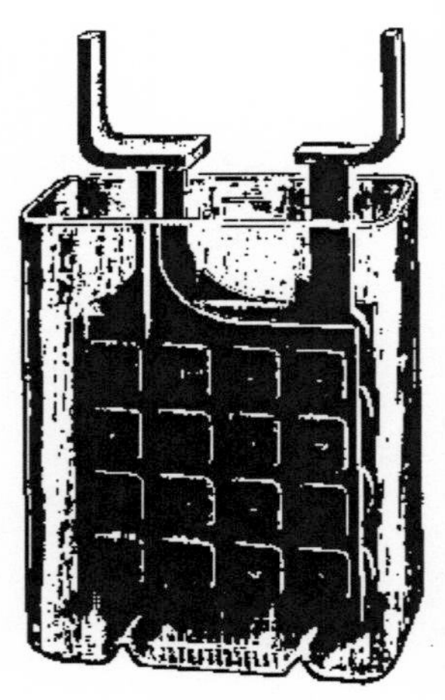

Fig. 246.

The storage battery stores energy and not electricity. The energy of the charging current is converted into the potential energy of chemical separation in the storage cell. When the circuit of the charged secondary cell is closed, the potential chemical energy is reconverted into the energy of an electric current in precisely the same way as in a primary cell.

Fig. 246 is a complete storage cell containing one positive and two negative plates.

C. MAGNETIC EFFECTS.

452. Magnetic Character of the Current. — **Experiment.** — Connect two or three chromic acid cells in parallel (§ 469). Close the circuit through a heavy wire, and then dip a portion of it into fine iron filings. A thick cluster of them will adhere to the wire (Fig. 247).

Fig. 247.

This experiment illustrates the fact that a conductor through which a current is passing possesses magnetic properties. The iron filings are magnetized by the current and set themselves at right angles to the wire. When the circuit is broken, they lose their magnetism and drop off.

453. Magnetic Field about Conductor. —Experiment. — Bend a wire *AB* and support it in the shallow vessel nearly full of water, as shown in Fig. 248. Magnetize a sewing-needle and oil it so that it will float on the water. It will turn with its north pole pointing north; but when a current is passed through the wire, it will set itself perpendicular to a line joining the needle and the wire. Reverse the current, and the needle will also reverse its direction.

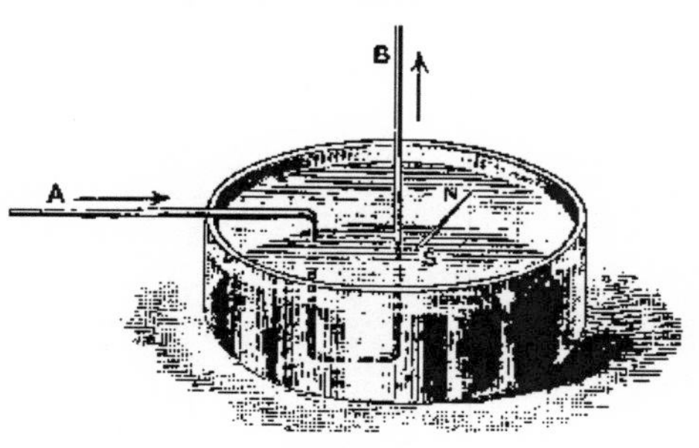

Fig. 248.

Experiment. — Support horizontally a sheet of stiff paper. Pass vertically through it a wire which connects the poles of a battery of two or more chromic acid cells. Close the circuit and sift a few very fine iron filings on the paper, jarring it slightly with a pencil as they fall. They arrange themselves in circles with the wire at the centre. Place a small mounted magnetic needle on the paper near the wire. The needle sets itself tangent to these circles, and points in the opposite direction to that traversed by the hands of a watch, when the current comes up through the paper from below. (What is the direction of the lines of force?)

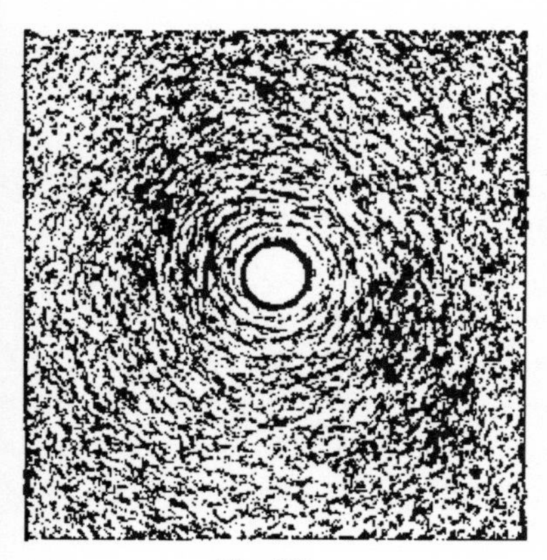
Fig. 249.

These experiments show that the lines of magnetic force about a wire through which an electric current is flowing, are concentric circles. Fig. 249 was made from a photograph of these circular lines of force as shown by iron filings on a plate of glass. Their direction relative to the current is given by the following rule: —

Grasp the wire by the right hand so that the extended thumb points in the direction of the current; then the fingers indicate the direction of the lines of force round the wire.

Fig. 250 is a sketch intended to show the direction of these circular lines of magnetic force (or magnetic whirl) which everywhere surround a wire conveying a current.

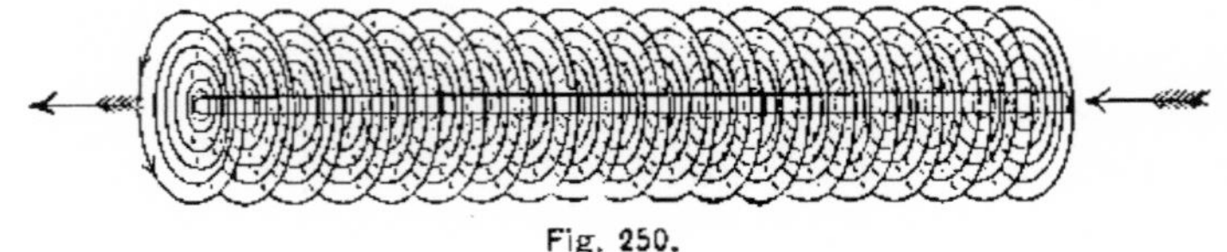

Fig. 250.

454. Properties of a Circular Conductor. — **Experiment.** — Bend a piece of No. 16 copper wire into the form shown in Fig. 251, the diameter of the circle being about 20 cm. Suspend it by a long thread, so that the ends dip into the mercury cups shown in section in the lower part of the figure. Send a current through the suspended wire by connecting a battery to the binding posts. A magnet brought near the face of the circular conductor will cause the latter to turn about a vertical axis and take up a position at right angles to the axis of the magnet.

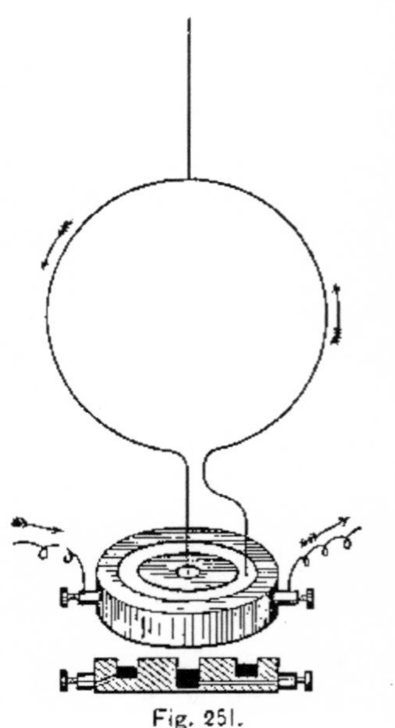

Fig. 251.

This experiment, due to Arago, shows that a circular current acts like a disk magnet, whose poles are its faces. The lines of force surrounding the conductor in this form pass through the circle and come round from one face to the other through the air outside the loop. The north-seeking side is the one from which the lines issue; and to an observer looking toward this side, the current flows round the loop counter-clockwise (Fig. 252). If instead of a single turn we take a long insulated wire and coil it into a number of parallel circles close together, the magnetic effect will be increased. Such a coil is called a *helix* or *solenoid*; and the passage of an electric current through it gives to it all the properties of a cylindrical bar magnet.

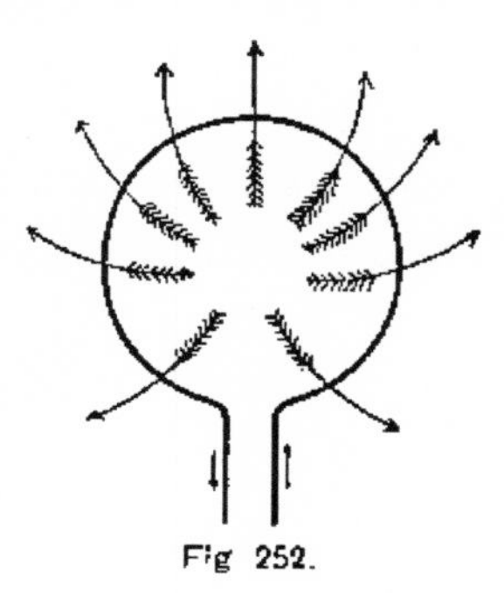
Fig 252.

455. The Electromagnet. — Experiment. — Wind neatly on a paper tube, about 2 cm. in diameter and 15 cm. long, three layers of No. 18 insulated copper wire. Pass an electric current through it and test its magnetic properties by bringing it near a mounted magnetic needle. Now fill the tube with straight pieces of soft iron wire and again bring it near the needle. Its magnetic effect will be greatly increased.[1]

This device, consisting of a helix encircling an iron core, is called an *electromagnet*. The presence of the iron

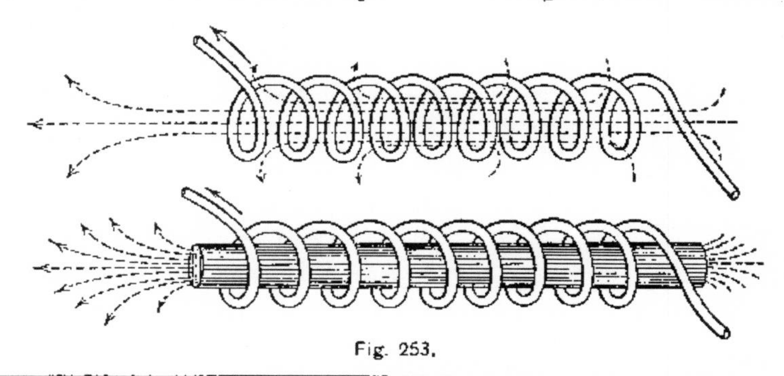
Fig. 253.

[1] The magnetic field in and around a helix may be shown by means of iron filings.

core greatly increases the number of lines of force running through the helix from end to end, by reason of its greater permeability (§ 377) as compared with air (Fig. 253).

When the iron core is not used, many of the lines leak out at the sides of the helix, and but few extend through from end to end. The core not only diminishes the leakage of the lines of force, but also adds many more to those previously running through the solenoid. Hence the magnetic strength of a helix is greatly increased by the iron core.

456. Mutual Relation of the Current and Lines of Force. — **Experiment.** — Support a long magnetized rod in a wooden clamp (Fig. 254); make a flexible conductor, by twisting together two or three strands of conducting tinsel (which can be bought in skeins), and connect as shown in the figure. When the current from two chromic acid cells, or from a single storage cell, is turned on, the flexible conductor will wind itself round the magnet in a helix. Use some convenient device to reverse the current, and the conductor will unwind and rewind itself in the opposite direction.

Fig. 254.

Here the conductor tends to wind itself in circles round the lines of force through the long magnet. If the magnet be grasped by the right hand, with the thumb pointing toward its north pole, the current will be found to flow round in the direction of the fingers (§ 453).

457. The Horseshoe Magnet. — The form given to an electromagnet depends on the use to which it is to be put. The *horseshoe* or *U-shape* (Fig. 255) is the most common. The advantage of this form becomes apparent when attention is directed to the fact that *all magnetic lines are closed curves;* that is, the lines of magnetic force or of magnetic induction are continuous,

Fig. 255.

passing through the core from the south to the north pole, and completing the circuit through the air from the north pole back to the south. Now, the shorter the air path of the magnetic lines, the larger their number and the stronger the magnet. The approach of the two poles in the U-shaped magnet shortens the air gap, and hence increases the number of lines of force. If then a bar of soft iron, called an *armature*, be placed across the poles, the air gap is further diminished when the armature is made to approach the poles (Fig. 256). When the armature is in contact with the poles, the magnetic circuit is all iron, and is said to be a closed circuit. The maximum number of magnetic lines then traverse it (Fig. 257).

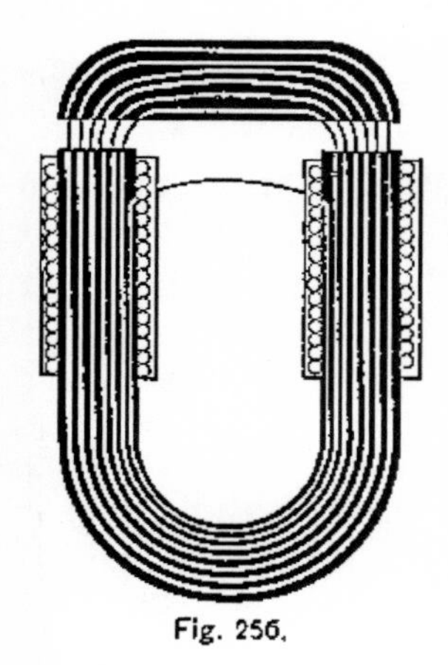
Fig. 256.

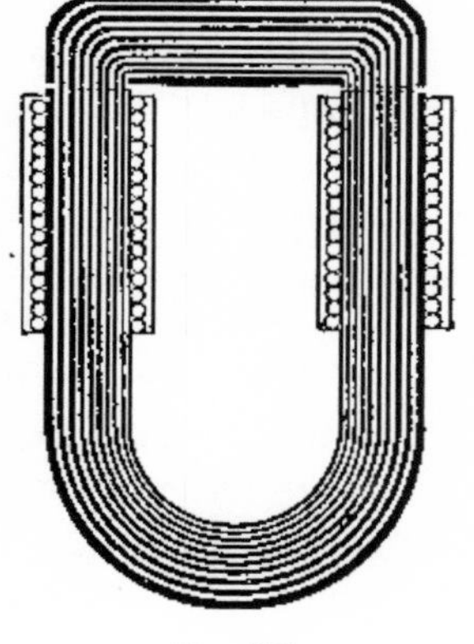
Fig. 257.

458. The Polarity of a Solenoid may be determined by the following rule:—

Grasp the coil with the right hand so that the fingers point in the direction of the current; the north pole will then be in the direction of the extended thumb.

In Fig. 258, if the fingers of the right hand are parallel to the arrows and point in the same direction, the extended thumb will point toward *N*, as the north pole of the coil. The converse of the rule enables one to ascertain the direction of the current when the polarity of the solenoid is known.

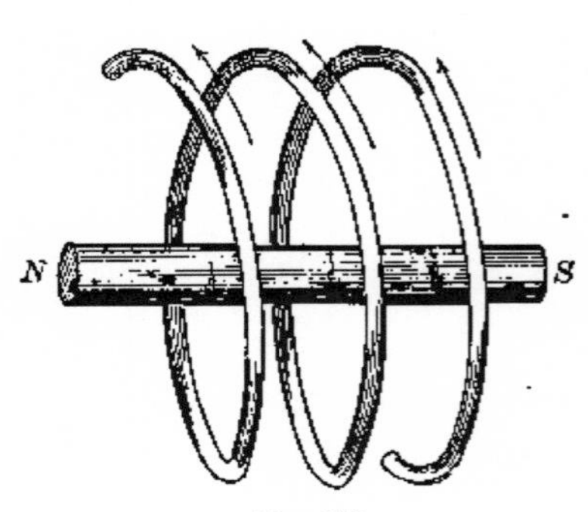

Fig. 258.

459. Mutual Action of Two Currents. — **Experiment.** — Construct a rectangular frame, about 25 cm. square, out of insulated copper wire No. 20, by winding four or five layers round the edge of a square board. Slip the wire off the board and tie the parts together in a number of places with thread. Bend the ends at right angles to the frame, remove the insulation, and give them the shape shown in Fig. 259. Suspend the wire frame by a long thread so that the ends dip into the mercury cups.

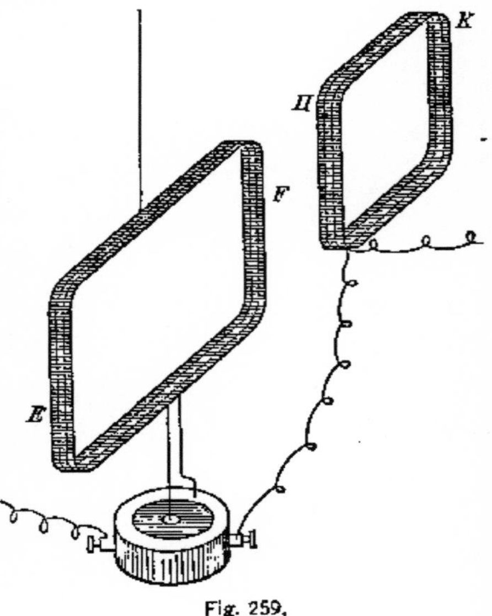

Fig. 259.

Wind several layers of the same size of insulated wire round a form about 10 cm. by 20 cm. Connect this coil in the same circuit with the rectangular coil and a battery of two or three cells joined in series.

First. Hold the coil *HK* with its plane perpendicular to the plane of the coil *EF*, with its edge *H* parallel to *F*, and with the currents in these two adjacent portions flowing in the same direction. The suspended coil will turn upon its axis, the edge *F* approaching *H*, showing attraction.

Second. Turn *HK* over so that the currents in the adjacent portions *H* and *F* flow in opposite directions. The edge *F* of the suspended coil will be repelled by *H*.

Third. Give *HK* a quarter turn around a vertical axis from the position shown in the figure. If it is near *F*, the suspended coil will turn in a direction to make the two parallel currents flow in the same direction round.

Fourth. Hold the coil *HK* within the rectangular coil *EF*, and with the edge *H* making an angle with the lower edge of *EF*. The coil *EF* will turn till the currents in its lower edge are parallel with those in *H*, and flowing in the same direction.

These results may be expressed by the following *laws of action between currents*: —

I. *Parallel currents flowing in the same direction attract.*

II. *Parallel currents flowing in opposite directions repel.*

III. *Currents making an angle with each other tend to become parallel and to flow in the same direction.*

Maxwell included the entire phenomenon under one law, viz.: *The two circuits tend to move so that the number of lines of force common to the two shall be a maximum.*

Figure 260 was made from a photograph of the magnetic figure or field of force around two wires passing through the holes and carrying parallel currents in the

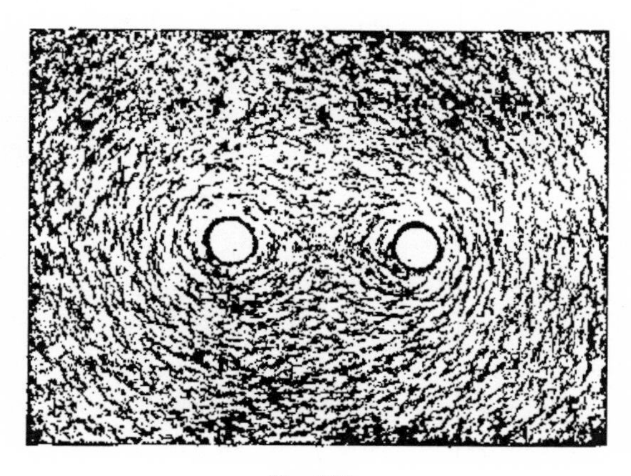

Fig. 260.

same direction. Many of the circular lines of force ar extended so as to include both wires, and the tension alon them tends to draw the two conductors together.

Figure 261 was made in the same manner from the mag netic field around the two wires when the parallel cur rents were in oppo site directions The circular line of force ar crowded togethe between the wires and their reactio to recover thei normal positio forces the wire apart.

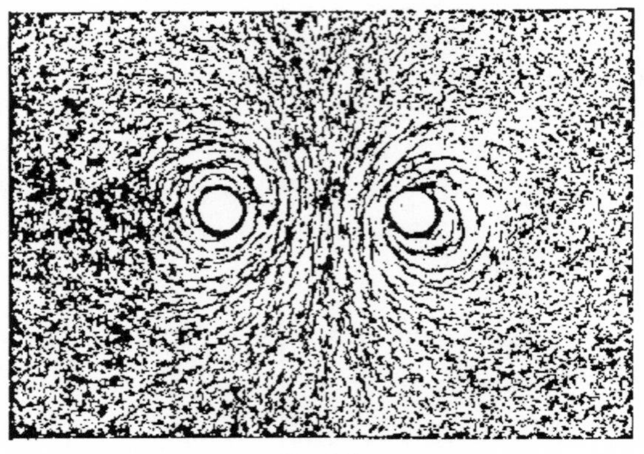

Fig. 261.

XIV. ELECTRICAL QUANTITIES.

a. OHM'S LAW.

460. Resistance. — Frequent mention has already bee made of *electrical resistance*, *electromotive force*, and *strengt of current.* We shall now consider these quantities mor closely, and shall study the relation between them whic is known as *Ohm's Law.* This law forms the basis o most electrical measurements of steady currents.

No conducting body possesses perfect conductivity, bu every conductor presents some obstruction to the passag of electricity. This obstruction is called its *electrica resistance.* It is the reciprocal of *conductivity.* Th greater the conductivity of a conductor the less it resistance, the one decreasing in the same ratio as th other increases.

461. Unit of Resistance. — The practical unit of resistance is the *ohm.* It was defined by the International Congress of Electricians in Chicago in 1893 as follows: "The international ohm is represented by the resistance offered to an unvarying electric current by a column of mercury at the temperature of melting ice, 14.4521 gm. in mass, of a constant cross-sectional area, and of the length 106.3 cm." The cross-sectional area of this thread of mercury is very nearly one square millimetre.

462. Laws of Resistance. — I. *The resistance of a conductor is proportional to its length.* For example, if 39 ft. of No. 24 copper wire (B. & S. gauge) have a resistance of 1 ohm, then 78 ft. of the same wire will have a resistance of 2 ohms.

II. *The resistance of a conductor is inversely proportional to its cross-sectional area, and in the case of round wire is inversely proportional to the square of its diameter.* For example, No. 24 wire has a diameter of .0201 inch, and No. 30 has a diameter of .01 inch, or nearly one-half of No. 24. 39 ft. of No. 24 has a resistance of 1 ohm, and 9.7 ft. of No. 30, which is nearly $\frac{39}{2^2}$, has a resistance also of 1 ohm, both at 22° C.

III. *The resistance of a conductor of a given length and cross-sectional area depends upon the material of which it is made, and is affected by any cause which modifies its molecular condition.* For example, the resistance of 2.2 ft. of No. 24 German-silver wire is 1 ohm, whereas it takes 39 ft. of copper wire of the same diameter to give the same resistance. Moreover, this is true only at a definite temperature; if these substances are heated, the percentage increase of resistance of the copper is nearly ten times as much per degree as that of German-silver, and twenty times as much as that of an alloy called platinoid. All metals have their resistance increased by increase of temperature. Carbon and most electrolytic conductors decrease in resistance as the temperature rises.

Experiment. — Connect the poles of a fresh chromic acid cell with a piece of fine platinum wire of such a length that the current heats it to a dull red color. Now apply a piece of ice to one portion and notice the rise in temperature of the remaining portion. Explain.

463. Formula for Resistance. — The above laws are conveniently expressed in the following formula for the resistance of a wire: —

$$r = k\frac{l}{C.M.},$$

in which k is a constant depending on the material, l the length of the wire in feet, and $C.M.$ denotes "circular mils." A "mil" is a thousandth of an inch, and circular mils are the square of the mils, that is, the square of the diameter of the wire in thousandths of an inch. The constant k becomes then the resistance in ohms of one mil-foot, that is, of a wire one foot long and one mil in diameter. The following are the values of k in ohms for several metals, at 20° C.: —

Silver	9.53	Iron	61.3	German-silver	181.3
Copper	10.19	Platinum	70.5	Mercury	574

464. The Strength of a Current is measured by the magnitude of the effects produced by it. Either the chemical, the electromagnetic, or the heating effects may be made the basis of a system of measurement. The quantity of an ion deposited in a second furnishes a convenient magnitude for the determination of unit strength of current. The unit of current strength is the *ampere*. It is that current which will deposit by electrolysis, under suitable conditions, 0.001118 gm. of silver, or 0.0003287 gm. of copper in one second. The ampere deposits 4.025 gm. of silver in one hour. A milliampere is a thousandth of an ampere. It is to be noted that the electrolytic method measures only the quantity of electricity passing through the decomposing cell, called a *voltameter*, in the given time.

465. Electromotive Force is the name given to the cause of an electric flow. It is often called *electric pressure* from its superficial analogy to water pressure. The unit of electromotive force (E.M.F.) is the *volt*. It is the E.M.F. which will cause a current of one ampere to flow through a resistance of one ohm. The E.M.F. of a battery depends upon the materials employed, and is entirely independent of the size and shape of the plates. The E.M.F. of a Daniell cell is about 1.1 volts; of a fresh chromic acid cell, 2 volts; and of a Leclanché cell, 1.5 volts. The E.M.F. of a Carhart-Clark standard cell is 1.44 volts at 15° C.

466. Ohm's Law. — The definite relation existing between strength of current, resistance, and E.M.F. is known as *Ohm's law*. It may be expressed as follows: —

The strength of a current equals the electromotive force divided by the resistance; or in symbols: —

$$I = \frac{E}{R}, \qquad (30)$$

where I is the current in amperes, E the E.M.F. in volts, and R the resistance in ohms. Applied to a battery, if R is the resistance external to the cell, and r the internal resistance of the cell itself, then

$$I = \frac{E}{r + R}.$$

Thus, if a chromic acid cell of 2 volts E.M.F. and half an ohm internal resistance is closed with a wire having a resistance of one and a half ohms, the current will be $\frac{2}{1.5+0.5}=1$ ampere.

From the equation $I=\frac{E}{R}$, we derive $E=IR$; that is, the effective E.M.F. equals the product of the current and resistance.

Again, $\frac{E}{I}=R$, or the resistance of a conductor is the constant ratio between the E.M.F. and the current which it produces through the conductor.

467. Methods of Varying Strength of Current. — It is evident from Ohm's law that the strength of the current furnished by an electric generator may be increased in two ways: 1. By increasing the E.M.F. 2. By reducing the internal resistance.

The E.M.F. may be increased by joining several cells in series, and the internal resistance may be diminished by connecting them in parallel. Enlarging the plates of a battery or bringing them closer together diminishes the internal resistance.

468. Connecting in Series. — If several cells are connected so that the positive pole of one is joined to the negative pole of the next and so on, then the total E.M.F. is the sum of the E.M.F.'s of the several cells. The cells are then said to be joined in *series*. Figure 262 is the conventional sign for a single cell. The short, thick line represents the

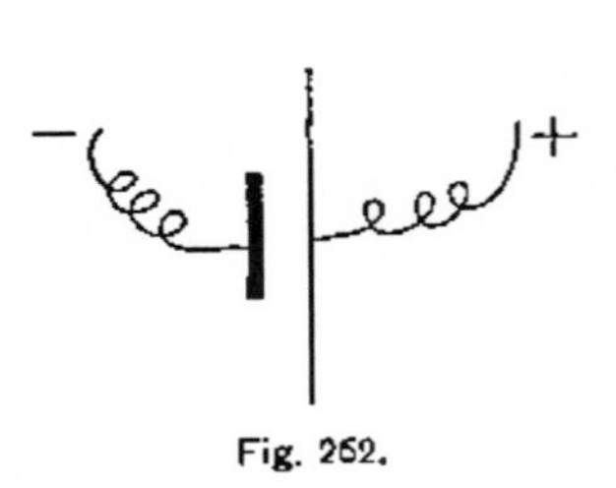

Fig. 262.

negative electrode, and the long, thin line the positive electrode. Figure 263 shows six cells joined in series. If each cell has an E.M.F. of 2 volts, then the total E.M.F. will be 6 × 2 or 12 volts. In connecting cells in this manner the internal resistance is also increased about six times, since the liquid conductor is six times as long as in one cell. The current for any external resistance R is then $I = \frac{12}{6r + R}$.

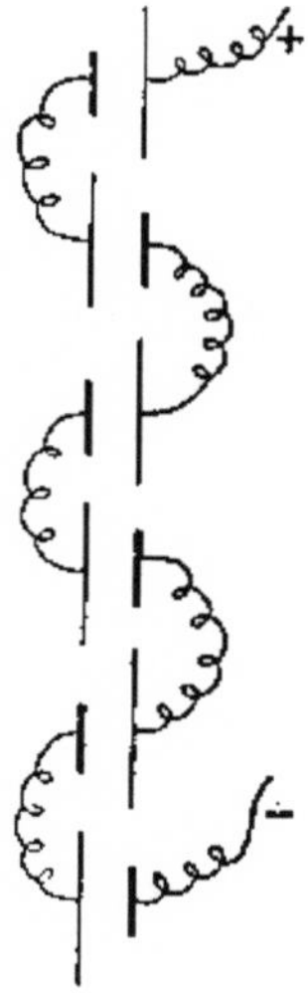

Fig. 263.

469. Connecting in Parallel. — When all the positive terminals are connected together on one side and the negative on the other, the cells are grouped in *parallel*. With n similar cells the effect of such a grouping (Fig. 264) is to reduce the internal resistance to $\frac{1}{n}$th that of a single cell. It is equivalent to increasing the area of the plates n times. All the cells side by side contribute equal shares to the output of the battery.

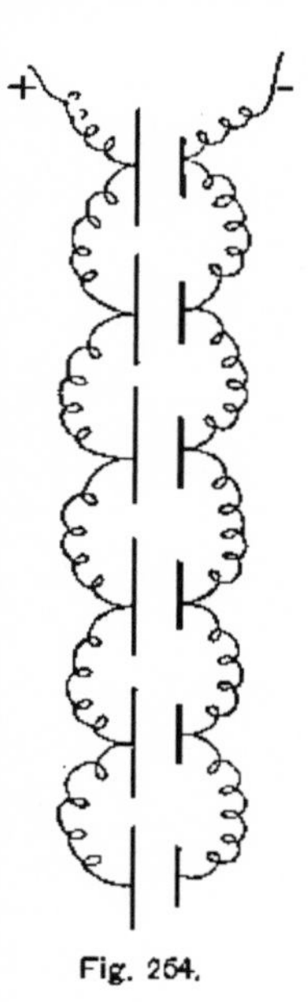

Fig. 264.

470. Relative Advantages of the Series and Parallel Grouping. — It is evident from Ohm's law that when the external resistance is small, there is nothing gained by increasing E and at the same time increasing r, since I then remains practically unchanged. But when R is large, the increase in r due to joining cells in series is more than counterbalanced by the increase in E. Consequently I is greater the larger the number of cells joined in series. For example: A battery of 6 cells, each having an E.M.F. of 2 volts and an internal resistance of 0.5 ohm, acts, first, through an external resistance of 0.1 ohm; and, secondly, through one of 500 ohms. If joined in parallel circuit, then when $R = 0.1$, $I = \frac{2}{0.1 + \frac{0.5}{6}} = 10.9$ amperes; and when $R = 500$, $I = \frac{2}{500 + \frac{0.5}{6}} = 0.004$ ampere. If joined in series, then when $R = 0.1$, $I = \frac{6 \times 2}{0.1 + 6 \times 0.5} = 3.87$ amperes; and when $R = 500$, $I = \frac{6 \times 2}{500 + 6 \times 0.5} = 0.024$ ampere. A comparison of these results shows that when the external resistance is small, the greater current is obtained by grouping in parallel; but when the external resistance is large, the series arrangement gives the greater current.

The largest current is obtained in any case by so grouping the cells that the external and internal resistances are equal to each other.

b. INSTRUMENTS FOR MEASUREMENT.

471. The Galvanometer. — If a comparison of currents is made by means of their magnetic effects, the instrument used for the purpose is called a *galvanometer*. If the galvanometer is calibrated, so as to read directly in amperes, it is called an *ammeter*. A galvanoscope becomes a galvanometer by providing it with a scale so that the deflections may be measured. In very sensitive instruments a small mirror is attached to the movable part of the instrument; it is then called a *mirror galvanometer*. Sometimes a beam of light from a lamp is reflected from this small mirror back to a scale, and sometimes the light from a scale is reflected back to a small telescope, by means of which the deflections are read. In either case the beam of light then becomes a long pointer without weight.

472. The Tangent Galvanometer consists of a vertical coil of wire (Fig. 265) from twenty-five to thirty centimetres in diameter, at the centre of which is supported a magnetized needle about two centimetres long, furnished with a long, light pointer moving over a graduated scale. Owing to the size of the coil the magnetic field at the centre is nearly uniform; that is, the lines of force there are sensibly parallel straight lines as in Fig. 261; and any movement of the short needle round a vertical axis will not carry its poles into a magnetic field of different strength. Under these conditions *the strength of the current is proportional to the tangent of the angle of deflection.* For example, if two different batteries, placed successively in circuit with a tangent galvanometer, give deflections of 55° and $35\frac{1}{2}°$ respectively, then the strengths of these currents are as $\frac{\tan 55°}{\tan 35\frac{1}{2}°} = \frac{1.428}{0.714} = \frac{2}{1}$; that is, the strength of one current is double that of the other. (See Appendix.)

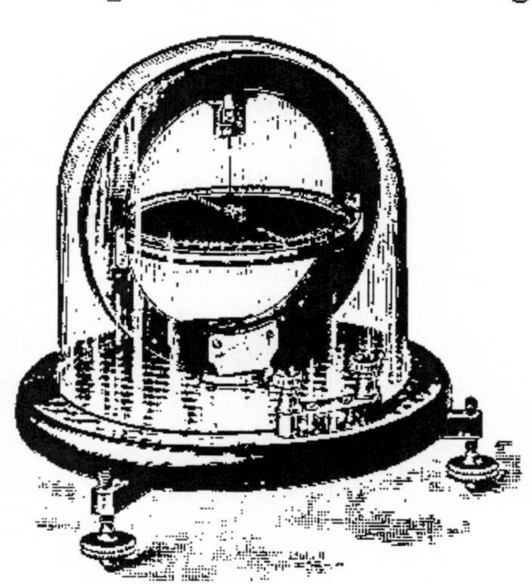

Fig. 265.

The meaning of the "tangent of an angle" may be made clear by consulting Fig. 266. The line *AH* is

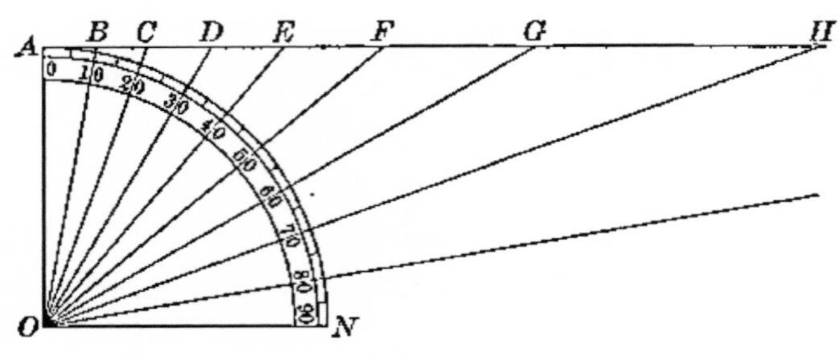

Fig. 266.

drawn perpendicular to *OA*. The tangents of 10°, 20°, 30°, etc., are equal to the quotients of *AB*, *AC*, etc., divided by *OA*; that is, they are proportional to the lengths of the lines *AB*, *AC*, *AD*, etc.

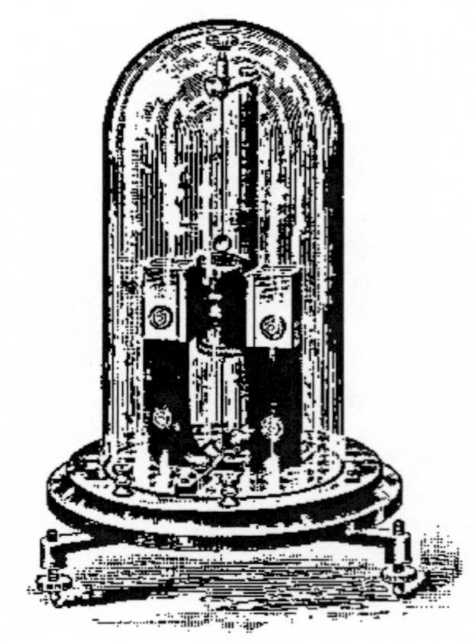

Fig. 267.

473. The d'Arsonval Galvanometer. — One of the most useful forms of galvanometer is the d'Arsonval, Fig. 267. Between the poles of a strong permanent magnet swings a coil suspended by a fine wire in such a way that the current is led in by the suspending wire and out by the wire connecting the coil to a spring at the bottom. A small mirror for reflecting a beam of light is attached to the coil. Inside the coil is a soft iron tube supported from the back. In some of the most recent forms of this instrument the coil is made narrower and the iron tube is omitted. Figure 268 illustrates a mirror galvanometer with horizontal magnets. The glass in front of the movable mirror is silvered halfway up; and the observer, looking through a hole in the upright strip, sees both a fixed and a movable image of the scale.

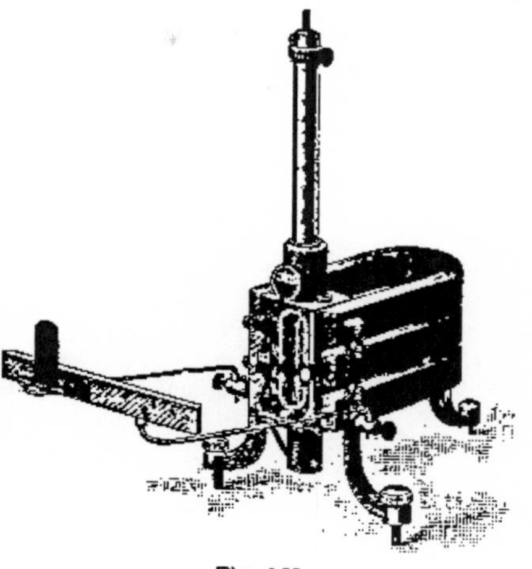

Fig. 268.

In the d'Arsonval galvanometer the coil is movable and the magnet is fixed. Its chief advantages are simplicity of construction, almost total independence of the direction and strength of the magnetic field at the place where it is used, and the quickness with which it comes to rest after the coil has been deflected by a current through it.

Fig. 269.

474. Voltmeters. — An instrument designed to measure the difference of potential in volts is called a *voltmeter*. For direct currents the most convenient portable voltmeter is made on the principle of the d'Arsonval galvanometer. The appearance of one of the best-known instruments of this class is shown in Fig. 269. The interior is represented by Fig. 270, where a portion of the instrument is cut away to show the coil and the springs. The current is led in by one spiral spring and out by the other. Attached to the coil is a very light aluminum pointer, which moves over the scale seen in Fig. 269, where it stands at zero. Soft iron pole-pieces are screwed fast to the poles of the permanent magnet, and they are so shaped that the divisions of the scale in volts are equal. Such an instrument needs to be frequently recalibrated because of slight changes in the strength of the magnet. A similar instrument, called an *ammeter*, is designed to measure currents in amperes. The resistance of a voltmeter should be high, so that it will take the smallest practicable current; the resistance of an ammeter should be as small as possible, so that it will not increase the resistance of the circuit in which it is placed.

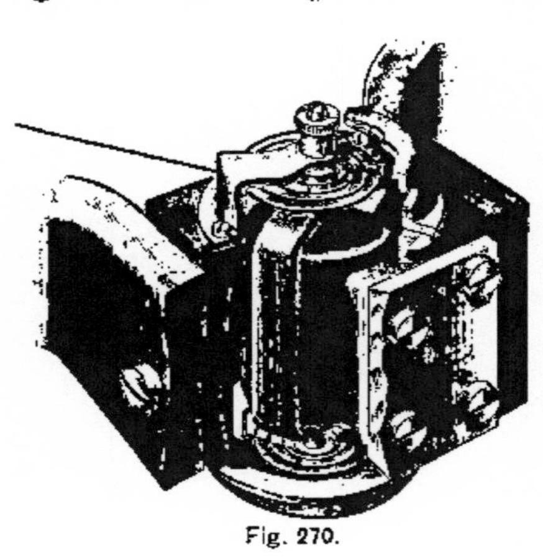
Fig. 270.

475. Resistance Coils. — Coils of wire of known resistance, for use with a galvanometer in measuring resistance, can be purchased of makers of electrical instruments. They should be mounted as shown in Fig. 271. Each coil is wound on its spool double, the inner ends being connected so that the current passes an equal number of times in both directions round the spool; the coil does not then produce a

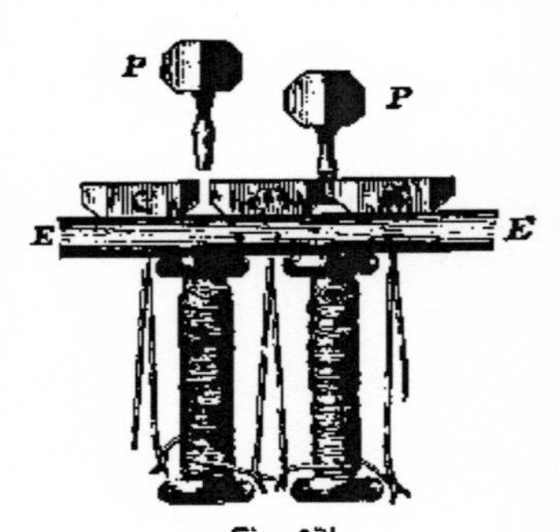

Fig. 271.

magnetic field. This method of winding diminishes what is known as self-induction (§ 483), which is an E.M.F. affecting the current on opening and closing the circuit. The ends of the coils are connected to the heavy brass blocks, C^1, C^2, C^3, on top of the box. When a brass plug, as P, is inserted between two blocks, as C^2 and C^3, the current goes through the plug; but when the plug is withdrawn, as between C^1 and C^2, the current must go through the corresponding coil.

The resistance of these coils is often arranged in ohms of the following numbers : —

1, 2, 2, 5, 10, 10, 20, 50, 100, 100, 200, 500, etc. In this way any number of ohms from one up to the full capacity of the resistance box can be thrown into circuit by withdrawing the proper plugs.

476. Divided Circuits. — When the wire leading from any electric generator is divided into two branches, as at B (Fig. 272), the current also divides, part flowing by one path and a part by the other. The sum of these two currents is always equal to the current in the undivided part of the circuit, since there is no accumulation of electricity at any point. Either of the branches between B and A is called a *shunt* to the other, and *the currents through them are inversely proportional to their resistances.*

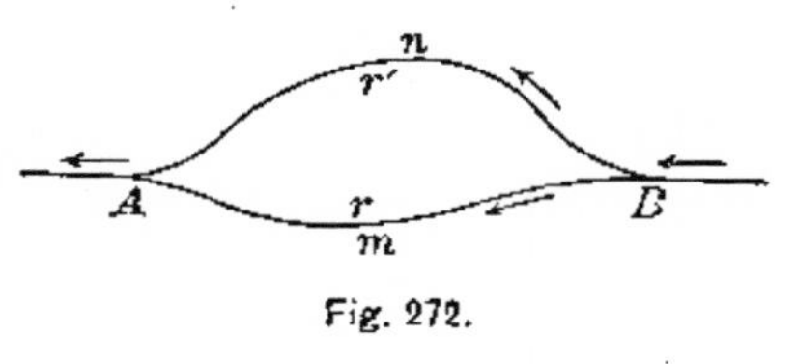

Fig. 272.

477. Resistance of a Divided Circuit. — Let the total resistance between the points A and B be represented by R, that of the branch BmA by r, and of BnA by r'. The conductivity of BA equals the sum of the conductivities of the two branches; and, as conductivity is the reciprocal of resistance, the conductivities of BA, BmA, and BnA are $\frac{1}{R}$, $\frac{1}{r}$, and $\frac{1}{r'}$ respectively, and $\frac{1}{R}=\frac{1}{r}+\frac{1}{r'}$. From this we easily derive $R=\frac{rr'}{r+r'}$. To illustrate, let a galvanometer whose resistance is 100 ohms have its binding posts connected by a shunt of 50 ohms resistance; then the total resistance of this divided circuit is $\frac{100 \times 50}{100+50}=33\frac{1}{3}$ ohms. The introduction of a shunt always lessens the resistance between the points connected, as A and B.

478. Fall of Potential along a Conductor. — When a current flows through a conductor a difference of potential exists, in general, between different points on that conductor. Let A, B, C be three points on a conductor conveying a current, and let there be *no source of E.M.F. between these points.* Then if the current flows from A to B, the potential at A is higher than at B, and the potential at B is higher than at C. If the potential difference between A and B and that between B and C be measured, the ratio of the two will be the same as the ratio of the resistances between the same points. This is only another statement of Ohm's law. For since $I=\frac{E}{R}$, and the current is the same through the two adjacent sections of the conductor, the ratio of the potential differences to the resistances of the two sections is the same. This important principle, of which great use is made in electrical measurements, may be expressed by saying that, when the current is constant, *the loss of potential along a conductor is proportional to the resistance passed over.*

479. Wheatstone's Bridge consists of four resistances connected as shown in Fig. 273. The four conductors R, R', R'', X constitute the *arms*, and the conductor CD the *bridge*. When the circuit is closed the current divides at B, the two parts reuniting at A. The loss of potential along BCA is the same as that along BDA. Now if no current flows through the galvanometer G, then the potentials of C and D must be the same. Under these conditions the fall of potential from B to C is the same as from B to D. We may get an expression for these two potential differences, and place them equal to each other. Let I' be the current through R'; it will also be the current through R, because none flows across through the galvanometer. Also let I'' be the current through the branch BDA. Then the difference of potential between B and C is the same as between B and D, and by Ohm's law (§ 466)

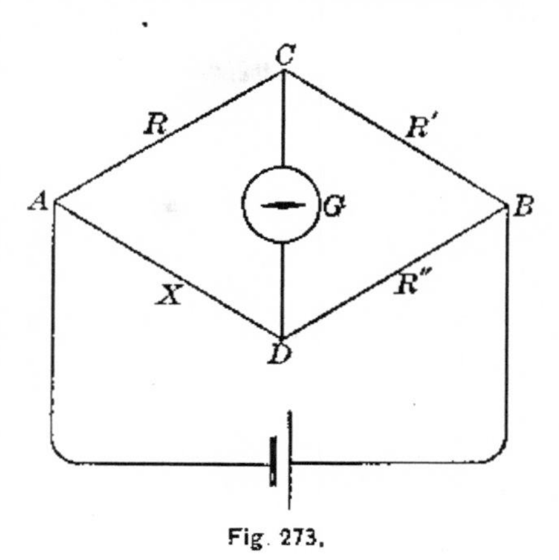

Fig. 273.

$$R'I' = R''I'' \quad . \quad . \quad . \quad . \quad . \quad . \quad (a)$$

In the same way, the equal potential differences between C and A and D and A give

$$RI' = XI'' \quad . \quad . \quad . \quad . \quad . \quad . \quad (b)$$

Dividing (a) by (b), we have

$$\frac{R'}{R} = \frac{R''}{X}. \qquad (31)$$

When no current passes through the galvanometer, the four resistances in the arms of the bridge form a proportion. If three of them are known the proportion gives the formula, $X = \frac{RR''}{R'}$.

Problems.

1. No. 30 wire has a diameter of 0.01 inch. How many feet of German-silver wire of this number will there be in a resistance of 5000 ohms?

2. A coil of copper wire has a resistance of 25.34 ohms. If 20 ft. of the same wire has a resistance of 1.75 ohms, how many feet are there in the coil?

3. How many feet of iron wire, having a diameter of 0.1014 in., will it take to make a resistance coil of 50 ohms?

4. What is the resistance of a cylindrical column of mercury 1 m. long and 1 cm. in diameter at a temperature of 0° C.?

5. A copper wire and an iron wire have the same length and the same resistance; the diameter of the iron wire is how many times that of the copper?

6. If copper wire 0.032 in. in diameter has a resistance of 54 ohms per mile, what will be the resistance of 1000 ft. of copper wire 0.096 in. in diameter?

7. If an electric current in 45 min. deposits by electrolysis 17.75 gm. of copper, what is the strength of the current? (§ 464).

8. How long will it require for a current of 2 amperes to deposit 50 gm. of silver?

9. Four Daniell cells, each having an E.M.F. of 1.1 volts and an internal resistance of 2.5 ohms, are joined in parallel. What will be the current through an external resistance of 1 ohm? Compute the current through the same external resistance when the cells are joined in series.

10. A current of 12 amperes flows through an electric lamp; the difference of potential between the terminals of the lamp is 45 volts. What is the apparent resistance of the lamp?

11. The fall of potential through a set of resistance coils is 450 volts. If the current is 10 amperes, what is the resistance of the coils?

12. What is the resistance of an incandescent lamp taking 0.5 ampere at a pressure of 110 volts.

13. A galvanometer whose resistance is 1000 ohms has its binding posts joined by a shunt wire of 10 ohms resistance. What is the combined resistance between the binding posts?

14. A galvanometer of 200 ohms resistance is to be provided with a shunt such that one-tenth of the whole current shall pass through the galvanometer. Compute the resistance of the shunt.

15. Three wires of 10, 20, and 30 ohms resistance respectively connect the poles of a battery. Compute their combined resistance.

16. A chromic acid cell has an E.M.F. of 2 volts and an internal resistance of 0.2 ohm. What external resistance must be placed in circuit to make the current one ampere?

17. Two equal masses of copper are drawn into wire, one 50 ft. long and the other 200 ft. If the resistance of the shorter wire is 5 ohms, what is the resistance of the longer?

18. A cell of 2 volts E.M.F. sends a current through a galvanometer of 50 ohms resistance, the terminals of the galvanometer being joined by a wire of 10 ohms resistance. If the resistance of the cell and connecting wires is 1 ohm, what is the current through the galvanometer?

19. A battery of two similar cells in series gives a current of 0.2 ampere through an external resistance of 10 ohms, and 0.3 ampere through 5 ohms. Calculate the E.M.F. and the internal resistance of one of the cells.

20. Twelve cells, each having an E.M.F. of 1.5 volts and an internal resistance of $\frac{1}{3}$ ohm, are joined in a mixed circuit of six cells in each of two parallel series. Compute the current through an external resistance of 3 ohms.

XV. ELECTROMAGNETIC INDUCTION.

480. Electromotive Forces induced by Magnets. — **Experiment.** — Connect a coil of insulated wire, consisting of a large number of turns, in circuit with a sensitive galvanometer (Fig. 274). Thrust quickly into the coil the north pole of a bar magnet. The needle of the galvanometer will be deflected, and the direction of the deflection will show that a momentary current has been set flowing round the coil counter-clockwise to one looking down upon it. The current ceases as soon as the magnet stops moving. When the magnet is removed, a current is produced in the opposite direction to the first one. If the south pole be thrust into the coil, and then withdrawn, the currents in both cases are the reverse of those produced by the north pole. If we substitute a helix of a smaller number of turns, or a weaker bar magnet, the deflection will be less, showing that a weaker current is set flowing, or a smaller E.M.F. has been generated.

Fig. 274.

The momentary electromotive forces generated in the helix are known as *induced electromotive forces*, and the currents as *induced currents*. The magnet carries with it into the coil its lines of force; and when the relative positions of a magnet and a closed conductor are so altered that a variation is produced in the number of the lines of force linked with the coil, an induced electromotive force is generated in the conductor in accordance with the following laws: —

I. *An increase in the number of lines of force linked with a coil produces an indirect electromotive force, while a decrease in the number of lines produces a direct electromotive force.* By direct electromotive force is meant one in the direction of the motion of watch-hands, and by indirect electromotive force, one in the opposite direction. The observer must always be looking along the lines of force. Thus, if in Fig. 275 the magnet be thrust into the coil in the direction of the arrow, the current will flow counter-clockwise, as shown by the arrows on the coil. The motion of the magnet into the coil carries its lines of force with it, and thus increases the number passing through the coil.

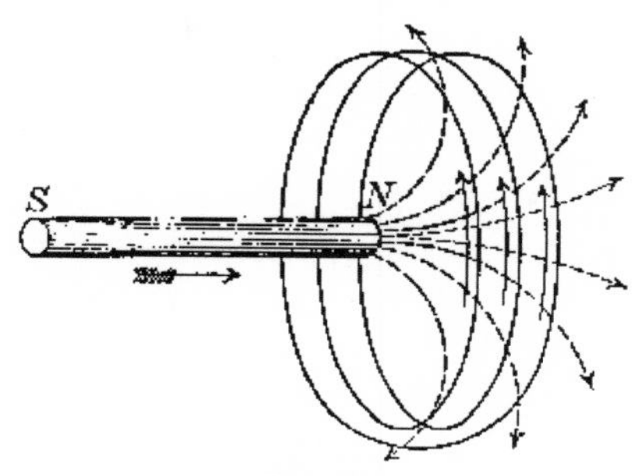

Fig. 275.

II. *The electromotive force produced is equal to the rate of increase or decrease in the number of lines of force linked with the coil.*

Experiment. — Wind a number of turns of fine insulated wire around the armature of a horseshoe magnet, leaving the ends of the iron free to come in contact with the poles of the permanent magnet. Connect the ends of the coil to a sensitive galvanometer, the armature being in contact with the magnetic poles, as shown in Fig. 276. Keeping the magnet fixed, suddenly pull off the armature. The galvanometer will show a momentary current. Suddenly bring the armature up to the poles of the magnet; another momentary current in the reverse direction will flow through the circuit.

Fig. 276.

When the armature is in contact with the poles of the magnet, the number of lines of force passing through the coil is a maximum. As the armature is pulled away, the number of magnetic lines threading through the coil rapidly diminishes. This experiment illustrates Faraday's method of obtaining electric currents by the agency of magnetism.

481. Currents induced by Other Currents. — Experiment. — Insert within a coil connected to a sensitive galvanometer a second coil in circuit with a battery (Fig. 277). The effects are exactly like those obtained by the use of the magnet, even to the direction of the galvanometer deflection, when the polarity of the coil (§ 458) corresponds with that of the magnet. Furthermore, the effects are the same if the coil is inserted before closing the circuit.

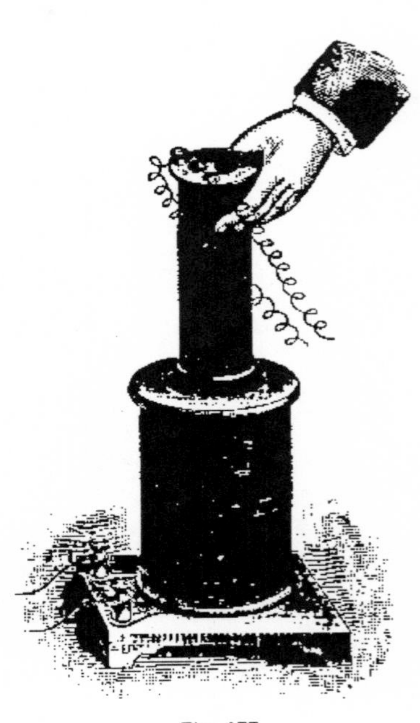

Fig. 277.

This experiment shows that a coil through which a current is passing has the same effect in producing an induced electromotive force as a magnet. This was to be expected, since we have seen that a coil carrying a current has properties identical with a magnet. The coil connected with the battery is called the *primary coil*, and the other the *secondary coil.*

482. Lenz's Law. — *The direction of an induced current is always such that it produces a magnetic field opposing the motion or change which induces the current.* This is known as *Lenz's law.* It is only the law of conservation of energy applied to electricity. To illustrate the law, consider Fig. 275. When the north pole of the magnet is thrust into the coil, the induced current flowing in the direction of the arrows produces lines of force running in the opposite direction to those from the magnet (§ 458). These lines of force tend to oppose the change in the magnetic field within the coil, or the magnetic field set up by the coil opposes the motion of the magnet.

Again, when the primary coil of Fig. 277 is inserted into the secondary, the induced current in the latter is opposite in direction to the primary current, and parallel currents in opposite directions repel each other. In every case of electromagnetic induction the change in the magnetic field which produces the induced current is always opposed by the magnetic field due to the induced current itself.

483. Self-induction. — **Experiment.** — Make a helix of thin brass wire with the turns close together and about 2 cm. in diameter. Solder to one end a small weight and to the weight a short piece of platinum wire. Support the other end of the helix in some convenient way so that its height can be readily adjusted, and let the platinum point just touch the surface of some mercury in a small cup. Connect one electrode of a battery to the upper end of the helix and the other electrode to the mercury. The current through the parallel turns of the helix will cause them to attract one another (§ 459), thus shortening the helix and lifting the platinum wire out of the mercury with a brilliant spark. The attraction then immediately ceases, and the platinum point again drops into the mercury. The operation is thus repeated, and the helix is set vibrating with a succession of bright sparks at the break. (If the helix is long enough and properly weighted, it will divide into vibrating segments separated by nodes, like a stretched cord (§ 214)).

Experiment. — Attach to one pole of a chromic acid battery a flat file and to the other a piece of iron wire. Draw the iron wire quickly across the ridges of the file. Only slight sparking will be visible. Put an electromagnet (§ 457) in the circuit and repeat the experiment. A series of bright sparks will now be seen as the circuit is broken at the ribs of the file.

The sparks in both cases are due to the relatively high E. M. F. induced by the mutual inductive action of adjacent turns of the helix. When there is only a single circuit, as in these cases, the action is called *self-induction.* When the circuit is opened, the self-induced E. M. F. tends to prolong the current, and a spark breaks over the opening. The current flowing across under the impulse of self-induction is often called the *extra current.* There is a similar self-induced E. M. F. when the circuit is closed, and this prevents the instantaneous rise of the current to its maximum value given by Ohm's law.

Let a coil be wound round a wooden cylinder (Fig. 278). Then some of the lines of magnetic force around one turn will thread through adjacent turns. If the cylinder PR is iron instead of wood, the magnetic flux through all the parallel turns of wire will be greatly increased (§ 455). At the instant when the circuit is closed, the increase of magnetic induction through the parallel turns of wire will give rise to a counter E.M.F. of self-induction; and when the circuit is opened, the decrease of magnetic induction through the coil causes a direct E. M. F.; that is, one in the same direction as the current itself (§ 481).

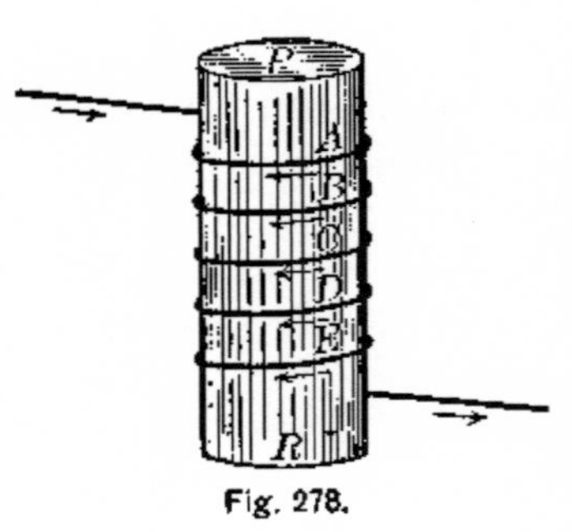

Fig. 278.

484. Self-induction with a Wheatstone's Bridge. — Experiment. — Connect an electromagnet in one of the arms of a Wheatstone's bridge (§ 479) as a resistance to be measured. Adjust for a balance so that no current flows through the galvanometer when the battery circuit is closed first, and the galvanometer circuit a few seconds later. Then, keeping the galvanometer circuit closed, close and open the battery circuit. The galvanometer will show a sharp deflection in one direction when the main circuit is closed, and in the other direction when it is opened.

The E. M. F. generated by self-induction in the electromagnet destroys the balance. The deflections of the galvanometer demonstrate that the induced E. M. F. is in one direction when the circuit through the electromagnet is closed, and in the other direction when it is opened.

485. The Induction Coil. — An apparatus for producing a high E. M. F. by means of induction is called an *induction coil* or a *transformer*. It consists of a coil of coarse insulated wire surrounding an iron core, and a second coil of a very large number of turns of fine wire surrounding the first (Fig. 279).

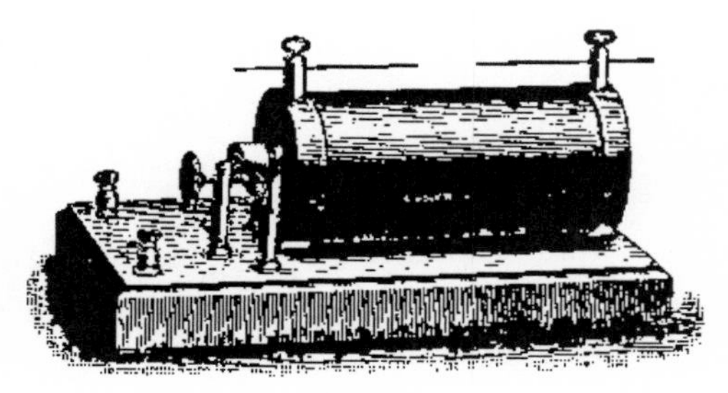

Fig. 279.

The inner or primary coil is connected to a battery through a current interrupter; also through a device called a *commutator* for changing the direction of the current. At the "make" and "break" of the circuit currents are induced in the secondary coil in accordance with the laws of current induction (§ 480). In coils designed to give sparks between the terminals of the secondary coil a *condenser* is added. It is placed in the supporting base of the coil, and consists of two sets of interlaid layers of tin-foil, separated by sheets of paper saturated with oil or paraffin. The two sets are connected with two points of the primary circuit on opposite sides of the current interrupter.

486. Action of the Coil. — Figure 280 shows the arrangement of the various parts of an induction coil. The current first passes through the heavy primary wire *PP*, thence through the spring *h*, which carries the soft iron block *F*, then across to the screw *b*, and so back to the negative pole of the battery. This current magnetizes the iron core of the coil, and the core attracts the soft iron block *F*, thus breaking the circuit at the point of the

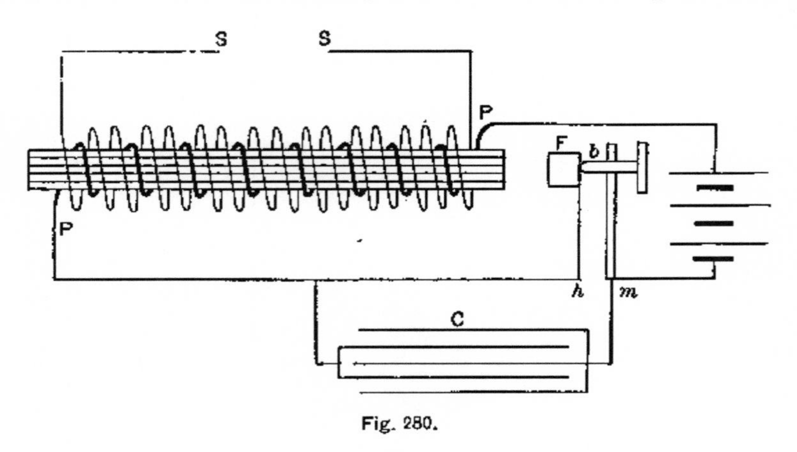

Fig. 280.

screw *b*. The core is then demagnetized, and the release of *F* again closes the circuit. Electromotive forces are thus induced in the secondary coil *SS*, both at the make and the break of the primary. Hence, if the secondary is closed, alternating currents are induced in it by the intermittent current in the primary. The high E. M. F. of the secondary is due to the large number of turns of wire in it and to the influence of the iron core in increasing the number of lines of force which pass through the entire coil.

The self-induction of the primary has a very important bearing on the action of the coil. At the instant the circuit is closed, the counter E. M. F. opposes the battery current, and prolongs the time of reaching its greatest strength. Consequently the E. M. F. of the secondary coil will be diminished by self-induction in the primary. The E. M. F. of self-induction at the "break" of the primary is direct, and this added to the E. M. F. of the battery produces a spark at the break points.

487. The Condenser. — The addition of a condenser increases the E.M.F. of the secondary coil in two ways: 1. It gives such an increase of capacity to the primary coil that at the moment of breaking the circuit the potential difference between the contact points does not rise high enough to cause a spark discharge across the air gap. The interruption of the primary is therefore more abrupt, and the E.M.F. of the secondary is increased. 2. After the break, the condenser *C*, which has been charged by the E.M.F. of self-induction, discharges back through the primary coil and the battery. The condenser causes an electric recoil in the current, and returns the stored charge as a current in the reverse direction through the primary, thus demagnetizing the core, increasing the rate of change of magnetic flux, and increasing the induced E.M.F. in the secondary. The condenser momentarily stores the energy represented by the spark when no condenser is used, and then returns it to the primary and by mutual induction to the secondary, as indicated by the longer spark or the greater current. When the secondary terminals are separated, the discharge is all in one direction and occurs when the primary current is broken.

488. Experiments with the Induction Coil. — 1. *Physiological Effects.* — Hold in the hands the electrodes of a very small induction coil, of the style used by physicians. When the coil is working, a peculiar muscular contraction is produced.

The "shock" from large coils is dangerous on account of the high E.M.F. The danger decreases with the increase in the rapidity of the impulses or alternations. Experiments with induction coils, worked by alternating currents of very high frequency, have demonstrated that the discharge of the secondary may be taken through the body without injury.

2. *Mechanical Effects.* — Hold a piece of cardboard between the electrodes of an induction coil giving a spark 3 cm. long. The card will be perforated, leaving a burr on each side. Thin plates of any non-conductor can be perforated in the same manner.

3. *Heating Effects.* — Make a torpedo (§ 419) and place it in circuit with an induction coil. Close the battery circuit and the torpedo will explode.

4. *Chemical Effects.* — Place on a plate of glass a strip of white blotting-paper moistened with a solution of potassium iodide and starch paste. Attach one of the electrodes of a small induction coil to the margin of the paper. Handle a wire leading to the other electrode with an insulator, and trace characters with the wire on the paper when the coil is in action. The discharge decomposes the chemical salt, as shown by the blue mark. This blue mark is due to the reaction between the iodine and the starch.

489. Discharges in Partial Vacua. — **Experiment.** — Cover the inside of a vase or glass goblet with tin-foil, a little over halfway up, and place the goblet on the table of the air-pump under a bell-jar provided with a brass sliding rod passing air-tight through the cap at the top (Fig. 281). Surround the part of the rod inside the jar with a glass tube and push the rod down till it touches the tin-foil. Connect the rod and the air-pump table to the terminals of the induction coil. When the air is exhausted a beautiful play of light will fill the bell-jar. If the vase is made of uranium glass, or if it stands on a block of that material, the effect is still more beautiful. This experiment is known as *Gassiot's cascade*

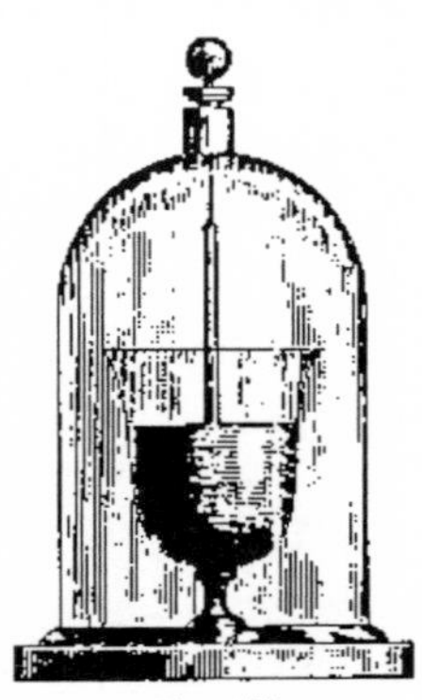

Fig. 281.

The best effects are obtained with discharges from the secondary of an induction coil in glass tubes when the exhaustion is carried to a pressure of about 2 mm. of mercury, and the tubes are permanently sealed. Platinum electrodes are melted into the glass at the two ends. Such tubes are known as *Geissler tubes.* They are made in a great variety of forms (Fig. 282), and the luminous effects are more intense in the narrow connecting tubes than in the large bulbs at the ends. The colors are determined by the nature of the residual gas. Hydrogen glows with a brilliant crimson; the vapor of water gives the same color, indicating that the vapor is dissociated by the discharge. An examination of this glow by the spectroscope gives the characteristic lines of the gas in the tube.

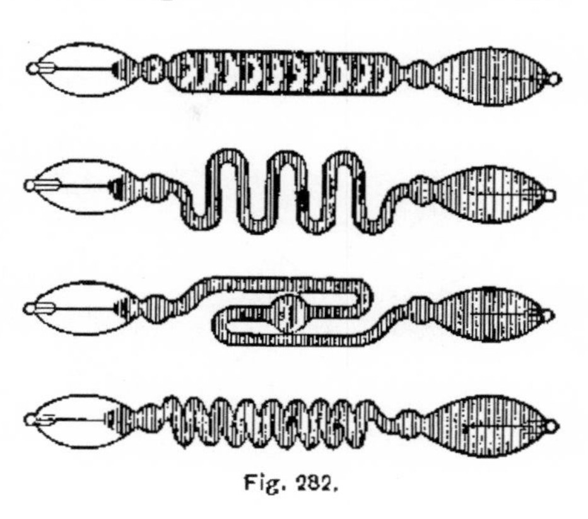

Fig. 282.

Geissler tubes often exhibit *stratifications*, which consist of portions of greater brightness separated by darker intervals. Stratifications have been produced throughout a tube 50 feet long. These striæ present an unstable flickering motion, resembling that sometimes observed during auroral displays.

490. Discharges in High Vacua. — When the exhaustion of a tube is carried to about a millionth of an atmosphere, the effects of an electric discharge through it are entirely changed in character. The light emanating from the residual gas nearly disappears. The dark space, which is present about the negative or cathode of a Geissler tube, broadens till it reaches the opposite wall of the bulb. Tubes of this kind were first investigated by Sir William Crookes; they are therefore known as *Crookes tubes*. A stream of the electrified particles, or of much more minute parts called *corpuscles*, is then projected from the cathode across to the opposite wall of the bulb. The impact excites remarkable luminous effects in the glass. Other substances, such as diamond, ruby, and various sulphides, under the impact of the cathode rays, as they are called, exhibit beautiful fluorescent colors (Fig. 283).

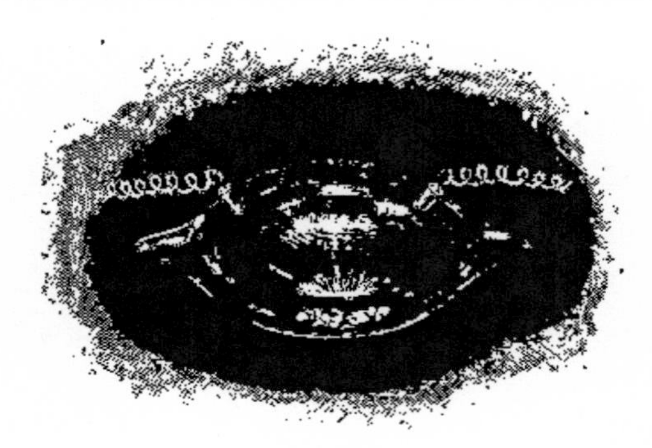

Fig. 283.

These cathode rays are straight and at right angles to the cathode surface, except when they are deflected by a magnet or by mutual repulsion. The cathode stream, when once deflected by a magnet, does not recover its former direc-

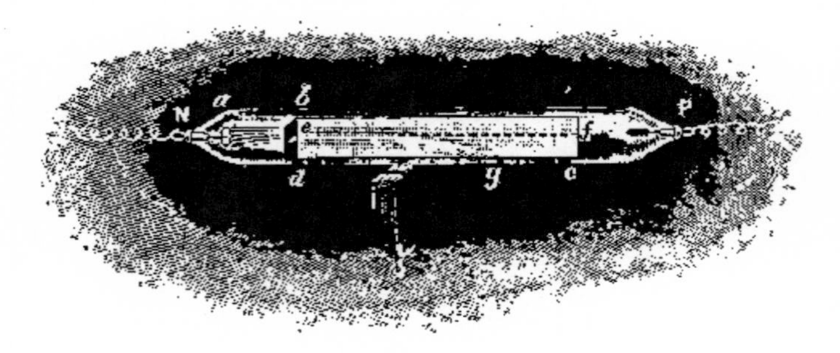

Fig. 284.

tion after passing the magnet (Fig. 284). A screen, *bd*, placed in the path of cathode rays stops them more or less completely, and appears to cast a shadow by protecting the wall of the tube from their impact. At the same time the obstruction is acted on mechanically, as if by a bombardment or a wind. If the cathode is concave, the rays cross at a focus, and at this focus platinum foil can be raised to a white heat (Fig. 285).

Fig. 285.

491. Roentgen Rays. — The rays of radiant matter, as Crookes called it, emanating from the cathode, give rise to another kind of rays when they strike the walls of the tube, or a piece of platinum placed in their path. These last rays, to which Roentgen, their discoverer, gave the name of "*X-rays*," can pass through glass, and so get out of the tube. They also pass through wood, paper, flesh, and many other substances opaque to light. They are stopped by bones, metals (except in very thin sheets), and by some other substances. Roentgen discovered that they affect a photographic plate like light. Hence, photographs can be taken of objects which are entirely invisible to the eye, such as the bones in a living body, or bullets embedded in the flesh.

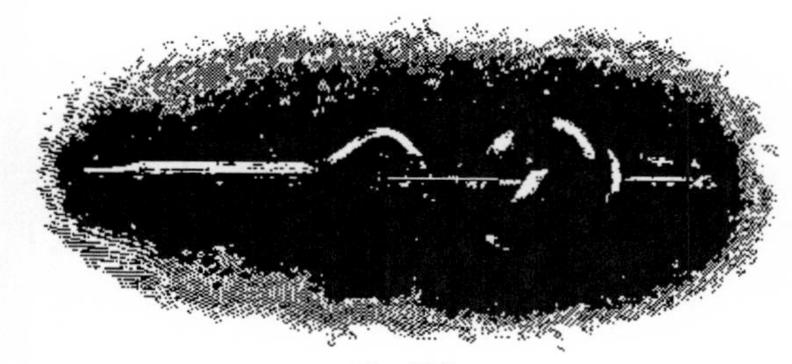

Fig. 286.

A Crookes tube adapted to the production of Roentgen rays (Fig. 286) has a concave cathode *K*, and at its focus an inclined piece of platinum

A, which serves as the anode. The X-rays originate at A and issue from the side of the tube. Figure 287 is a photograph taken by the aid of one of these tubes. The sensitized plate was enclosed in an ordinary plate-holder, the hand was laid on the holder next to the sensitized side of the plate, and the X-ray tube was held about a foot above the hand.

Fig. 287.

The tube of Fig. 286 is called a *focus tube*, because the cathode is so shaped that all the cathode rays reach the inclined platinum anode at nearly the same point. The X-rays themselves cannot be focussed; all photographs taken by them are only shadow pictures; but, as in the case of light, the shadow will be sharper when the source of light is of small area than when it is larger (§ 232). When the tube is working well, the anode A becomes red-hot near its centre, and is then a source of copious X-rays.

492. The Fluoroscope. — Soon after the discovery of X-rays it was found that certain fluorescent substances, like platino-barium-cyanide, and calcium tungstate, become luminous under the action of X-rays. This fact has been turned to account in the construction of a *fluoroscope* (Fig. 288), by means of which shadow pictures of concealed objects become visible. An opaque screen is covered on one side with the fluorescent substance; this screen fits into the larger end of a box blackened inside, and having at the other end an opening adapted to fit closely around the eyes, so as to exclude all outside light. When an object, such as the hand, is held against the fluorescent screen and the

Fig. 288.

fluoroscope is turned toward the Roentgen tube, the bones are plainly visible as darker objects than the flesh because they are more opaque to X-rays. The beating heart may be made visible in a similar manner.

XVI. DYNAMO-ELECTRIC MACHINES.

493. A Dynamo-electric Machine is a device designed to produce electric currents by the expenditure of mechanical work. The so-called generation of electric currents consists always in the generation of electromotive forces or electric pressures. Every dynamo-electric machine has three essential parts: 1. The *inductor* or *magnet* by which the magnetic field is produced. 2. The *armature*, or system of conductors in which electromotive forces are generated by a change in the magnetic field or flux through it. 3. The *brushes* or *collecting rings.* When the field is produced by a permanent magnet, the machine is called a *magneto;* when the electromagnet is used, the machine is a *dynamo.*

494. The Ideal Simple Dynamo. — Suppose a single loop of wire to revolve between the poles of a magnet *NS* (Fig. 289) in the direction of the arrow and round a horizontal axis. The magnetic flux runs across from N to S, a indicated by the light lines. The loop encloses in th position shown the largest possible magnetic flux. Whe it has rotated through a quarter of a turn, the magneti flux will be parallel to its plane, and none will then pas through it. During this quarter turn, the decrease in th magnetic flux through the loop generates a direct E.M.F looking in the directio N to S, as indicated b the arrows on the loop During the next quarte turn the magnetic flu through the rotating loo will increase again, bu it will run through th loop in the opposite direction. This is equivalent to continuous decrease in the original direction, and therefor the direction of the induced E.M.F. round the loop re mains the same for the entire half turn. If the loop i part of a closed circuit, a current will flow in the directio of the E.M.F. induced. During the next half revolu tion, the current will flow in the opposite direction roun the loop. Hence the E.M.F. and the current throug the loop reverse twice during every revolution.

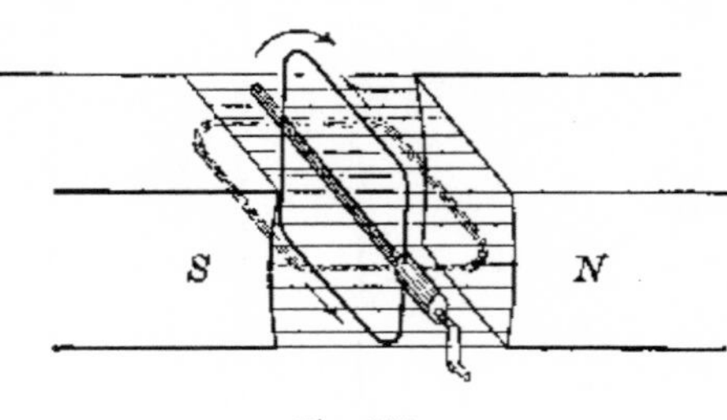

Fig. 289.

495. The Shuttle Armature. — If the loop is composed o several turns of wire instead of one, the E.M.F. generate by the rotation of the coil is increased in the same ratio as the number of turns of wire. The E.M.F. can also be increased by winding the coil on iron, for the iron increases the magnetic flux through the coil (§ 377). Fig. 290 is a cross-section of a

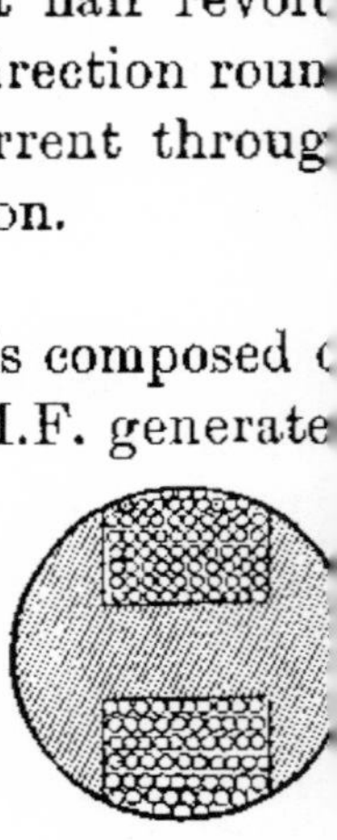

Fig. 290.

coil with a large number of turns of wire wound in grooves ploughed in an iron cylinder. It is called the ***shuttle armature.*** Such an armature is very inefficient and not suitable for large currents. Currents are produced in the solid iron core, as well as in the insulated wire, and these absorb energy and heat the iron. It is used extensively in "magnetos" for ringing bells on telephone lines.

496. The Commutator. — When it is desired to convert the alternating currents flowing in the armature into a current in one direction through the external circuit, a special device called a *commutator* is employed. For a single coil in the armature, the commutator consists of two parts only. It is a split tube with the two halves insulated from each other and from the shaft on which they are mounted (Fig. 291). The two ends of the coil are connected with the two halves of the tube. Two brushes, with which the external circuit is connected, bear on the commutator, and they are so placed that they exchange contact with the two commutator segments at the same time that the current reverses in the coil. In this way one of the brushes is always positive and the other negative, and the current flows in the external circuit from the positive brush back to the negative, and thence through the armature to the positive again.

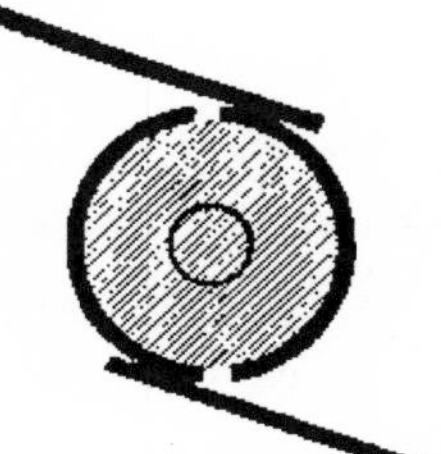

Fig. 291.

497. The Gramme Ring. — The use of a commutator with more than two parts is conveniently illustrated in connection with the *Gramme ring.* This armature has a core made either of iron wire, or of thin disks at right angles to the axis of rotation. The iron is divided for the purpose of preventing induction currents in it, which waste energy. The relation of the several parts of the machine is illustrated by Fig. 292. A number of coils are wound in one direction and are all joined in series. Each junction between coils is connected with a commutator bar. When a coil is in the highest position in the figure, the maximum flux passes through it; as the ring rotates, the flux through the coil decreases, and after a quarter of a revolution there is no flux through it. The current

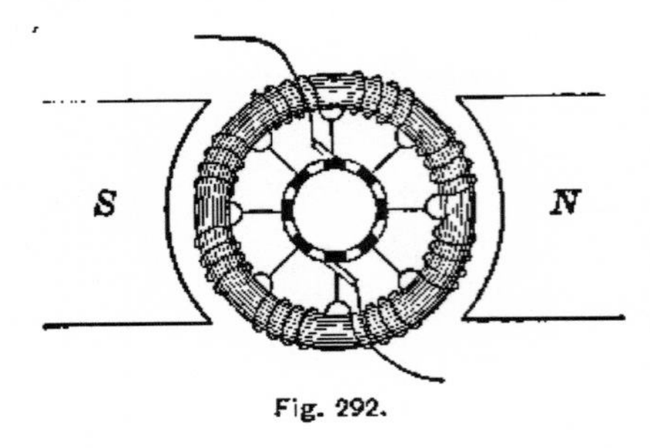

Fig. 292.

through each coil reverses twice during each revolution, exactly as in the case of the single loop. No current flows entirely round the armature, because the E.M.F. generated in one coil at any instant is exactly counterbalanced by the E.M.F. generated by the coil opposite. But when the external circuit connecting the brushes is closed, a current flows up on both sides of the armature. The current has then two paths through the armature, and one brush is constantly positive and the other negative. By increasing the number of coils, the potential difference between the brushes never drops to zero, as it does with a single coil twice during each revolution, but it is kept nearly constant and at the highest value given by half the coils in series. The brushes must bear on the commutator near the part of the field where the E.M.F. in any coil passes through zero and reverses.

498. The Field-magnet. — In dynamos the magnetic field is produced by a large electromagnet excited by the current flowing from the armature, which is led either wholly or in part round the field-magnet cores. When the entire current is carried round the coils of the field-magnet, the dynamo is said to be *series wound.* When the field-magnet is excited by coils of many turns of fine wire connected as a shunt to the external circuit, the dynamo is said to be *shunt wound.* The connections of the field and external circuit of the two kinds of winding are shown in Fig. 293. A combination of these two methods of exciting the field-magnet is called *compound*

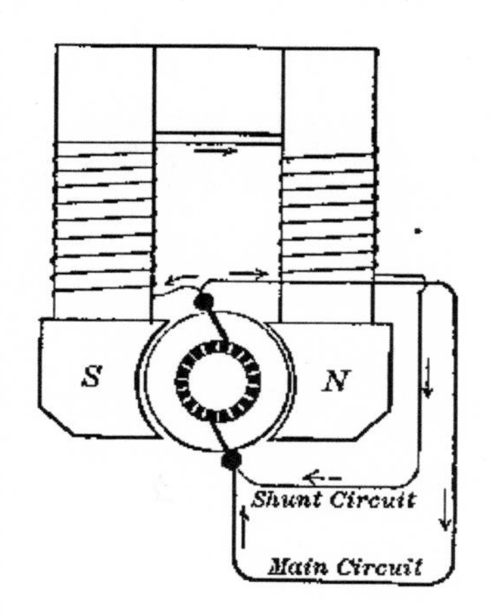

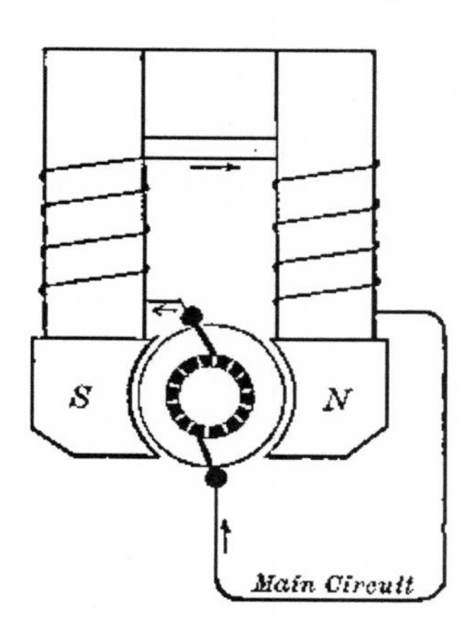

Fig. 293.

winding. The residual magnetism remaining in the cores is sufficient to start the machine. The current thus produced increases the magnetic flux through the armature and so increases the E.M.F.

499. The Drum Armature.— The modern *drum armature* for direct currents is an evolution from the one-coil shuttle armature. It is shown at *A*, Fig. 294, which represents the parts of a four-pole machine. Fig. 295 is the same machine ready to run.

In the drum armature the iron core, which is made up

of laminated disks, contains a series of grooves, parallel to the shaft, and coils are wound in them at equal angular

Fig. 294.

distances round the circumference. These sections of the armature may all be joined in series, and the junctions between them are then connected with the commutator bars *C*, as in the Gramme ring. If there are only two poles to the field-magnet, there are only two brushes and two circuits through the armature. The ma-

Fig. 295.

chine of Fig. 294 has four poles P, four armature circuits, and four brushes. A four-pole machine may be wound so as to require only two brushes. When there are four sets of brushes, two of them are positive and two are negative ; the two positives are connected in parallel as the positive terminal, and the two negatives as the negative terminal. With such a winding, divided into numerous sections with corresponding commutator bars, the current in the external circuit is almost perfectly continuous and free from small fluctuations.

500. The Electric Motor. — An *electric motor* for direct currents is exactly like a generator. In fact, any direct current generator may be used as a motor. A study of the magnetic field across which a current is flowing aids greatly in understanding both a generator and a motor. Fig. 296 is the magnetic field (shown by iron filings) distorted by the magnetic influence of a current in a wire looped through the two holes between unlike magnetic poles. These lines of force are under tension, and the loop has therefore a magnetic stress acting on it and tending to turn it counter-clockwise. If the loop is rotated clockwise by mechanical means, it turns against the magnetic torque on it, and work must be done against the resistance of this magnetic drag. When, therefore, the armature of a dynamo is rotated by mechanical

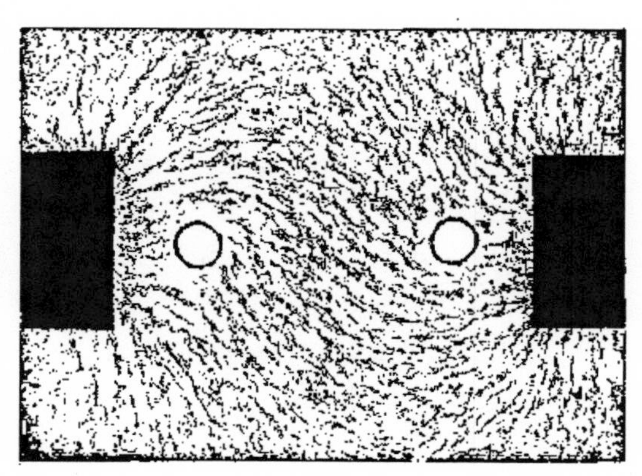

Fig. 296.

means against the internal magnetic action between th field and the armature, mechanical power is converte into electrical energy, and the machine is a generator If, on the other hand, the rotation is in the directio of the magnetic effort between the field and the arma ture, work is done by the machine as a motor. Wit the current through the field and the armature in th same direction in both cases, the armature rotates i one direction as a generator and in the other directio as a motor; but if the field be reversed when the machin is used as a motor, the armature will turn in the same direc tion as when acting as a generator without reversing th field. A series-wound machine turns one way as a gen erator and the other as a motor; a shunt-wound machin turns in the same direction whether operating as a gener ator or as a motor.

501. Alternators. — If the brushes of a dynamo bear o two continuous rings on the shaft instead of on a commu tator, the current in the external circuit will alternate o reverse every time the armature turns through the angu lar distance from one field pole to the next. A complet series of changes in the current and E.M.F. in both direc tions takes place while the armature is turning the dis tance from one pole to the next one of the same name This is called a *cycle*. The *frequency* is the product o the number of *pairs of poles* and the number of rotation per second. In two-pole machines the frequency is the same as the number of rotations per second. Frequencie are now restricted between the limits of about twenty-fiv and sixty cycles per second. Multipolar machines ar used to avoid excessive speed of rotation.

Figure 297 is a diagram of an alternator with a stationar

field and a rotating armature. In large machines the armature is generally the stationary member and the field revolves. The field is excited by a direct current machine. The armature coils are reversed in winding from pole to pole; they are then all joined in series, and the terminals are brought out to the two slip rings at *B*. The brushes bearing on these slip rings lead to the external circuit.

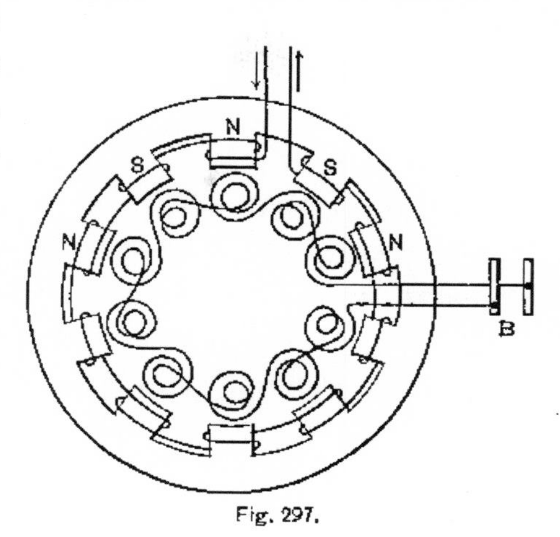

Fig. 297.

502. Transformers. — A *transformer* is an induction coil with a primary of many turns of wire and a secondary of a smaller number, both wound round a divided iron core forming a closed magnetic circuit; that is, the magnetic flux is entirely through iron (Fig. 298). A transformer is employed with alternating currents either to step down from a high E.M.F. to a low one, or the reverse. The two electromotive forces are directly proportional to the number of turns of wire in the two coils. While the primary and the secondary are wound round the same iron core, they are as perfectly insulated from each other as possible. The iron serves as a path for the flux of mag-

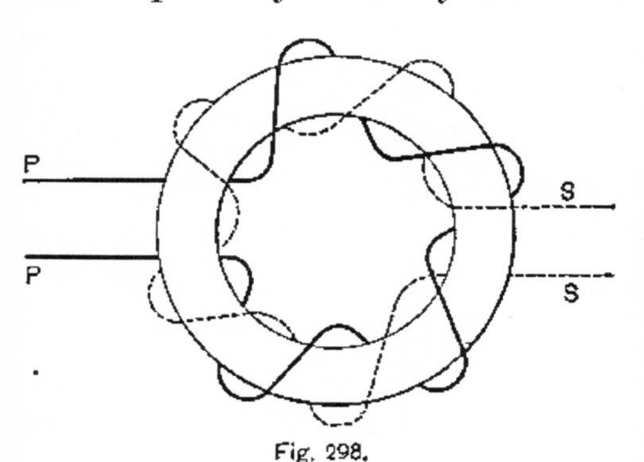

Fig. 298.

netic induction, and all the lines of force produced by the primary also pass through the secondary, and *vice versa.* When the secondary is open, the transformer acts simply as a "choke coil"; the current in the primary is then only the very small one required to magnetize the iron and to furnish the small amount of energy lost in it, for the counter E.M.F. of self-induction is then nearly equal to the impressed E.M.F. When the secondary is closed, the energy of the current supplied by it lacks only a few per cent of the energy absorbed by the primary; for, while the secondary E.M.F. is less than the primary, the secondary current is greater in the same ratio.

XVII. THE ELECTRIC LIGHT.

503. Electric Lights are of two classes, *arc* and *incandescent.* The former is produced by a current of a few amperes passing across from one carbon rod to another, the highly heated vapor between the two rods acting as a conductor. The carbon rods are placed in electrical connection with the two poles of a dynamo, and the current is established by bringing the carbon points together. Upon separation they are heated to an exceedingly high temperature and the current continues to pass across through the heated vapor. The ends of the carbon rods in the open air are disintegrated, a depression or crater forming in the positive (Fig. 299) and a cone on the

Fig. 299.

negative carbon. Most of the light of the open electric arc comes from the bottom of the crater, which is the hottest part of the carbon. Sir Humphry Davy, who first produced the electric arc light, obtained an arc four inches in length between two charcoal points. For this purpose he used a great battery of 2000 cells. The pores of the charcoal sticks had been previously filled with mercury. Despretz, with 600 Bunsen cells, obtained an arc 7.8 inches long. The arc light may be produced in a vacuum. The intense heat is not due, therefore, to combustion, but to the conversion of the energy of the current into heat in the arc. The electric arc light may be produced even in water, but its brilliancy is much reduced and the water is rapidly decomposed.

504. The Open and the Enclosed Arc. — When the electric arc is maintained between rods of hard retort carbon in the open air, the carbon burns away rather rapidly, the positive about twice as fast as the negative. The E.M.F. required is then from 45 to 55 volts for a 10-ampere current. Modern arc lamps are mostly "enclosed arcs"; that is, the lower carbon and a part of the upper one are enclosed in a small glass globe, which is air-tight at the bottom, but allows the upper carbon to slip through a check-valve at the top. Soon after the arc begins to burn, the oxygen is absorbed and the arc is then enclosed in an atmosphere of nitrogen and carbon monoxide. The enclosed arc is longer than the open arc, and the E.M.F. required is about 80 volts, while the current for the same consumption of energy is smaller than the open arc requires. The carbons for the enclosed arc last about ten times as long as in the open air.

505. The Arc Lamp. — The wasting away of the carbons necessitates some automatic devices to make them approach each other. Most arc lamps contain one electromagnet for separating the carbons when the current is turned on, and a second one to make the upper carbon feed down when the arc requires it. Fig. 300 represents one form of enclosed arc lamp with automatic feed of the upper carbon. *M* is a lifting magnet to start the arc by separating *C* from *C′* in the small inner globe. The magnet *E* has a circular or disk armature *D*, to which is attached a pinion engaging the rack *R*. The upper electromagnet is a solenoid containing an iron core, which is drawn into it by the magnetism induced in the iron. The lower magnet has both a series and a shunt winding, and the two are so connected that they oppose each other magnetically; that is, they act differentially. The ends of the fine wire winding composing the shunt are attached to the two carbon holders. When the arc increases in length, the potential difference between the carbons increases and a larger current flows through the shunt winding. The magnetism due to the series winding, which is connected in series with the carbons themselves, is then neutralized by the effect of the shunt coil. This differential magnet, which is quick acting, then releases its armature, and the upper carbon drops slightly by its own weight. It does

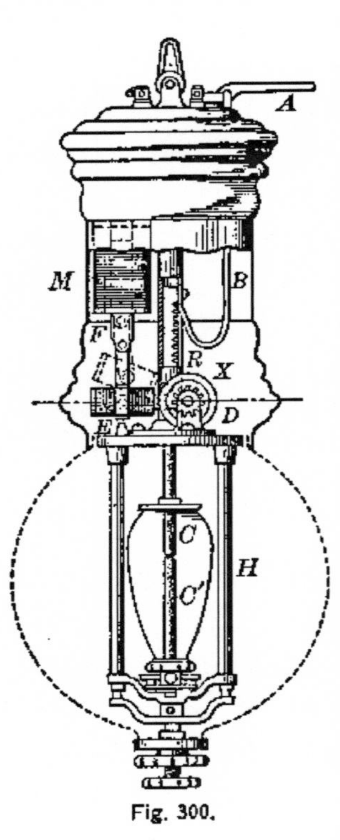

Fig. 300.

not reach the lower carbon, because the differential magnet again grips its armature and stops the descent before the arc is extinguished. The upper carbon is thus fed down as needed till the carbon holder touches the check-valve.

506. Incandescent Lamps.—The smaller subdivision of the electric light is made by means of carbon filaments enclosed in exhausted glass bulbs (Fig. 301). These small lamps are usually placed in parallel across from one main conductor to the other (Fig. 302). The carbon filaments have a resistance of from 25 to 800 or 900 ohms when hot, according to the voltage used and the candle power of the lamp. On account of this resistance, the current heats the filament to incandescence. No combustion takes place, but the carbon gradually deteriorates, is thrown off, and blackens the bulb. The lamp then gives less light, and if burned long enough, the carbon filament will break and interrupt the current.

Fig. 301.

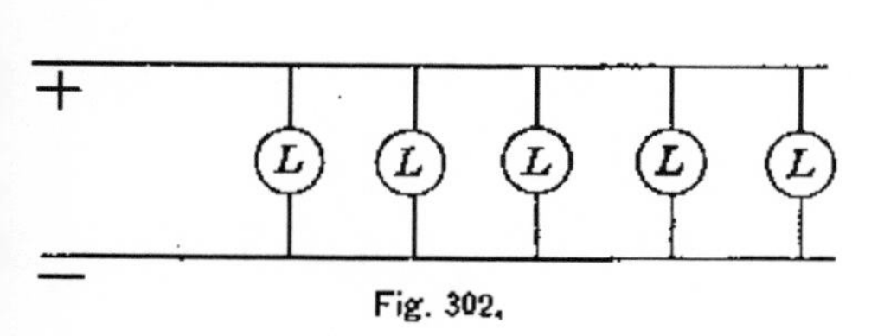

Fig. 302.

The Nernst incandescent lamp employs a refractory earth as the luminescent body. This material is not conducting till it is preheated by a special device for the purpose. It is then traversed by a current which heats it to brilliant incandescence. The Nernst lamp gives light of an intensity intermediate between a carbon filament and an electric arc. Commercial glow lamps take from $2\frac{1}{2}$ to 4 watts for each candle power.

XVIII. THE ELECTRIC TELEGRAPH.

507. The Electric Telegraph is a system of transmitting messages by means of simple signals through the agency of an electric current. Its essential parts are the *line*, the *transmitter* or *key*, the *receiver* or *sounder*, and the *battery*.

508. The Line is an iron or copper wire, insulated from the earth except at its ends, and serving to connect the signaling apparatus. The ends of this conductor are connected with large metallic plates, or with gas or water pipes, buried in the earth. By this means the earth becomes a part of the electric circuit containing the signaling apparatus.

Fig. 303.

509. The Transmitter or Key (Fig. 303) is merely a current interrupter, and usually consists of a brass lever *A*, turning about pivots at *B*. It is connected with the line by the screws *C* and *D*. When the lever is pressed down, a platinum point projecting under the lever is brought in contact with another platinum point *E*, thus closing the circuit. When not in use, the circuit is left closed, the switch *F* being used for that purpose.

Fig. 304.

510. The Receiver or Sounder (Fig. 304) consists of an electromagnet *A* with a pivoted armature *B*. When the circuit is closed through the terminals D and E, the armature is attracted to the magnet, producing a sharp click. When the circuit is broken, a spring C causes the lever to rise and strike the back stop with a lighter click.

511. The Relay. — When the resistance of the line is large, the current is not likely to be strong enough to operate the sounder with sufficient energy to render the signals distinctly audible. To remedy this defect, an electromagnet,

Fig. 305.

called a *relay* (Fig. 305), whose helix A is composed of many turns of fine wire, is placed in the circuit by means of its terminals C and D. As its armature H moves to and fro between the points at K, it opens and closes a second and shorter circuit through E and F, in which the sounder is placed. Thus the weak current, through the agency of the relay, brings into action a current strong enough to do the necessary work.

512. The Battery consists of a large number of cells, usually of the gravity type, connected in series. (Why?) It is generally divided into two sections, one placed at each terminal station, these sections being connected in series through the line. The principal circuits of the great telegraph companies are now worked by means of currents from dynamo machines.

513. The Signals are a series of sharp and light clicks separated by intervals of silence of greater or less duration, a short interval between the clicks being known as a "dot," and a long one as a "dash." By a combination of "dots" and "dashes" letters are represented and words are spelled out.

514. The Telegraph System described in the preceding sections is known as Morse's, from its inventor. Fig. 306 illustrates diagrammatically the instruments necessary for one terminal station, together with the mode of connection. The arrangement at the other end of the line is an exact duplicate of this one, the two sections of the battery being placed in the line, so that the negative pole of one and the positive of the other are connected with the earth. At intermediate stations the relay and the local circuit are connected with the line in the same manner as at a terminal station.

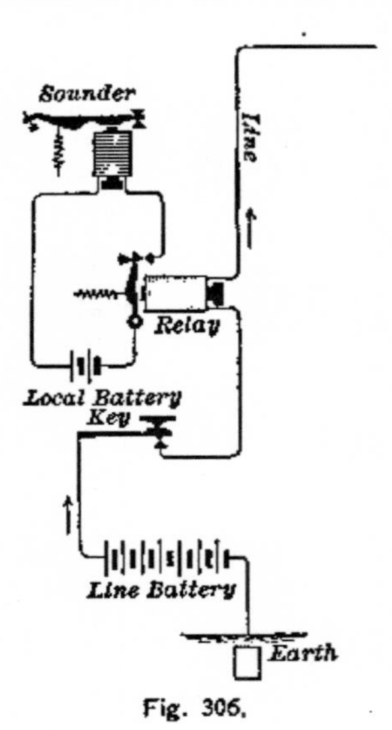

Fig. 306.

515. The Electric Bell (Fig. 307) is used for sending signals as distinguished from messages. Besides the gong, it contains an electromagnet, having one terminal connected directly with a binding-post, and the other, through a light spring attached to the armature (shown on the left of the figure) and a contact screw, with another binding post. One end of the armature is supported by a

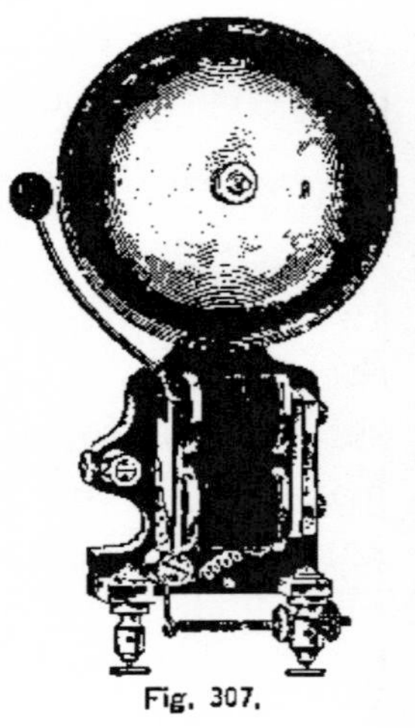
Fig. 307.

stout spring, or on pivots, and the other carries the bent arm and hammer to strike the bell. Included in the circuit are a battery and a push-button *B*, shown in section in Fig. 308.

When the springs *S* are brought into contact by pushing *B*, the circuit is closed, the electromagnet attracts the armature, and the hammer strikes the gong. The movement of the armature opens the circuit by breaking contact between the spring and the point of the screw; the armature is then released, the retractile spring at the bottom carries it back, and contact is again established between the spring and the screw. The whole operation is repeated automatically as long as the circuit is kept closed at the push-button.

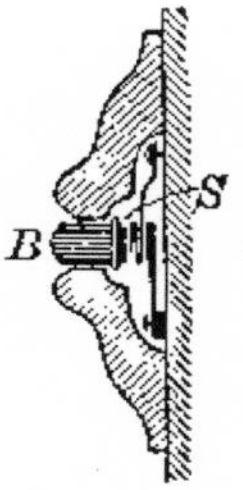

B S

Fig. 308.

XIX. WIRELESS TELEGRAPHY.

516. Electric Waves. — The fundamental discovery which has led to telegraphy through space without the use of wires was made by Hertz in 1887–88. Joseph Henry discovered long ago that the discharge of a Leyden jar is oscillatory (§ 426). Later it was found that the spark discharge of an induction coil is likewise oscillatory. Hertz discovered that this oscillatory spark gives rise to electromagnetic waves in the ether, which appear to be the same as waves of light, except that they are very much longer, or of lower frequency. They travel in straight lines, and are reflected and refracted in accordance with the same laws as light.

Evidence of these waves may be readily obtained by setting up an induction coil, with two sheets of tin-foil on glass, Q and Q', connected with the secondary terminals, and with two discharge balls, F and F', as shown in Fig. 309. The receiving apparatus should be tuned so as to have the same period of oscillation as this transmitter; but even without this, so simple a device as a large picture-frame with a conducting gilt border, may be used to detect electric waves in the neighborhood of the induction coil. If the frame has shrunken so as to leave very narrow gaps in the mitre at the corners, then minute sparks may be seen, in a dark room, breaking across these

Fig. 309.

gaps when the induction coil produces vivid sparks between the polished balls, F and F'. The plane of the frame should be held parallel with the sheets of tin-foil. The passage of electromagnetic waves through a conducting circuit produces electric oscillations in it, and these oscillations cause electric surges across a minute air gap. An apparatus tuned to the same frequency as the transmitter or oscillator is called a receiver or resonator.

517. The Coherer.— A very sensitive device for the detection of electromagnetic waves was discovered by Branly. He found that when metallic filings are placed loosely between solid electrodes in a glass tube they offer a high resistance to the passage of an electric current; but when electric oscillations are produced in the neighborhood of the tube, the resistance of the filings falls to so small a value that a single voltaic cell sends through them a current strong enough to work a relay (§ 511). The tube containing the metallic filings is called a *coherer*. It is shown at *C* in Fig. 310. If the tube is slightly jarred, the filings resume their state of high resistance. A slight discharge from the cover of an electrophorus (§ 413) through

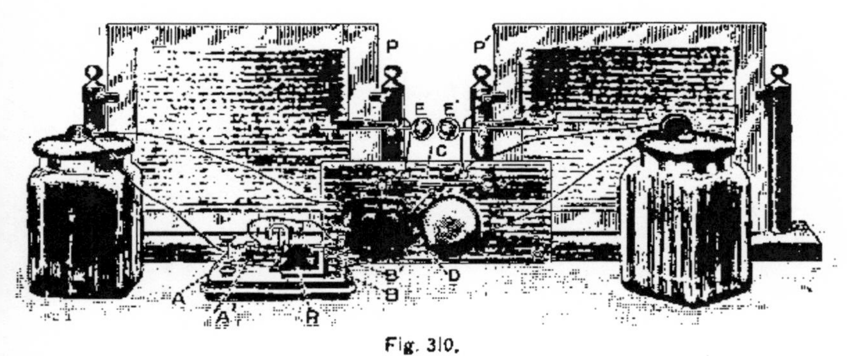

Fig. 310.

the filings lowers the resistance just as electric oscillations do. It is thought that minute sparks between the filings partially weld them together and make them conducting.

518. Receiver for Electromagnetic Waves. — The transmitter of electric oscillations for laboratory experiments is represented by the apparatus of Fig. 309. The *receiver* is somewhat more complicated, but the different parts and their connection may be easily made out with the help of Fig. 310. The metallic plates, *P* and *P′*, are duplicates of those belonging with the transmitter. The two balls, *E* and *E′*, are connected both with the plates and with the ends of the coherer *C*. A circuit is formed including the voltaic cell (shown on the right), the coherer, and the fine wire coil *R* of the relay through the binding posts, *B* and *B'*. Another circuit is formed through the contact points of the relay, the voltaic cell at the left, and the vibrating electric bell. The bell is so placed that the ball or hammer strikes the coherer when the bell rings. The transmitter and receiver may then be placed at some distance apart, with the metallic plates parallel to each other. When the key *K* of the transmitter is firmly pressed for a moment, a spark passes across the gap between the balls; at the same instant the coherer allows a current to pass through, and this current operates the relay and closes the circuit through the electric bell. The blow of the bell hammer on the coherer restores it to its sensitive condition of high resistance, ready for the reception of another short succession of electric waves.

519. Marconi's System. — Marconi, who has succeeded in sending messages without wires over very long distances, employs a powerful induction coil in connection with his transmitter, and a coherer as the sensitive detector of electric waves. One terminal of the induction coil is connected with the earth, and the other with a wire running up a tall mast and ending in a ball or a plate of metal. The receiver differs from Fig. 310 in the same way as the transmitter differs from Fig. 309. Another tall mast supports a wire ending in a ball or a plate. The lower end of the wire is connected with one end of the coherer; the other end of the coherer goes to earth. Instead of an electric bell is a sounder (§ 510). Auxiliary devices are used which cannot be described here. Ships fitted with Marconi's system communicate successfully when one hundred fifty miles or more apart at sea.

XX. THE TELEPHONE AND THE MICROPHONE.

520. The Telephone (Fig. 311) consists of a permanent magnet O, one end of which is surrounded by a coil of many turns of fine copper wire b, whose ends are connected with the binding-posts t and t. At right angles to the magnet, and not quite touching the pole within the coil is an elastic diaphragm or disk a of soft sheet-iron, kept in place by the conical mouthpiece d. If the instrument is placed in an electric circuit when the current is unsteady, or alternating in direction, the magnetic field due to the helix, when combined with that due to the magnet, alters intermittently the number of lines of force which branch out from the pole, thus varying the attraction of the magnet for the disk. The result is that the disk vibrates in exact keeping with the changes in the current.

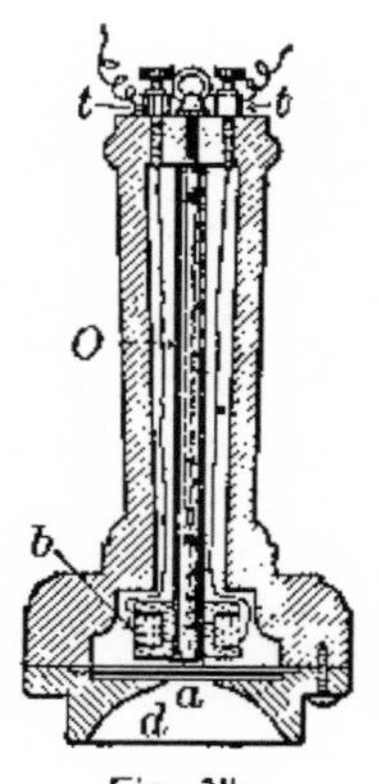

Fig. 311.

521. The Microphone is a device for varying an electric current by means of a variable resistance in the circuit. One of its simplest forms is shown in Fig. 312. It consists of a rod of gas-carbon A, whose tapering ends rest loosely in conical depressions made in blocks C, C, of the same material attached to a sounding-board. These blocks are placed in circuit with a battery and a telephone, by means of the wires X and Y. While the current is

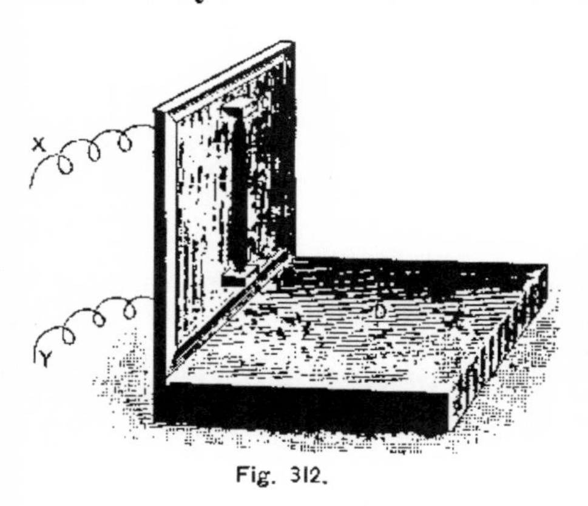

Fig. 312.

passing, the least motion of the sounding-board, caused either by sound waves or by any other means, moves the loose carbon pencil and varies the pressure between its ends and the supporting bars. A slight increase of pressure between two conductors resting loosely one on the other lessens the resistance of the contact, and conversely. Hence the vibrations of the sounding-board cause variations in the pressure at the points of contact of the carbons, and consequently make corresponding fluctuations in the current and vibrations of the telephone disk.

522. The Blake Transmitter is a form of microphone used in connection with the telephone in the telephone service of our cities and towns. An idea of its construction and mode of action can be obtained from an examination of Fig. 313, which illustrates diagrammatically the apparatus at a subscriber's station. A larger section of the microphone itself is represented by Fig. 314. A conical mouthpiece A has back of it a diaphragm of elastic sheet-iron, secured by its edges to the supporting frame. Back of this disk are two springs E and C, insulated from each other at the top. The spring C has a carbon button D attached to its lower end, against which rests the light hammer-shaped end of the other spring. These springs are connected

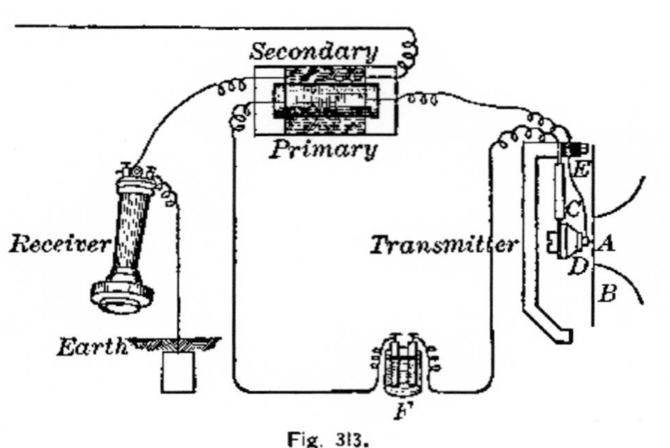

Fig. 313.

respectively with the poles of a battery, through the primary of a small induction coil. One of the terminals of the secondary coil is connected with the earth; the other is connected with the line wire leading to the other station. In circuit with this secondary coil is a telephone. When two subscribers are connected for conversation, the line wire, starting from the secondary of the induction coil, runs through the central exchange to the second subscriber's station, where the instruments shown in Fig. 313 are duplicated.

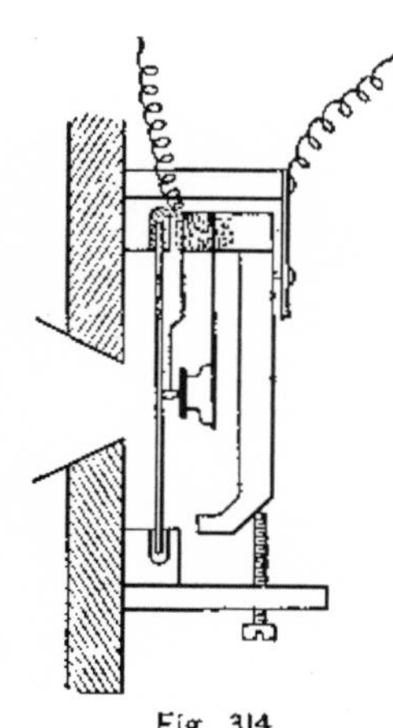
Fig. 314.

Sound waves striking against the diaphragm *A* cause a varying pressure between the free ends of the springs *E* and *C*, which act as a microphone. The battery current is consequently unsteady, and capable of inducing currents in the secondary coil. These induced currents have all the peculiarities of the inducing current. The telephone in the secondary circuit will be affected by them, and will accordingly send off sound waves similar to those which disturb the disk of the transmitter.

When a telephone is used as a transmitter, its action is as follows: Sound waves falling on the diaphragm set it vibrating; the vibrations vary the magnetic field near the end of the magnet, and consequently produce induced currents in the coil surrounding it. When these alternating currents pass through a similar instrument as a receiver, they cause the attraction of the magnet for the iron disk to vary. The disk is thus set vibrating with frequencies and amplitudes corresponding to those of the transmitting diaphragm.

APPENDIX.

I. GEOMETRICAL CONSTRUCTIONS.

The principal instruments required for the accurate construction of diagrams on paper are the *compasses* and the *ruler*. For the construction of angles of any definite size the *protractor* (Fig. 315) can be used. There are, however, a number of angles, as 90°, 60°, and those which can be obtained from these by bisecting them and combining their parts, that can be constructed by the compasses and ruler alone.

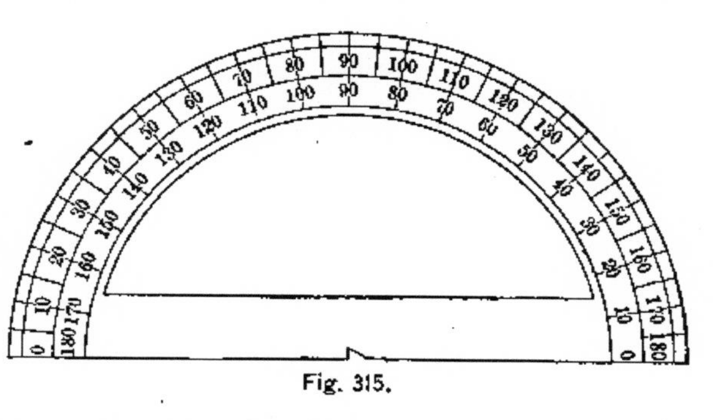

Fig. 315.

A convenient instrument for the rapid construction of the angles, 90°, 60°, and 30°, is a triangle made of wood, horn, hard rubber, or cardboard, whose angles are these respectively. Such a triangle may be easily made from a postal card as follows: Lay off on the short side of the card (Fig. 316) a distance a little less than the width, as AB. Separate the points of the compasses a distance equal to twice this distance. Place one point of the compasses at B, and draw an arc cutting the adjacent side at C.

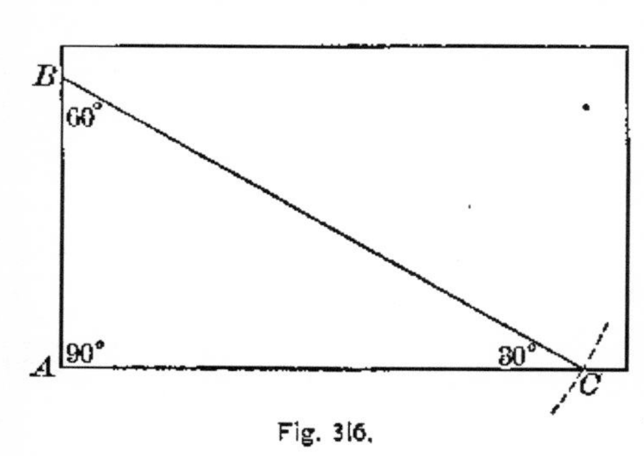

Fig. 316.

Cut the card into two parts along the straight line BC. The part ABC will be a right-angled triangle, having the longest side twice as long as the shortest side, with the larger acute angle 60° and the smaller 30°. With this triangle and a straight edge the majority of the constructions required in elementary physics can be made.

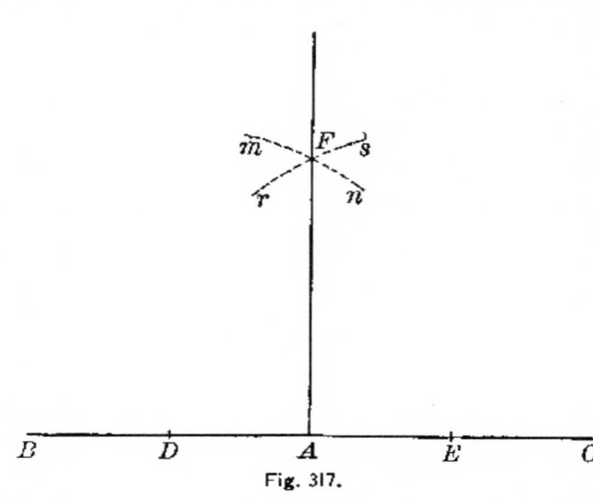

Fig. 317.

PROB. 1. — *To construct an angle of 90°.*

Let A be the vertex of the required angle (Fig. 317). Through A draw the straight line BC. Measure off AD, any convenient distance; also make $AE = AD$. With a pair of compasses, using D as a centre, and a radius longer than AD, draw the arc mn; with E as a centre and the same radius, draw the arc rs, intersecting mn at F. Join A and F. The angles at A are right angles.

PROB. 2. — *To construct an angle of 60°.*

Let A be the vertex of the required angle (Fig. 318), and AB one of the sides. On AB take some convenient distance as AC. With a pair of compasses, using A as a centre and AC as a radius, draw the arc CD. With C as a centre and the same radius, draw the arc mn, intersecting CD at E. Through A and E draw the straight line AE; this line will make an angle of 60° with AB.

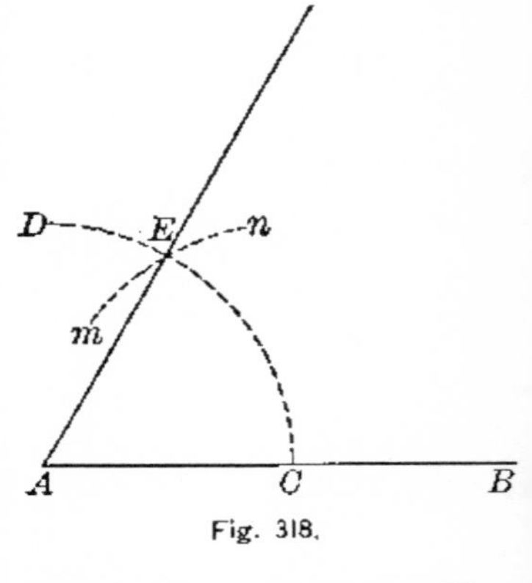

Fig. 318.

PROB. 3. — *To bisect an angle.*

Let BAC be an angle that it is required to bisect (Fig. 319). Measure off on the sides of the angle equal distances, AD and AE. With D and E as centres and with the same radius, draw the arcs mn and rs, intersecting at F. Draw AF. This line will bisect the angle BAC.

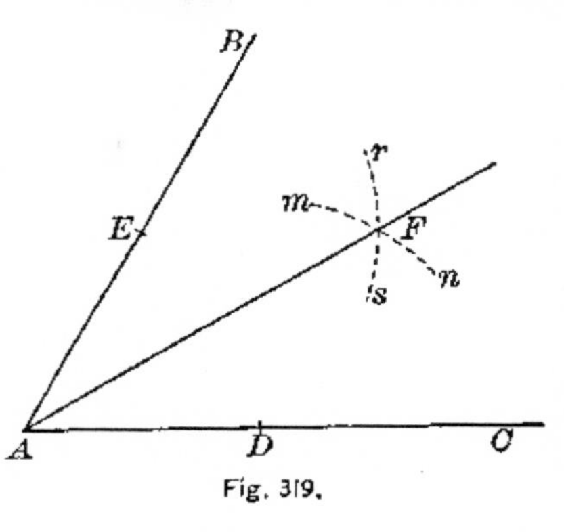

Fig. 319.

PROB. 4. — *To make an angle equal to a given angle.*

Let BAC be a given angle; it is required to make a second angle equal to it (Fig. 320). Draw DE, one side of the required angle. With A as a centre and any convenient radius, draw the arc mn across the given angle. With D as a centre and the same radius, draw the arc rs. With s as a centre and a radius equal to the chord of

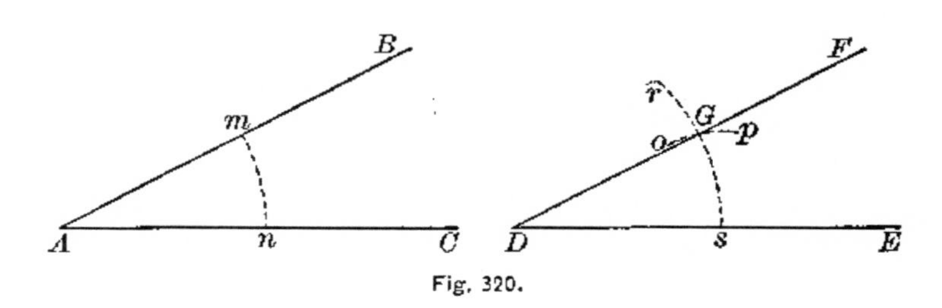

Fig. 320.

mn, draw the arc op, cutting rs at G. Through D and G draw the line DF. This line will form with DE the required angle, as FDE.

PROB. 5. — *To draw a line through a point parallel to a given line.*

Let A be the point through which it is required to draw a line parallel to BC (Fig. 321). Through A draw ED, cutting BC at D. At A make the angle EAG equal to EDC. Then AG or FG is parallel to BC.

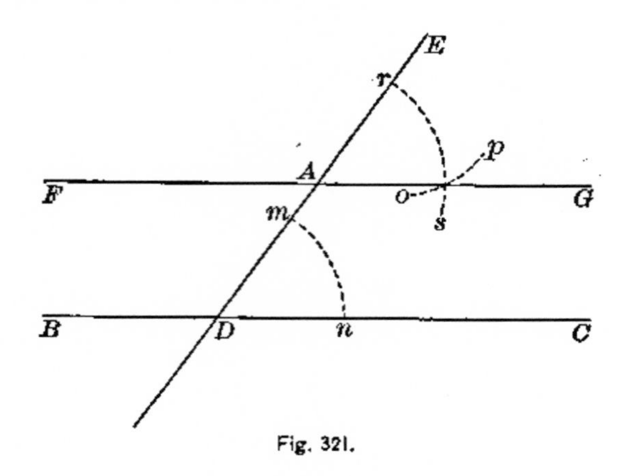

Fig. 321.

PROB. 6. — *Given two adjacent sides of a parallelogram to complete the figure.*

Let AB and AC be two adjacent sides of the parallelogram (Fig. 322). With C as a centre and a radius equal to AB,

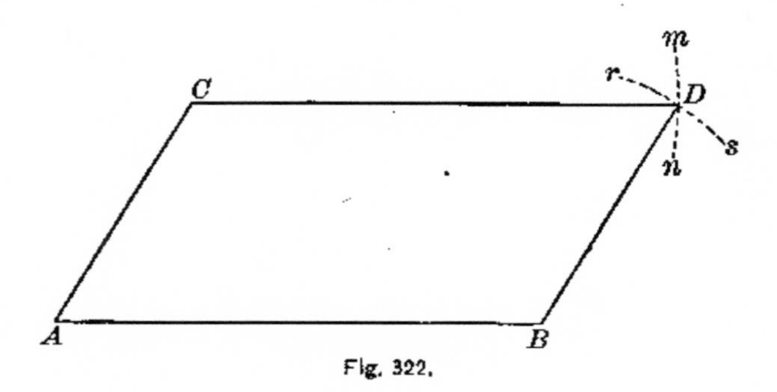

Fig. 322.

draw the arc mn. With B as a centre and a radius equal to AC, draw the arc rs, cutting mn at D. Draw CD and BD. Then $ABDC$ is the required parallelogram.

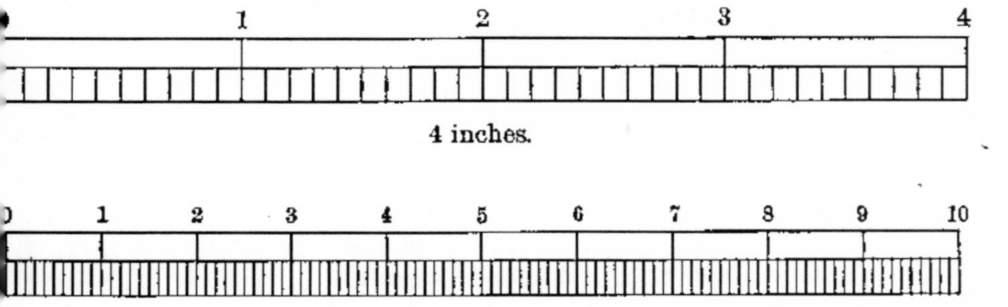

4 inches.

1 decimetre = 10 centimetres = 100 millimetres.

1 sq. decim. = 15.5 sq. in.
1 sq. in. = 6.45 sq. cm.

A cube of water at 4° C., one of whose faces is this square, has a mass of one kilogramme. The volume is a litre. A cube of water at 4° C., with each of its faces one sq. cm., has a mass of one gramme. A cubic inch of water at 4° C. has a mass of 0.03611 lb. or 0.58 oz.

1 sq.in.

1 sq.cm.

III. CONVERSION TABLES.

1. Length.

To reduce	Multiply by	To reduce	Multiply by
Miles to km.	1.60935	Kilometres to mi. . .	0.62137
Miles to m.	1609.347	Metres to mi. . . .	0.0006214
Yards to m.	0.91440	Metres to yds. . . .	1.09361
Feet to m.	0.30480	Metres to ft.	3.28083
Inches to cm.	2.54000	Centimetres to in. . .	0.39370
Inches to mm.	25.40005	Millimetres to in. . .	0.03937

2. Surface.

To reduce	Multiply by	To reduce	Multiply by
Sq. yards to m.^{2} . . .	0.83613	Sq. metres to sq. yds. .	1.19599
Sq. feet to m.^{2}	0.09290	Sq. metres to sq. ft. . .	10.76387
Sq. inches to cm.^{2} . . .	6.45163	Sq. centimetres to sq. in. .	0.15500
Sq. inches to mm.^{2} .	645.163	Sq. millimetres to sq. in. .	0.00155

3. Volume.

To reduce	Multiply by	To reduce	Multiply by
Cu. yards to m.^{3} . . .	0.76456	Cu. metres to cu. yds. .	1.30802
Cu. feet to m.^{3}	0.02832	Cu. metres to cu. ft. . .	35.31661
Cu. inches to cm.^{3} . . .	16.38716	Cu. centimetres to cu. in.	0.06102
Cu. feet to litres . . .	28.31701	Litres to cu. ft.	0.03532
Cu. inches to litres . .	0.01639	Litres to cu. in.	61.02337
Gallons to litres	3.78543	Litres to gallons . . .	0.26417
Pounds of water to litres .	0.45359	Litres of water to lbs. . .	2.20462

4. Weight.

To reduce	Multiply by	To reduce	Multiply by
Tons to kgm. . . .	907.18486	Kilogrammes to tons . .	0.001102
Pounds to kgm. . . .	0.45359	Kilogrammes to lbs. . .	2.20462
Ounces to gm. . . .	28.34953	Grammes to oz.	0.03527
Grains to gm. . . .	0.064799	Grammes to grains .	15.43236

5. Force, Work, Activity, Pressure.

To reduce	Multiply by	To reduce	Multiply by
Lbs.-weight to dynes,	444520.58	Dynes to lbs.-weight,	22496×10^{-10}
Ft.-lbs. to kgm.-m. . . .	0.138255	Kgm.-m. to ft.-lbs. . . .	7.233
Ft.-lbs. to ergs	13549×10^{3}	Ergs to ft.-lbs. . . .	$.0.7381 \times 10^{-7}$
Ft.-lbs. to joules . . .	1.3549	Joules to ft.-lbs. . . .	0.7381
Ft.-lbs. per sec. to H.P.	18182×10^{-7}	H.P. to ft.-lbs. per sec. .	550
H.P. to watts	745.196	Watts to H.P.	0.001342
Lbs. per sq. ft. to kgm. per m.^{2}	4.8824	Kgm. per m.^{2} to lbs. per sq. ft.	0.2048
Lbs. per sq. in. to gm. per cm.^{2}	70.3068	Gm. per cm.^{2} to lbs. per sq. in.	0.01422

Calculated for g = 980 cm., or 32.15 ft. per sec. per sec.

6. Miscellaneous.

To reduce	Multiply by	To reduce	Multiply by
Lbs. of water to U.S. gal.	0.11983	U.S. gal. to lbs. of water,	8.345
Cu. ft. to U.S. gal. . .	7.48052	U.S. gal. to cu. ft. . . .	0.13368
Lbs. of water to cu. ft. at 4° C.	0.01602	Cu. ft. of water at 4° C. to lbs.	62.425
Cu. in. to U.S. gal. . .	0.004329	U.S. gal. to cu. in. . .	231
Atmospheres to lbs. per sq. in.	14.69640	Lbs. per sq. in. to atmospheres	0.06737
Atmospheres to gm. per cm.^{2}	1033.296	Gm. per cm.^{2} to atmospheres	0.000968
Lb.-degrees F. to calories,	252	Calories to lb.-degrees F.	0.003968
Calories to joules . . .	4.18936	Joules to calories . . .	0.2387
Miles per hour to ft. per sec.	1.46667	Ft. per sec. to miles per hour	0.68182
Miles per hour to cm. per sec.	44.704	Cm. per sec. to miles per hour	0.02237

IV. MENSURATION RULES.

Area of triangle	$= \frac{1}{2}$ (base × altitude).
Area of triangle	$= \sqrt{s(s-a)(s-b)(s-c)}$ where $s = \frac{1}{2}(a+b+c$
Area of parallelogram	= base × altitude.
Area of trapezoid	= altitude × $\frac{1}{2}$ sum of parallel sides.
Circumference of circle	= diameter × 3.1416.
Diameter of circle	= { circumference ÷ 3.1416. circumference × 0.3183.
Area of circle	= { diameter squared × 0.7854. radius squared × 3.1416.
Area of ellipse	= product of diameters × 0.7854.
Area of regular polygon	$= \frac{1}{2}$ (sum of sides × apothem).
Lateral surface of cylinder	= circumference of base × altitude.
Volume of cylinder	= area of base × altitude.
Surface of sphere	= { diameter × circumference. 4 × 3.1416 × square of radius.
Volume of sphere	= { diameter cubed × 0.5236. $\frac{4}{3}$ of radius cubed × 3.1416.
Surface of pyramid } Surface of cone }	$= \frac{1}{2}$ (circumference of base × slant height).
Volume of cone	$= \frac{1}{3}$ (area of base × altitude).

V. TABLE OF DENSITIES.

The following table gives the mass in grammes of 1 $cm.^3$ of the substance: —

Substance	Density
Agate	2.615
Air, at 0° C. and 76 cm. pressure	0.00129
Alcohol, ethyl, 90%, 20° C.	0.818
Alcohol, methyl	0.814
Alum, common	1.724
Aluminum, wrought	2.670
Antimony, cast	6.720
Beeswax	0.964
Bismuth, cast	9.822
Brass, cast	8.400
Brass, hard drawn	8.700
Carbon, gas	1.89
Carbon disulphide	1.293
Charcoal	1.6
Coal, anthracite	1.26 to 1.800
Coal, bituminous	1.27 to 1.423
Copper, cast	8.830
Copper, sheet	8.878
Cork	0.14 to 0.24
Diamond	3.530
Ebony	1.187
Emery	3.900
Ether	0.736
Galena	7.580
German-silver	8.432
Glass, crown	2.520
Glass, flint	3.0 to 3.600
Glass, plate	2.760
Glycerine	1.260
Gold	19.360
Granite	2.650
Graphite	2.500
Gypsum, crys.	2.310
Human body	0.890
Hydrogen, at 0° C. and 76 cm. pressure	0.0000896
Ice	0.917
Iceland spar	2.723
India-rubber	0.930
Iron, white cast	7.655
Iron, wrought	7.698
Ivory	1.820
Lead, cast	11.360
Magnesium	1.750
Marble	2.720
Mercury, at 0° C.	13.596
Mercury, at 20° C.	13.558
Milk	1.032
Nitrogen, at 0° C. and 76 cm. pressure	0.001255
Oil, olive	0.915
Oxygen, at 0° C. and 76 cm. pressure	0.00143
Paraffin	0.824 to 0.940
Platinum	21.531
Potassium	0.865
Silver, wrought	10.56
Sodium	0.970
Steel	7.816
Sulphuric Acid	1.84
Sulphur	2.033
Sugar, cane	1.593
Tin, cast	7.290
Water, at 0° C.	0.999
Water, at 20° C.	0.998
Water, sea	1.207
Zinc, cast	7.000

VI. TABLE OF NATURAL SINES AND TANGENTS.

Angle.	Sine.	Tangent.	Angle.	Sine.	Tangent.	Angle.	Sine.	Tangent.
0	0.000	0.000	31	0.515	0.601	62	0.883	1.881
1	0.017	0.017	32	0.530	0.625	63	0.891	1.963
2	0.035	0.035	33	0.545	0.649	64	0.899	2.050
3	0.052	0.052	34	0.559	0.675	65	0.906	2.145
4	0.070	0.070	35	0.574	0.700	66	0.914	2.246
5	0.087	0.087	36	0.588	0.727	67	0.921	2.356
6	0.105	0.105	37	0.602	0.754	68	0.927	2.475
7	0.122	0.123	38	0.616	0.781	69	0.934	2.605
8	0.139	0.141	39	0.629	0.810	70	0.940	2.747
9	0.156	0.158	40	0.643	0.839	71	0.946	2.904
10	0.174	0.176	41	0.656	0.869	72	0.951	3.078
11	0.191	0.194	42	0.669	0.900	73	0.956	3.271
12	0.208	0.213	43	0.682	0.933	74	0.961	3.487
13	0.225	0.231	44	0.695	0.966	75	0.966	3.732
14	0.242	0.249	45	0.707	1.000	76	0.970	4.011
15	0.259	0.268	46	0.719	1.036	77	0.974	4.331
16	0.276	0.287	47	0.731	1.072	78	0.978	4.705
17	0.292	0.306	48	0.743	1.111	79	0.982	5.145
18	0.309	0.325	49	0.755	1.150	80	0.985	5.671
19	0.326	0.344	50	0.766	1.192	81	0.988	6.314
20	0.342	0.364	51	0.777	1.235	82	0.990	7.115
21	0.358	0.384	52	0.788	1.280	83	0.993	8.144
22	0.375	0.404	53	0.799	1.327	84	0.995	9.514
23	0.391	0.424	54	0.809	1.376	85	0.996	11.43
24	0.407	0.445	55	0.819	1.428	86	0.998	14.30
25	0.423	0.466	56	0.829	1.483	87	0.999	19.08
26	0.438	0.488	57	0.839	1.540	88	0.999	28.64
27	0.454	0.510	58	0.848	1.600	89	1.000	57.29
28	0.469	0.532	59	0.857	1.664	90	1.000	Infinity
29	0.485	0.554	60	0.866	1.732			
30	0.500	0.577	61	0.875	1.804			

INDEX.

Numbers refer to pages.

Breinigsville, PA USA
06 March 2011
257011BV00001B/105/P

9 781408 690918